Experimental Brain Research Supplementum 13

Processes of Recovery from Neural Trauma

Edited by
G. M. Gilad A. Gorio G. W. Kreutzberg

With 135 Figures and 16 Tables

Springer-Verlag
Berlin Heidelberg New York Tokyo

Dr. Gad M. Gilad
Center for Neuroscience and
Behavioral Research
The Weizmann Institute of Science
Rehovot 76100, Israel

Dr. Alfredo Gorio
Department of Cytopharmacology
Fidia Research Laboratories
Via Ponte della Fabbrica 3a
35031 Abano Terme, Italy

Prof. Dr. Georg W. Kreutzberg
Max-Planck-Institut für Psychiatrie
Am Klopferspitz 18A
8033 Planegg-Martinsried, FRG

ISBN-13: 978-3-642-70701-8 e-ISBN-13: 978-3-642-70699-8
DOI: 10.1007/978-3-642-70699-8

Library of Congress Cataloging-in-Publication Data. Main entry under title: Processes of recovery from neural trauma. (Experimental brain research. Supplementum; 13) Includes bibliographies and index. 1. Central nervous system-Wounds and injuries. 2. Nervous system-Regeneration. 3. Neurophysiology. 4. Surgery, Experimental. I. Gilad, G. (Gad) II. Gorio, Alfredo. III. Kreutzberg, George W. IV. Series. [DNLM: 1. Central Nervous System-injuries. 2. Peripheral Nerves-injuries. 3. Wound and Injuries-rehabilitation. W1 EX485B v. 13 / WL 300 P963] RD593.P76 1986 617'.48044 85-27858

Softcover reprint of the hardcover 1st edition 1986

Offsetprinting: Beltz Offsetdruck, Hemsbach/Bergstr.
Binding: J. Schäffer OHG, Grünstadt

2125/3140-543210

It is for the science of the future to change, if possible, this harsh decree. Inspired with high ideals, it must work to impede or moderate the gradual decay of the neurones, to overcome the almost invincible rigidity of their connections, and to reestablish normal nerve paths, when disease has severed centers that were intimately associated.

Santiago Ramon Y Cajal (1928)
Degeneration and Regeneration of the Nervous System
(Vol. 2, Oxford University Press, London)

Foreword

The book on Processes of Recovery from Neural Trauma contains the proceedings of an international symposium which was sponsored by the Weizmann Institute of Science and was held in June 1984 in Israel.

One of the central problems in neurobiology and clinical neurology are the processes which underlie the ability of the nervous system to recover from damage. Neurons of the mammalian nervous system cannot be replaced once the generative embryonic period has ended. Thus, recovery after trauma is limited by the degree to which regeneration of damaged neural processes is possible and by the capacity of uninjured neurons to form new synaptic contacts or to use their existing ones more effectively. The lack of successful axon regeneration in the central nervous system is the reason for the permanent damage and the very limited recovery of function seen in victims of traumatic injury. Over the past few years, research on neural trauma has witnessed significant progress, resulting from interdisciplinary research carried out in an increasing number of laboratories throughout the world. Accordingly, the book includes reports on the progress made in the forefront of research in this field.

We hope that in addition to being a source of new information, readers of the book will be stimulated to work and to generate new ideas and research strategies in this field of research.

We would like to thank Frau Marthe Bellstedt and Frau Ursula Qreini from the secretarial staff of the Max Planck Institute for Psychiatry in Martinsried, and Mr. Menashe Dornay from the Weizmann Institute of Science. Their expertise in computer assisted word processing was invaluable in the production of this volume. We also would like to express our thanks to Dr. Thiekoetter and his staff from Springer Verlag for their help and advice in publishing these proceedings.

January 1986 The Editors

Contents

List of Contributors*

* The address of the authors is given on the first page of each contribution
[1] Page, on which contribution commences

Neuronal Response to Nerve Trauma

Reaction on Motoneurons and Their Microenvironment to Axotomy

W. Tetzlaff, M. B. Graeber, and G. W. Kreutzberg

Abteilung für Neuromorphologie, Max-Planck-Institut für Psychiatrie, Am Klopferspitz 18A, 8033 Martinsried, FRG

Interruption of the continuity of an axon induces a number of morphological, metabolic and physiological changes in the nerve cell body that are referred to as chromatolysis, "axonal response" or "retrograde reaction" (reviewed in Lieberman 1971; Grafstein and McQuarrie, 1978; Kreutzberg 1982; Anderson et al. 1982). It is assumed that these changes are part of an inherent neuronal program that is a prerequisite for the axonal growth process leading to regeneration and finally functional restitution of the motor or sensory neuronal unit. Many of the retrograde reactions in neurons of the peripheral nervous system capable of regeneration are lacking in axotomized central neurons, which do not regenerate. However, a number of recent observations have proven that also central neurons can grow new axons provided they are supplied with the right stimulus which can come, for example, from a peripheral nerve graft (Richardson et al. 1980; Benfey and Aguayo 1982). We believe a better understanding of the changes occurring in regenerating peripheral neurons might also help in understanding principles of suppressed or stimulated regeneration in CNS neurons.

We have studied these events in a variety of motoneuron systems but mainly in the facial nucleus of the rat. In this case we have developed a routine technique for microdissection and isolation of the facial nucleus which allows the biochemical investigation of the axotomized motoneurons. Our interest has been focused on enzyme activities concerned in basal metabolism and the metabolism of neurotransmitters as well as on the cellular microenvironment in the regenerating facial nucleus. Little is known how the neuron gets informed about the traumatization of its axon and how it triggers the onset of the regeneration program. The earliest change so far known is an increase in ornithine decarboxylase (ODC) activity (Tetzlaff and Kreutzberg 1985). Already 8 hours after the traumatization a level of 200% over control is reached and the maximal activity was found 24 hours post lesion. Thereafter the activity declines to subnormal levels where it stays during the regeneration period. Histochemistry demonstrated this enzyme to be localized in the perikarya of the motoneurons. ODC is the rate limiting enzyme of polyamine synthesis. Polyamines are suspected to be involved in gene regulation at the level of DNA and RNA synthesis (Canellakis et al. 1979; Tabor and Tabor 1976). It is likely that this early polyamine synthesis in the axotomized motoneurons is related to the early changes in messenger RNA and ribosomal RNA that are also seen early after axotomy (reviewed in Austin 1985). In another neuronal system, the superior cervical ganglion of the rat, this ODC increase occurs even earlier (Gilad and Gilad 1983), probably due to the shorter distance from the lesion to the nerve cell body.

This initiation of the neuronal regeneration response on the first day is accompanied by an increased uptake of glucose that can be demonstrated autoradiographically already 24 hours after the lesion (Singer and Mehler 1980; Kreutzberg and Emmert 1980). Despite the limited resolution of the technique, our original finding that this increase takes place in the perikarya and in the neuropil has been recently confirmed (Erhart et al. 1984). At least a part of the glucose consumed will be directed to the pentose phosphate shunt which is also activated after axotomy (Kreutzberg 1963; Härkönen and Kauffmann 1974). This activation is measurable already 24 hours after axotomy and reaches a maximum 7 days later (Tetzlaff and Kreutzberg 1984). As known from measurements in other neuronal systems the activity of other glucose consuming pathways such as glycolysis is unchanged (reviewed in Thomas 1969; Watson 1976). The pentose phosphate shunt contributes ribose and NADPH. Ribose might be needed for the increased RNA synthesis. NADPH furnishes proton equivalents for lipid synthesis, probably necessary for membrane formation in the regenerating axons. An increase in activity in this pathway is one of the most regular findings in axotomized neurons of the CNS and PNS. To test the significance of this pathway for the regeneration of the axonal process we were able to inhibit the increase in activity with caffeine and theophylline by about 60% (Tetzlaff and Kreutzberg, in preparation). Such an inhibition did not impair the regeneration of crushed facial or hypoglossal axons indicating that the products of this pathway are not a limiting factor for axonal elongation. However, the amount of cellular material needed after a crush lesion is probably less than after a transection as in the latter case a substantial number of searching fibers are formed.

In contrast to these enzymes related to the basal metabolism of the neuron there is another group of substances which in general decrease in response to axotomy. These are the enzymes involved in transmitter metabolism such as dopamine hydroxylase, tyrosine hydroxylase and in the facial nucleus, cholineacetyltransferase and acetylcholinesterase (AChE) (Hoover and Hancock 1985). The decrease in AchE becomes measurable already 24h after axotomy and reaches a plateau 7 days following the traumatization (Tetzlaff and Kreutzberg 1984). Biochemical and histochemical quantifications revealed a decrease to 60% of normal values. The reappearance of the enzyme is related to the type of traumatization and may thus be correlated with the degree of regeneration (Kreutzberg et al. 1984).

The situation is somewhat different in the motor nucleus of the vagus where AChE and butyrylcholinesterase (BChE) activity disappear (light microscopically) from the neurons within a few days after axotomy. The recovery of BChE occurs later than the reappearance of AChE. While AChE reappears after 2 weeks, BChE does not become visible in light microscopic histochemistry before the end of the third week. (Kreutzberg et al. 1984). This suggests that the synthesis of the two cholinesterases is coded by two different genes which are regulated by different mechanisms.

The retrograde changes after axotomy are not limited to the neurons but also include the surrounding microenvironment of the glial cells and blood vessels. As shown in earlier studies (Kreutzberg 1968; Blinzinger and Kreutzberg 1968) microglial cells proliferate and reach a perineuronal position thus isolating the neurons from their synaptic inputs. Electron microscopic histochemistry revealed high activity of 5'-nucleotidase on the membranes of these perineuronal microglial cells (Kreutzberg and Barron 1978). The increase in this enzyme activity does not begin before the second day after axotomy when biochemically and histochemically measured and it reaches a maximum during the second week after axotomy (150% of control) (Tetzlaff and Kreutzberg 1984). Thus the glial reaction to axotomy starts with a certain delay as compared to the neuronal response which points out the secondary nature of these glial changes. 5'-nucleotidase produces adenosine as well as other nucleosides. Adenosine might act on the blood vessels as a potent vasodilator (Berne et al. 1974) and it might act on the neurons themselves via adenosine receptors that occur in the neuropil of the facial nucleus. Whether or not adenosine plays a role in neuronal regeneration remains to be established.

In order to study the molecular cytology of the satellite glial cells during regeneration of the facial nerve we have applied a number of antibodies against various constituents of the cytoskeleton and other elements of these cells. Some of the perineuronal glial cells surrounding chromatolytic neurons were clearly shown to have cross reactivity against two different monoclonal antibodies and one polyclonal antiserum against glial fibrillary acidic protein (GFAP). Anti S-100 also stained the cells. No decoration has been found by applying antibodies against fibronectin, laminin and the cell surface markers L2, O4 and Cl.

The presence of GFAP-like immunoreactivity in satellite glial cells led us to postulate that apart from the perineuronally proliferating microglial cells a second population of glial cells reacts to facial nerve injury. As the glial fibrillary acidic (GFA) protein has been shown to be the major protein constituent of glial filaments these cells can be classified as neuroglia, belonging to the astrocyte family. A reinvestigation of the ultrastructure of the GFAP-positive cells using pre-embedding (PAP/DAB) and post-embedding (immuno gold) immunocytochemical methods confirmed the presence of labelling of glial filament bundles in perineuronal fibrous astrocytes. Compared with the unoperated nucleus where GFAP-immunoreactivity is very low the regenerating facial nucleus contains a large number of fibrous astrocytes. Since astrocytes do not proliferate in the operated nucleus, it seems likely that protoplasmic astrocytes are transformed into the fibrous type during axonal reaction. Microglial cells possess a very dense cytoplasm in which a reticular net of intermediate filaments not stained by anti-GFAP is also present although this has proved difficult to recognize.

Motoneuron regeneration is of great clinical relevance as a substantial amount of accidents are associated with traumatization of peripheral nerves. A popular surgical approach is a delayed suture of the nerve two or three weeks after the trauma which neccessarily includes a trimming of the nerve stumps i.e. a second axonal lesion. The rationale for this procedure is based on the experimental observation that a second lesion of a nerve two or three weeks after the initial trauma leads to an enhanced growth of the axonal sprouts from the proximal stump (reviewed by McQuarrie 1984). It is assumed that while the first lesion conditions the neuron and triggers the regeneration program the second lesion could provide an additional stimulus to boost the response of the motoneurons. An alternative or additional reason for this conditioning lesion effect could be based on stimulatory factors in the peripheral nerve stump that undergoes Wallerian degeneration. Because of the obvious importance of this phenomenon we have investigated the neuronal response of the facial nerve after a conditioning lesion.

The earliest parameter reacting to axotomy, ornithine decarboxylase, does not increase again when the second axotomy (cut) is performed 14 days after the first lesion (crush) e.g. during the regeneration period (Tetzlaff and Kreutzberg 1985). There is a second increase of ODC activity if the second lesion is performed when axonal regeneration and maturation is accomplished, which is the case 20 days after the initial trauma. This observation supports the view that ODC is somehow involved in triggering the regeneration program of the neuron. The ODC-reaction-products, the polyamines, may have a regulatory function during RNA and protein synthesis. The missing reincrease during the regeneration period might reflect that in the facial nucleus no further stimulation of the neuronal regeneration program takes place after a conditioning lesion. There is evidence that the second lesion is also signalled to the nerve cell body and its microenvironment. In the facial nucleus the second axotomy is followed by a further decrease of AChE activity and by a second increase of 5'nucleotidase (Tetzlaff and Kreutzberg 1984). However, a simple amplification of the original response to axotomy does not take place. The neuronal pentose phosphate shunt enzymes are not activated further by a preceding axotomy of the facial nerve; on the contrary they were found to be at a lower level.

Nevertheless in the facial nerve a conditoning lesion reduces the period of functional recovery by 10 to 15 %. This can only be achieved by an earlier and/or faster outgrowth of the regenerating axons. If no further stimulation of the neuronal metabolism itself takes place, this conditioning lesion effect would be due to a shortening of the initial delay, that is the time before axonal outgrowth begins. Another possibility is that stimulatory factors in the distal nerve stump (Politis et al.1982; Skene and Shooter 1983; see also Freemann, this volume) which is undergoing Wallerian degeneration enhance the fibre outgrowth velocity as recently shown for the sciatic nerve after conditioning lesion (Bisby and Pollock 1983).

The complex response of the facial motoneurons to a conditioning lesion might reflect our observation that the neuronal vulnerability is enhanced since an additional 10% of the neurons disappear if they are axotomized again during the regeneration period. Thus the conditioning lesion may not be beneficial to all neuronal populations (see also McQuarrie et al. 1978; Arvidsson and Aldskogius 1982). These observations must be taken into account when performing a secondary suture in clinical practice, which includes a second axotomy by trimming the nerve stumps.

Conclusion

We have reported here a number of new observations related to enzymatic, structural and cellular changes of regenerating neurons and their environment. Correlation of these changes with the regeneration process is, however, problematic. Nevertheless, a few trends emerge which correlate relatively frequently: (1) Decrease of synthesis of transmitter-related compounds, e.g. enzymes or receptor proteins; (2) loss of synaptic input e.g. by synaptic stripping; (3) increase in glucose consumption as well as in enzymes involved in the hexose monophosphate shunt; (4) general hypertrophy of neurons expressed as volume increase and augmentation in organelles. There seems also to be an increase in materials axonally transported, among which are special proteins not normally seen or only present in extremely tiny amounts (Willard and Skene 1982). The function of these so-called growth associated proteins (GAPs) might be another clue to understanding the regeneration process.

References

Anderson HJ, Aguayo AJ, Blackshaw SE, Bray D, Gilliatt RW, Grinnell AD, Kreutzberg GW, Parnas I, Purves D, Rotshenker S, Schwab ME, Willard MB (1982) Early responses to neural injury. In: Nicholls JG (ed) Repair and regeneration of the nervous system, Springer, Berlin, p 315

Arvidsson J, Aldskogius H (1982) Effect of repeated hypoglossal nerve lesions on the number of neurons in the hypoglossal nucleus of adult rats. Exp Neurol 75:520-524

Austin L (1985) Molecular aspects of nerve regeneration. In: Lajtha A (ed) Alterations of metabolites in the nervous system, 2nd ed, Plenum, New York, p 1

Benfey M, Aguayo AJ (1982) Extensive elongation of axons from rat brain into peripheral nerve grafts. Nature 296:150-152

Berne RM, Rubio R, Curnish R (1974) Release of adenosine from ischaemic brain. Circ Res 35:262-272

Bisby MA, Pollock B (1983) Increased regeneration rate in peripheral nerve axons following double lesions: enhancement of the conditioning lesion phenomenon. J Neurobiol 14:467-472

Blinzinger K, Kreutzberg GW (1968) Displacement of synaptic terminals from regenerating motoneurons by microglial cells. Z Zellforsch Mikrosk Anat 85:145-157

Canellakis ES, Viceps-Madore D, Kyriakidis DA, Heller JS (1979) The regulation and function of ornithine decarboxylase and of the polyamines. In: Horecker BL, Stadtman ER (eds) Current Topics in Cellular Regulation, Academic, New York, p 155

Erhart G, Hochsattel H, Pilgrim C, Wagner HJ (1984) Localization of functional changes in rat brain stem nuclei during axon regeneration: a deoxyglucose study with improved autoradiographic resolution. Mikroskopie 41:138-144

Freeman JA, Bock S, Deaton M, McGuire B, Norden JJ, Jackson Snipes G (1985) Axonal and glial proteins associated with development and response to injury in the rat and goldfish optic nerve. (this volume)

Gilad GM, Gilad VH (1983) Early rapid and transient increase in ornithine decarboxylase activity within sympathetic neurons after axonal injury. Exp Neurol 81:158-166

Grafstein B, McQuarrie IG (1978) The role of the nerve cell body in axonal regeneration. In: Cotman CW (ed) Neuronal plasticity, Raven, New York, p 155

Härkönen MHA, Kauffman FC (1974) Metabolic alterations in the axotomized superior cervical ganglion of the rat. II. The pentose phosphate pathway. Brain Res 65:141-157

Hoover DB, Hancock JC (1985) Effect of facial nerve transection on acetylcholinesterase and (^{3}H)quinuclidinyl benzilate binding in rat facial nuclei. Neuroscience 15:481-487

Kreutzberg GW (1963) Changes of coenzyme (TPN) diaphorase and TPN-linked dehydrogenase during axonal reaction of the nerve cell. Nature 199:393-394

Kreutzberg GW (1968) DNA metabolism in glial cells during retrograde changes. In: Lodin Z, Rose SPR (eds) Macromolecules and the function of the neuron. Excerpta Medica, Amsterdam, p 51

Kreutzberg GW (1982) Acute neural reaction to injury. In: Nicholls JG (ed) Repair and Regeneration of the Nervous System, Springer, Berlin, p 57

Kreutzberg GW, Barron KD (1978) 5'Nucleotidase of microglial cells in the facial nucleus during axonal reaction. J Neurocytol 7:601-610

Kreutzberg GW, Emmert H (1980) Glucose utilization of motor nuclei during regeneration: a (^{14}C)2-deoxyglucose study. Exp Neurol 70:712-716

Kreutzberg GW, Tetzlaff W, Tóth L (1984) Cytochemical changes of cholinesterases in motor neurons during regeneration. In: Brzin M, Kiauta T, Barnard EA (eds) Cholinesterases, Walter de Gruyter, Berlin New York, p 273

Lieberman AR (1971) The axon reaction: a review of the principal features of perikaryal responses to axon injury, Int Rev Neurobiol 14:49-124

McQuarrie IG (1984) Effect of a conditioning lesion on axonal transport during regeneration. In: Elam JS, Cancalon P (eds) Axonal transport in neuronal growth and regeneration, Plenum, New York, p 185

McQuarrie IG, Grafstein B, Dreyfus CF, Gershon MD (1978) Regeneration of adrenergic axons in rat sciatic nerve: effect of a conditioning lesion. Brain Res 141:21-34

Politis MJ, Ederle K, Spencer PS (1982) Tropism in nerve regeneration in vivo. Attraction of regenerating axons by diffusible factors derived from cells in distal nerve stumps of transected peripheral nerves. Brain Res 253:1-12

Richardson PM, Guinness UM, Aguayo AJ (1980) Axons from CNS neurons regenerate into PNS grafts. Nature 284:264-265

Singer P, Mehler S (1980) 2-Deoxy(^{14}C)glucose uptake in the rat hypoglossal nucleus after nerve transection. Exp Neurol 69:617-626

Skene JHP, Shooter EM (1983) Denervated sheath cells secrete a new protein after nerve injury. Proc Natl Acad Sci USA 80:4169-4173

Tabor CW, Tabor H (1976) 1,4-Diaminobutane (Putrescine, Spermidine, and Spermine. In: Snell ES, Boyer PD, Richardson A, Meister CC (eds) Annual Review of Biochemistry, Vol 45, Annual Reviews Inc, Palo Alto, pp 288-305

Tetzlaff W, Kreutzberg GW (1984) Enzyme changes in the facial nucleus following a conditioning lesion. Exp Neurol 85:547-564

Tetzlaff W, Kreutzberg GW (1985) Ornithine decarboxylase in motoneurons during regeneration. Exp Neurol 89, in press

Thomas E (1969) Histotopochemie und Histopathochemie des peripheren Nervensystems bei Verletzungen und Tumoren. Gustav Fischer, Stuttgart

Watson WE (1976) Cell biology of brain. Chapman and Hall, London

Willard M, Skene JHP (1982) Molecular events in axonal regeneration. In: Nicholls JG (ed) Repair and regeneration of the nervous system, Springer, Berlin, p 71

Coupling Among Neurons in Neocortical Slices: A Specific Response to Dendrotomy

M. J. Gutnick and R. Lobel-Yaakov

Unit of Physiology, Corob Center for Medical Research, Faculty of Health Sciences, Ben-Gurion University of Negev, Beer Sheva, Israel

Introduction

Although it is generally agreed that chemical synaptic transmission is the dominant mechanism of neuronal interaction in the mammalian central nervous system, there is strong evidence that electrotonic junctions between neurons are also present in several brain regions (cf. Bennett and Goodenough 1978). In neocortex, evidence for electrical synapses comes primarily from brain slice studies in which it was demonstrated that intracellular injection of the fluorescent dye, Lucifer Yellow CH (LY), into a single neuron often leads to staining of several neurons (Gutnick and Prince 1981; Gutnick and Lobel-Yaakov 1983; Connors et al. 1983). Electrophysiological observations (see below) suggest that these dye-coupled neuronal aggregates are electrotonically coupled as well. Similar observations in hippocampal slices have recently been summarized in a comprehensive review by Dudek et al. (1983).

One possible interpretation of these data is that neuronal interaction through electrotonic synapses plays a normal role in cortical function. However, it is noteworthy that in in vivo studies of neocortex, neuronal dye-coupling is not a prominent consequence of intracellular LY injection (Takato and Goldring 1979; Greenwood et al. 1983). This raises the possibility that the pathways for cell-to-cell dye movement may form in response to trauma caused by the slice procedure itself. Indeed, recent evidence suggests that traumatic injury to the nervous system can induce changes in local neuronal connectivity. For example, studies of identified neurons in the mollusc Heliosoma indicate that axonal damage can induce formation of new electrical synapses (Bulloch and Kater 1981) and strengthen existing ones (Bulloch et al. 1980). Murphy et al. (1983) found that this axotomy-induced increase in electrotonic coupling is associated with a parallel increase in dye-coupling. Our data suggest that a similar phenomenon may occur in neocortex. Thus, dye-coupling incidence in neocortical slices varies as a function of the plane in which the slices are sectioned, and is significantly lowered when dendritic damage is minimized.

In the present paper, we summarize the characteristics of dye-coupling in neocortical slices and the experimental manipulations which modify its incidence. The data are consistent with the hypothesis that the phenomenon reflects movement of LY through junctions which form as part of a specific neuronal response to dendritic injury. Because increased electrotonic coupling would enhance synchrony of neuronal activities, the findings may be relevant to the pathophysiology of trauma-induced, focal, neocortical epilepsy.

Experimental Brain Research, Suppl. 13

Methods

Intracellular LY injections were made in slices of guinea pig somatosensory cortex or visual cortex, as previously described (Gutnick and Lobel-Yaakov 1983). Slices were maintained at 36°C in a medium (pH 7.4) containing 124 mM NaCl, 5 mM KCl, 1.25 mM NaH_2PO_4, 2 mM $MgSO_4$, 26 mM $NaHCO_3$, 2 mM $CaCl_2$, and 10 mM glucose. The solution was bubbled with a gas mixture of 95% O_2 and 5% CO_2. For dye injection, glass micropipettes were filled with a 5% solution of the lithium salt of LY in water; resistances were 125-250 Mohms. After stable impalement of a neuron (resting potential > 60 mV), LY was injected iontophoretically for 5-10 min with 1-2 nA, 200 ms, hyperpolarizing pulses at 3 Hz. The slice was removed from the chamber 10-30 min following injection, and was then fixed in 4%, phosphate-buffered formalin, dehydrated in alcohols, cleared in xylene, whole mounted, and examined and photographed with a fluorescence microscope.

Slices were either cut in a plane normal to the pial surface (radial slices), or in a plane parallel to the pial surface (tangential slices) (Fig. 1). Radial slices (450-500 μm thick) were either parasagittal or coronal, and therefore included all of the cortical layers. Injections were only made in the uppermost of tangential slices, which included the pial surface and the superficial cortical layers. For both angles of section, neuron impalement was not attempted until at least 1.5 h after sectioning. No more than one intracellular injection was made in each slice.

Results and Discussion

Dye-coupling in Radial Slices

Characteristics of Dye-Coupled Aggregates. In visual cortex, dye-coupling resulted from 35 of 79 intracellular LY injections in superficial cortical layers (< 500 μm beneath the pia), and from 12 of 35 injections in deeper layers (Fig. 2). Dye-coupled aggregates consisted of two to five neurons, all of which were confined to the same relative depth beneath the pia. Thus, coupling between superficial and deep neurons was never observed. Most dye-coupled neurons were pyramidal in shape. However, coupling between stellate and pyramidal cells or bipolar and pyramidal cells was also revealed. Often,

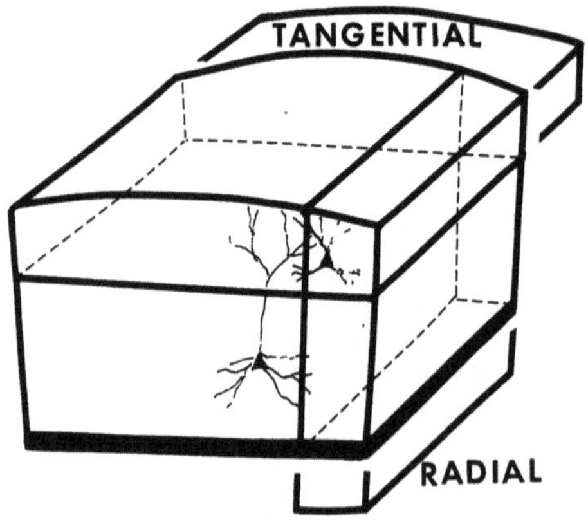

Fig. 1. Planes of section of neocortical slices. Radial slices were cut in the coronal or parasagittal plane. Tangential slices were cut in a plane parallel to the pia, and only the most superficial slice was used. Note that dendrites of superficial neurons were more likely to be damaged in radial slices than in tangential slices

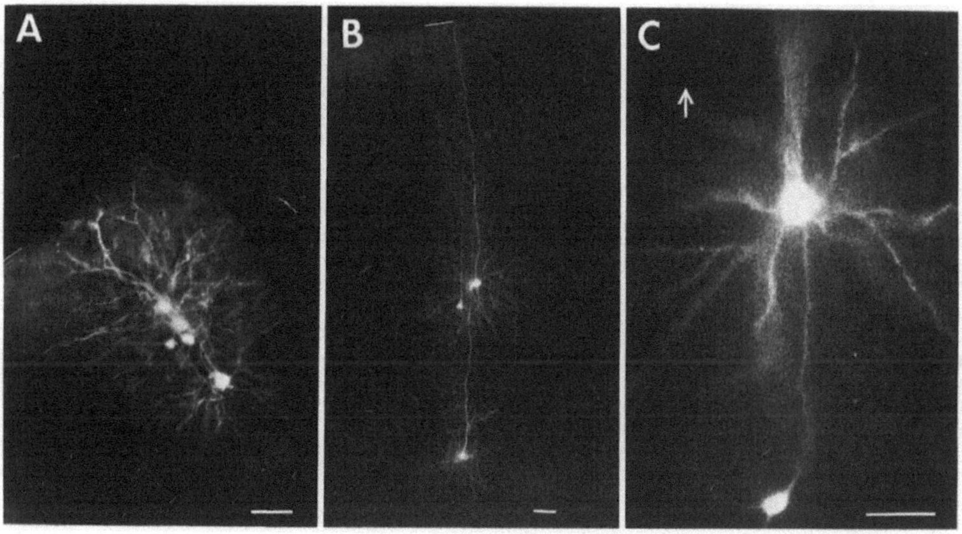

Fig. 2. Dye-coupling among neocortical neurons in radial slices following intracellular LY injections. (A) Injection of one layer 3 neuron caused staining of five superficial cells (Gutnick and Prince 1981). (B) Injection of one layer 5 neuron caused staining of three deep cells. Note the vertically oriented columnar organization of both aggregates. In both, dendrites extend to the pial surface (indicated by line). (C) Injection of one layer 3 cell caused staining of a pyramidal neuron and a deeper, multipolar neuron, which was located 350 µm beneath the pia. Direction to the pial surface indicated by an arrow (Gutnick and Lobel-Yaakov 1983). Scale bars, 50 µm

coupled neurons were arranged in a vertical row oriented perpendicular to the pia, suggestive of the columnar organization of the cortex (Fig. 2 A,B). In all dye-coupled aggregates there was considerable overlap of dendritic arborizations and of dendrites and somata, suggesting that the pathways for dye movement were dendro-dendritic and/or dendro-somatic. Although dye-coupling between neocortical glial cells has also been reported (Gutnick et al. 1981), neurons and glia were never coupled to each other. It is unlikely that dye-coupled aggregates consisted of more neurons than were actually stained, since there was no correlation between the number of cells stained per injection and the duration of the injection, the amount of iontophoretic current used, or the amount of time that passed between the end of the injection period and tissue fixation. However, the intensity of staining of the neurons coupled to the impaled cell did seem to be correlated with the duration of the injection period.

<u>Dye-coupling Is Not an Artifact of Impalement or Injection Procedures</u>. Several authors have criticized dye-coupling experiments, pointing out that multiple staining could be due to nonspecific uptake of extracellular dye into damaged neurons (Powell and Westerfield 1984) or penetration of more than one cell and consequent creation of artificial pathways for LY movement (Knowles et al. 1982; Alger et al. 1983). Although the latter problem is particularly critical in the hippocampal slice, where somata of pyramidal cells are closely packed into a single stratum, Andrew et al. (1982) convincingly demonstrated that such impalement artifacts could not entirely

account for the hippocampal dye-coupling they observed. In neocortex, neuronal cell bodies are distributed throughout the slice, and the problem of somatic packing density does not exist. Our results indicate that the dye-coupling phenomenon is not an artifact of cell damage due to impalement, dye uptake from the extracellular space, or other such factors. This contention is based on the following evidence: (1) In all dye-coupled aggregates, neuronal somata were clearly separated by a space of at least 10 μm (Fig. 2). Voltage at the tip of the LY-filled microelectrode was closely monitored throughout each experiment, and on the rare occasions when potential shifts suggested that the microelectrode might be passing directly from cell to cell, the data were discarded. It was quite common for dye-coupling to result from injection of the first neuron encountered in a slice. (2) Prolonged ejection of LY directly into the extracellular space did not result in staining of neuronal elements. (3) There was no correlation between dye-coupling and common electrophysiological determinants of cell health, such as membrane potential, membrane resistance, action potential amplitude, etc. For example, mean action potential amplitude at the conclusion of a dye injection period was 62.4 mV (SD = 15.3) for dye-coupled aggregates and 60.7 mV (SD = 8) for noncoupled neurons. (4) It is unlikely that dye-coupling was caused by damage to a tangle of dendrites and subsequent backfilling of several neurons; were this to happen, we would expect coupling between superficial and deep cells, because the apical dendrites of deeper neurons ramify extensively in superficial layers (Fig. 2B).

Evidence that Dye-Coupled Neuronal Aggregates are also Electrotonically Coupled. The most rigorous electrophysiological test for the presence of direct electrical connections between two neurons entails simultaneously impaling both cells, injecting current into one, and recording its electrotonic manifestation in the other. Although this technically difficult procedure has been successfully employed in the hippocampal slice (cf. Dudek et al. 1983), it has not revealed electrotonic coupling in neocortical slices (Gutnick and Lobel-Yaakov, unpublished observations). This may be because coupled aggregates contain so few cells that the probability of simultaneously impaling two related neurons is very low. It is important to note that although this test is definitive in most systems, results may be ambiguous in cortical slices, where dual intracellular recording entails blind impalement of neurons whose dendrites project for long distances. Unless there is other evidence to confirm that the microelectrodes have indeed penetrated two different neurons, it remains possible that observed electrotonic coupling is between two distant parts of the same neuron (Dudek et al. 1983).

In vertebrate nervous systems, another widely used test for electrotonic coupling between neurons entails demonstration of intracellularly recorded, subthreshold depolarizations (SDs) following antidromic stimulation (Korn and Faber 1979). In order to determine that the SD reflects antidromic invasion of an adjacent, electrotonically coupled neuron, it is necessary first to show that neither is it a chemical synaptic potential nor does it reflect blocked antidromic invasion from the axon of the impaled neuron itself. In brain slices, the test can be performed under conditions in which chemical synaptic transmission is blocked with calcium current inhibitors. The intracellular recordings in Fig. 3 were made with an LY-filled microelectrode in a slice of neocortex that was bathed in a solution containing 0.5 mM Ca and 3 mM Mn. Low intensity stimuli to the white matter evoked an early antidromic spike followed by a fixed-latency SD

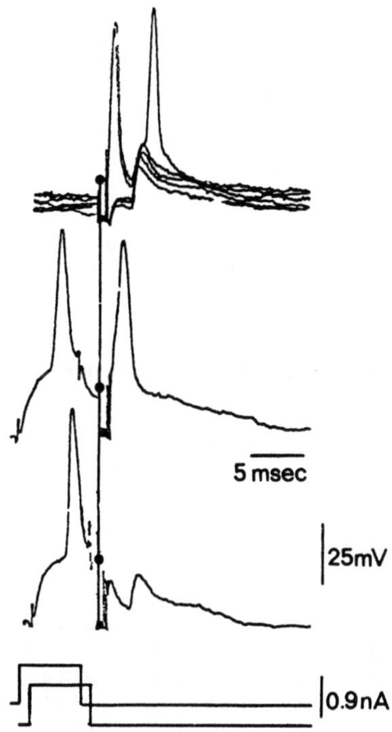

Fig. 3. Electrotonic coupling of dye-coupled neuronal aggregate. Intracellular recording from neuronal aggregate shown in Fig. 2C. Bathing medium contained 0.5 mM Ca and 3 mM Mn, which blocked chemical synapses. Upper frame: White matter stimulation (dots) evoked antidromic spikes followed by SDs which occasionally achieved threshold. Antidromic invasion was delayed (middle frame) or blocked (lower frame) when the white matter stimulus was preceded by orthodromic spikes which were evoked by appropriately timed intracellular depolarizing current pulses. Note that the SD did not collide with orthodromic spikes. Timing of intracellular pulses shown in bottom traces

which, on occasion, achieved threshold. Appropriately timed intracellular depolarizations through a bridge circuit evoked orthodromic spikes that collided with the antidromic action potential but not with the SD. It is thus apparent that the SD was an electrotonic potential generated by antidromic invasion of a coupled neuron. Subsequent LY injection (Fig. 2C) revealed that these recordings were from a dye-coupled aggregate which consisted of a layer 3 pyramidal neuron and a less intensely stained multipolar cell. Presumably, the pyramidal cell was the impaled neuron; the small size of the other neuron may account for the relatively long latency of the SD.

Uncoupling of Dye-Coupled Neuronal Aggregates

Because a decrease in intracellular pH is known to uncouple electrotonic synapses (Spray et al. 1981), it was interesting to determine whether procedures which bring about such a decrease also modify the incidence of neocortical dye-coupling. Results of experiments on snail neurons (Thomas 1977) indicate that increased extracellularly CO_2 lowers intraneuronal pH, and that the ability of a neuron to regulate its pH depends on extracellular HCO_2.

Gutnick and Lobel-Yaakov (1983) found that of 24 LY injections in neocortical slices exposed for 30 min to a medium containing Tris buffer (rather than carbonate) and exposed to high CO_2, only two resulted in dye-coupling. Under these conditions, the pH of the bathing fluid fell to between 5 and 6. Incidence of dye-coupling under these conditions (8.3%) was significantly lower than the control incidence (44.4%) (Chi-square=8.7, df=1, $p<0.005$). This uncoupling, which was not associated with an obvious change in the electrophysiological characteristics of the neurons, was apparently caused by a disruption of cellular pH regulation. Thus, in control slices that were not exposed to CO_2, but were incubated in Tris-buffered medium whose pH was lowered to 5.5 by adding impermient acid (HCl), dye-coupling incidence was 50%. In similar experiments, Connors et al. (1984) found that exposure of neocortical slices to elevated CO_2 completely eliminated glial dye-coupling, and significantly lowered neuronal dye-coupling. These results indicate that an experimental procedure which is known to decrease the permeability of gap junctions also decreases the incidence of dye-coupling.

Dye-coupling Incidence Varies with the Plane of Section During Slicing

In sharp contrast to results in radial slices, none of the 16 intraneuronal LY injections in tangential slices (Fig. 1) from as many animals resulted in dye-coupling. Each of the injected neurons was a superficial pyramidal cell (Fig. 4), which was also the most frequently encountered cell type in the dye-coupled aggregates of radial slices.

Two alternative explanations might account for the absence of dye-coupling in tangential slices: (1) Dye-coupling might be a normal feature of neuronal organization in cortex which is somehow disturbed by a tangential cut. This would occur, for example, if the tangential cut mechanically disrupted deep pathways for LY movement between superficial neurons. (2) Sectioning in the parasagittal or coronal planes might induce coupling between normally noncoupled superficial neurons. We distinguished between these possibilities by determining the incidence of dye-coupling in radial slices that had been further sectioned at the time of slice preparation so that the deeper layers were removed and only the superficial 500 µm of cortex remained. Under these conditions, dye-coupling was observed following 9 of 25 injections (36%), demonstrating that separation of deep from superficial layers did not cause uncoupling.

These data lead us to conclude that slicing the cortex in a plane normal to the pial surface somehow induced dye-coupling. This may be because dendrites of neurons in radial slices are more likely to have been damaged by the slicing procedure itself. In tangential slices, intact dendrites were observed to extend from the cell body for 150-200 µm in a plane parallel to the pial surface (Fig. 4).

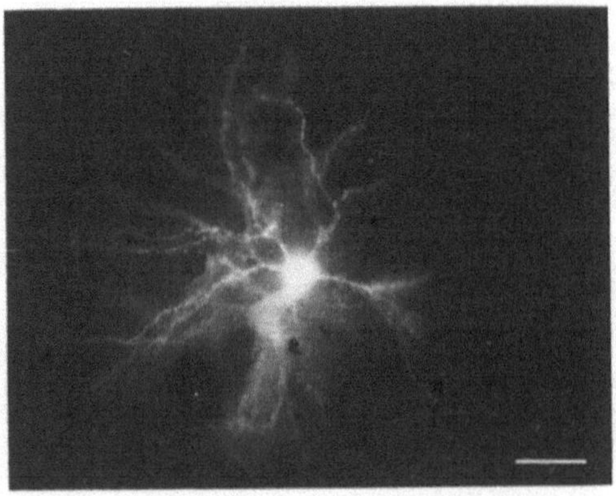

Fig. 4. Intracellularly injected, layer 3, pyramidal-shaped neuron in tangential slice, viewed from above. Scale bar, 50 µm

Thus, pyramidal cells occupied a cylindrical space with a diameter of 300-400 µm. Since the thickness of radial slices was 450-500 µm, it is almost certain that most injected neurons had sustained a significant degree of dendrotomy at the time of slicing. In contrast, neurons encountered in tangential slices had probably suffered far less dendritic damage, since we found that the basilar dendrites of most superficial pyramidal neurons in radial slices did not extend as deep as 500 µm beneath the pia.

Conclusions

We suggest that the partial dendrotomy which occurs during preparation of radial slices of neocortex triggers a specific neuronal process that results in formation of dye-coupling pathways. That this process does not simply entail generalized fusion of cut dendritic stumps is indicated by the finding that dye-coupling between superficial and deep neurons did not occur, even though the apical dendrites of deep cells, which were undoubtedly also damaged during slicing, overlapped considerably with those of superficial cells (Fig. 2B). It is important to note that all injected neurons appeared to be quite healthy, as determined by usual electrophysiological criteria, suggesting that they had successfully coped with the dendritic damage. Formation of junctions between neurons might be one manifestation of this coping process. This view is consistent with much recent evidence which underscores the plasticity of electrical synapses in a variety of preparations. It has recently been shown that an increase in intracellular cyclic AMP can rapidly lead to junction formation (Flagg-Newton et al. 1981; Radu et al. 1982), as can axotomy (Bulloch et al. 1980; Bulloch and Kater 1981; Murphy et al. 1983).

The functional significance of dye-coupling between neocortical neurons is still unclear. Several studies have implicated enhanced electrotonic coupling as a factor in the pathophysiology of certain clinical conditions which entail hypersynchronous neuronal activities (Traub and Pedley 1981; Rayport and Kandel 1981), and the possibility has been raised that traumatic brain injury might lead to formation of electrotonic junctions between neurons and thus cause increased susceptibility to epileptic seizures (Dudek et al. 1983). The present data are consistent with this notion.

Connors et al. (1983) demonstrated that dye-coupling incidence is high in neocortical slices from newborn rats and decreases significantly during the first postnatal weeks, suggesting that coupling may play a role in cortical development and then recede as the cortical circuitry matures. Our results raise the intriguing possibility that dendrotomy in mature cortex evokes the strengthening and/or reestablishment of direct connections between neurons that had previously been coupled during brain development. Since the neuronal migrations which underlie cortical development progress layer by layer along radially oriented glial processes (Rakic 1978), such a mechanism could explain our finding that superficial and deep cells were not dye-coupled to each other, and would also account for the apparent columnar organization of many dye-coupled aggregates.

Acknowledgments. This research was supported by a grant from the U.S.Israel Binational Science Foundation.

References

Alger BE, McCarren M, Fisher RS (1983) On the possibility of simultaneously recording from two cells with a single microelectrode in the hippocampal slice. Brain Res 270:137-141

Andrew RD, Taylor CP, Snow RW, Dudek FE (1982) Coupling in rat hippocampal slices: dye transfer between Cal pyramidal cells. Brain Res Bull 8:211-222

Bennett MVL, Goodenough DA (1978) Gap junctions, electrotonic coupling, and intercellular communication. Neurosci Res Program Bull 16:373-486

Bulloch AGM, Kater SB (1981) Selection of a novel connection by adult molluscan neurons. Science 212:79-81

Bulloch AGM, Kater SB, Murphy AD (1980) Connectivity changes in an isolated molluscan ganglion during in vivo culture. J Neurobiol 11:531-546

Connors BW, Benardo LS, Prince DA (1983) Coupling between neurons of developing rat neocortex. J Neurosci 3:773-782

Connors BW, Benardo LS, Prince DA (1984) Carbon dioxide sensitivity of dye-coupling among glia and neurons of the neocortex. J Neurosci 4:1324-1330

Dudek FE, Andrew RD, MacVicar BA, Snow RW, Taylor CP (1983) Recent evidence for and possible significance of gap junctions and electrotonic synapses in the mammalian brain. In: Jaspe HH, van Gelder NM (eds) Basic mechanisms of neuronal hyperexcitability. Liss, New York, p 32

Flagg-Newton JL, Dahl G, Loewenstein WR (1981) Cell junction and cyclic AMP. I. Upregulation of junctional membrane permeability and junctional membrane particles by administration of cyclic nucleotide or phosphodiesterase inhibitor. J Membr Biol 63:105-121

Greenwood RS, Godar SE, Winstead KK (1983) Focal penicillin seizures - motor activity and cellular physiology and morphology. Brain Res Bull 11:91-101

Gutnick MJ, Lobel-Yaakov (1983) Carbon dioxide uncouples dye-coupled neuronal aggregates in neocortical slices. Neurosci Lett 42:197-200

Gutnick MJ, Prince DA (1981) Dye coupling and possible electrotonic coupling in the guinea pig neocortical slice. Science 211:67-70

Gutnick MJ, Connors BW, Ransom BR (1981) Dye-coupling between glial cells in the guinea pig neocortical slice. Brain Res 213:486-492

Knowles WD, Funch PG, Schwartzkroin PA (1982) Electrotonic and dye coupling in the hippocampal slice. Neuroscience 7:1713-1722

Korn H, Faber DS (1979) Electrical interactions between vertebrate neurons: field effects and electrotonic coupling. In: Schmitt FO, Worden FG (eds) Neurosciences: fourth study program. MIT Press, Cambridge, p 333

Murphy AD, Hadley RD, Kater SB (1983) Axotomy-induced parallel increases in electrical and dye coupling between identified neurons of Heliosoma. J Neursosci 3:1422-1429

Powell SL, Westerfield M (1984) The absence of specific dye-coupling among frog neurons. Brain Res 294:9-14

Radu A, Dahl G, Loewenstein WR (1982) Hormonal regulation of cell junction permeability: upregulation by catecholamine and prostaglandin E1. J Membr Biol 70:239-251

Rakic P (1978) Neuronal migration and contact guidance in the primate telencephalon. Postgrad Med J 54:25-37

Rayport SG, Kandel ER (1981) Epileptogenic agents enhance transmission at an identified weak electrical synapse in aplysia. Science 213:462-464

Spray DC, Harris AL, Bennett MVL (1981) Gap junctional conductance is a simple and sensitive function of intracellular pH. Science 211:712-715

Takato M, Goldring S (1979) Intracellular marking with Lucifer Yellow CH and horseradish peroxidase of cells electrophysiologically characterized as glia in the cerebral cortex of the cat. J Comp Neurol 186:173-188

Thomas RC (1977) The role of bicarbonate, chloride and sodium ions in the regulation of intracellular pH in snail neurons. J Physiol 273:317-338

Traub RD, Pedley TA (1981) Virus-induced electrotonic coupling: hypothesis on the mechanism of periodic EEG discharges in Creutzfeldt-Jakob disease. Ann Neurol 10:405-410

Regeneration of Axons in the Mammalian Visual System

M. Berry[1], L. Rees[1], and J. Sievers[2]

1 Anatomy Department, Guy's Hospital, Medical School, London SE1 9RT, Great Britain
2 Anatomisches Institut, Universität Kiel, Olsshaussenstrasse 40, 2300 Kiel, FRG

Introduction

Mammalian axons regrow if severed in a peripheral nerve, but these same axons fail to penetrate the central nervous system (CNS) at the root/cord entry point (Lugaro 1906; Tower 1931; Ten Cate 1932; Paskind 1936; Westbrook and Tower 1940; Kimmel and Moyer 1947; Moyer et al. 1953; McCouch 1955; Gamble 1976; Stensaas et al. 1979), and do not regenerate if damaged in the CNS. Moreover, peripheral nerves will not grow into CNS parenchyma when embedded into the brain or spinal cord (Clark 1943; Benes 1968; Kiernan 1971; Knowles and Berry 1978), but central axons will grow into a peripheral nerve (Sugar and Gerard 1940; Brown and McCouch 1949; Feigin et al. 1951; Sager and Marcovic 1972; Kao 1974; Kao et al. 1970; Kao et al. 1977a,b; Chi et al. 1980). All these observations suggest that the regrowth of axons is inhibited in the environs of damaged CNS tissue. More recently, Aguayo and his co-workers have used horseradish peroxidase (HRP) and isotope tracing methods in more detailed nerve transplantation studies and confirmed that central axons enter a peripheral nerve segment implanted into the brain or spinal cord. Both those CNS axons projecting into the peripheral nervous system (PNS), and those which normally ramify exclusively intrinsically, are found in the peripheral nerve graft (Aguayo et al. 1979, 1982; Benfey and Aguayo 1982; Richardson et al. 1980). Such fibres grow in the graft for much longer distances than the length of the CNS tracts they originally constituted (Benfey and Aguayo 1982; David and Agauayo 1981; Richardson et al. 1980, 1981). However, interpretation of the findings of Aguayo and co-workers is limited because it is impossible to ascertain if the central axons found in a peripheral nerve graft are either regenerating fibres (i.e., outgrowth from the cut stumps of CNS axons), or collateral sprouts originating from uninjured axons coursing in the region of the PNS implant. We have set out to overcome this difficulty by anastomosing a segment of sciatic nerve to the cut end of the completely transected optic nerve in the rat. Using this model, all ganglion cell axons found in the sciatic nerve have regenerated into the graft, since all optic nerve axons are severed at the site of anastomosis. The model also offers a means of discovering the relative importance of tropic and trophic factors, produced by the peripheral nerve graft, which enhance both CNS axon outgrowth and neuronal survival.

Experimental Brain Research, Suppl. 13

Materials and Methods

Animals

WAG inbred albino rats (250 g) of either sex were used throughout. Comparisons were made of the regenerative response of ganglion cell axons after complete optic nerve section (ONS), and after a segment of sciatic nerve had been anastomosed to the cut end of the completely transected optic nerve (ONA rats). The retinae of 12 ONS and 18 ONA animals were whole mounted. In both groups, six rats were used for ganglion cell counting and six for axon counts. A further six ONA retinae were needed to estimate the number of HRP filled ganglion cells after injecting the distal part of the sciatic nerve. The controls for these latter animals were normal, intact WAG rats in which HRP was injected into the optic nerve, whilst controls for possible artifact due to diffusion of HRP were provided by six rats in which HRP was injected into the sciatic nerve transplant, 10 mm from the point of anastomosis with the optic nerve, immediately after operation and killed 36 h later. The normal, unoperated eye (always a right eye), mounted on the same slide as the treated eye of the same animal, acted as an additional control for shrinkage artefact and systemic uptake of HRP. Another group of 12 rats contained six with retinal lesions and six with implants of a sciatic nerve into a retinal lesion. All animals were killed 30 days post-lesioning (dpl).

Operations

The sheath of the optic nerve was exposed (under tribromoethanol anaesthesia) through the upper lid by evulsing the lid levator, superior oblique, superior rectus and the superior fibres of the retractor bulbi muscles. The dura and arachnoid membranes were slit longitudinally, usually on the lateral or superior aspect, avoiding subarachnoid branches of the ophthalmic artery and vein. The optic nerve was completely sectioned 2 mm from the bulb and a 2 mm proximal segment of optic nerve removed. The integrity of blood flow in the ophthalmic artery, which normally lies medially in the dural sheath and is not damaged, was confirmed by applanation ophthalmoscopy (Wise et al. 1971). In ONA animals, a segment of sciatic nerve 15 mm in

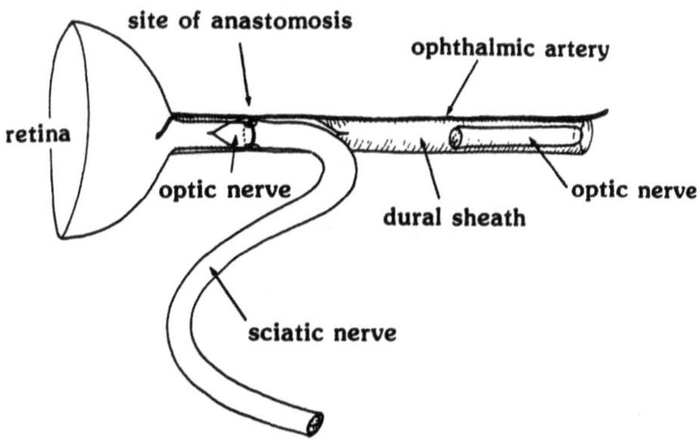

Fig.1. Diagram of the operative technique for anastomosing a segment of sciatic nerve to the cut retinal end of the optic nerve - see text for further details

length was removed from a donor WAG rat and the proximal end anastomosed to the retinal end of the sectioned optic nerve using two, 10/0 nylon sutures, after feeding the sciatic nerve into the most posterolateral part of the orbit through the temporalis muscle (Fig.1). The distal end of the sciatic nerve rested on the subcutaneous surface of this muscle. In the six rats in which the sciatic nerve was inserted into a retinal lesion, the epineurium of the sciatic nerve was sutured to the margins of a 1 mm^2, dorsal, scleral window, 1.0-1.5 mm from the optic nerve head, with two, 10/0 nylon sutures. Animals receiving retinal lesions only were prepared by the method of Goldberg and Frank (1979), passing a needle through the pars plana of the eye and scratching the vitreal retinal surface using applanation ophthalmoscopy.

Staining Procedures

The site of anastomosis in ONA rats was stained by the method of Nonidez (1939) and examined in serial 5-μm, wax-embedded sections. Retinal whole mounts were prepared by drying retinae, fixed in 10% neutral formalin for 3-4 h, to gelatin subbed slides. Ganglion cells were stained with cresyl violet whilst axons were silver impregnated using a physical developer technique (Kiernan 1981).

HRP-WGA (Triticium vulgaris) conjugate (Sigma, L2384) was injected into sciatic nerve transplants to assess regeneration. The sciatic nerve was crushed 10 mm from the site of anastomosis and 5 μl 4% HRP-WGA in sterile water was injected into the crush site. Animals were left for 36 h and perfused with buffered saline. Both retinae, from each animal, were dissected from the globe in cold saline, fixed for 2-3 min, by immersion in 1.0% paraformaldehyde and 1.25% glutaraldehyde in 0.1 M phosphate buffer (pH 7.4), whole mounted and air dried onto a subbed glass slide. The HRP was then developed using the TBM reaction as described by Mesulam (1982).

Quantitation

An estimate of the total number of axons in the ganglion cell fibre layer of the retina was obtained by counting all fibres about the optic nerve head under oil immersion. The frequencies of different sizes of ganglion cells were estimated using a square graticule with a total sampling area of 22500 μm^2, and an MOP videoplan (Kontron) imagine analysis system. The numbers of ganglion cells contained in the graticule area were sampled at standard points on the retina representing regions near the optic disc, at the periphery, and in mid-retinal regions in all four quadrants. The mean of 12 counts of the normal and treated retina in each animal were used to estimate percentage reductions and accumulations of retinal ganglion cells over all size ranges, at intervals of 6.6 μm^2. The total number of HRP-filled ganglion cells in the ONA group was estimated by counting all the labelled cells present in the retina by scanning the entire surface of whole mounts at x100 magnification.

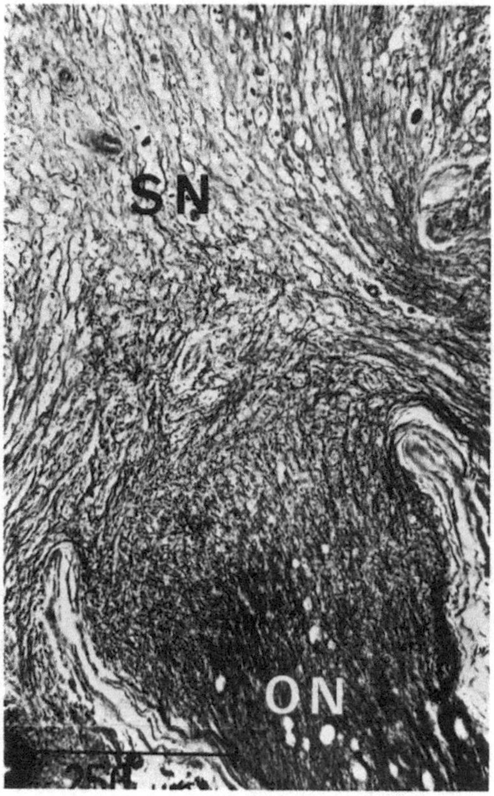

Fig. 2. Example of the site of anastomosis, 30 dpl - note the absence of scar tissue and the many axons which bridge the two nerves (ON - optic nerve; SN - sciatic nerve; Nonidez silver method; marker = 250 μm)

Results

ONS vs ONA Experiments

Site of Anastomosis. Silver preparations of the anastomotic site of all ONA animals showed that the optic nerve and sciatic nerve remained well approximated without the intervention of connective scar tissue. Axons were clearly seen passing between the two nerves (Fig.2).

Numbers of Axons. Figure 3 illustrates the ganglion cell fibre layer in representative retinae of normal, ONS and ONA rats, 30 dpl. The ganglion cell fibre layer appears equally reduced in density in both ONS and ONA animals. Fibre counts about the optic disc showed there to be a mean (±SE) of 2650+255 axons in ONS and 2362±412 in ONA retinae. However, it was quite clear that axon numbers were grossly underestimated because of fasciculation and thus such light microscopic counts show only that at least some 3000 fibres survive in both ONS and ONA retinae and that this method is too inaccurate to detect any differences between the two groups.

HRP Findings. In the ONA group of rats, HRP-filled cells were found in all retinae, 30 dpl. Both large and small cells were filled and distributed equally within the retinal whole mounts (Fig.4).

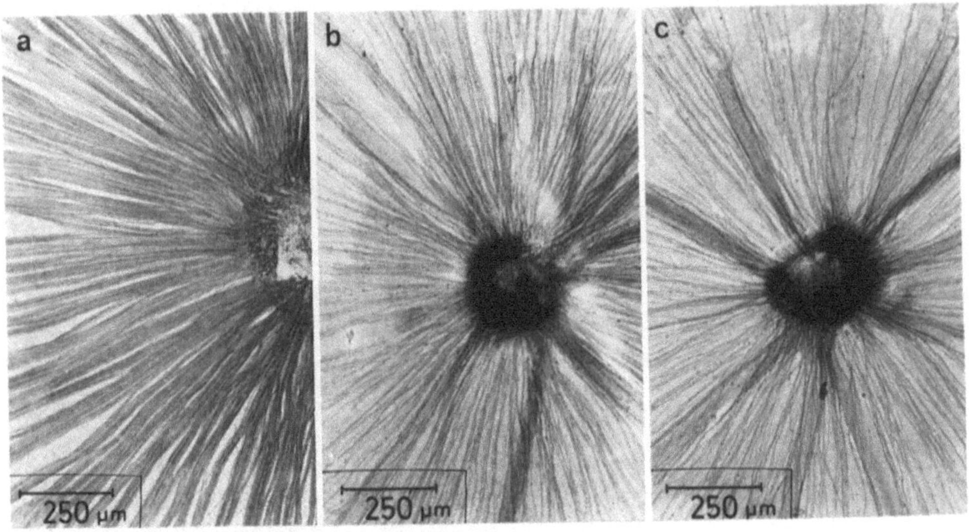

Fig.3. Representative examples of axons about the optic nerve head in (a) normal retina, (b) ONS retina, (c) ONA retina, 30 dpl. Note that the density of fibres in the ganglion cell fibre layer in ONS and ONA retinae appears equally reduced

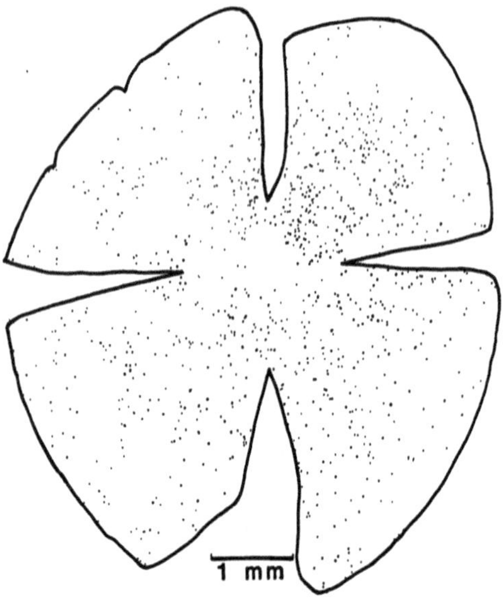

Fig.4. Distribution of HRP-labelled ganglion cells in the retina of rat 4 (see Table 1) after injection of HRP into a crushed region of the sciatic nerve 10 mm from the site of anastomosis at 30 dpl and killed 36 h later. Note that large through to small cells are all represented and equally distributed throughout the retina (camera lucida drawing of the retinal whole mount

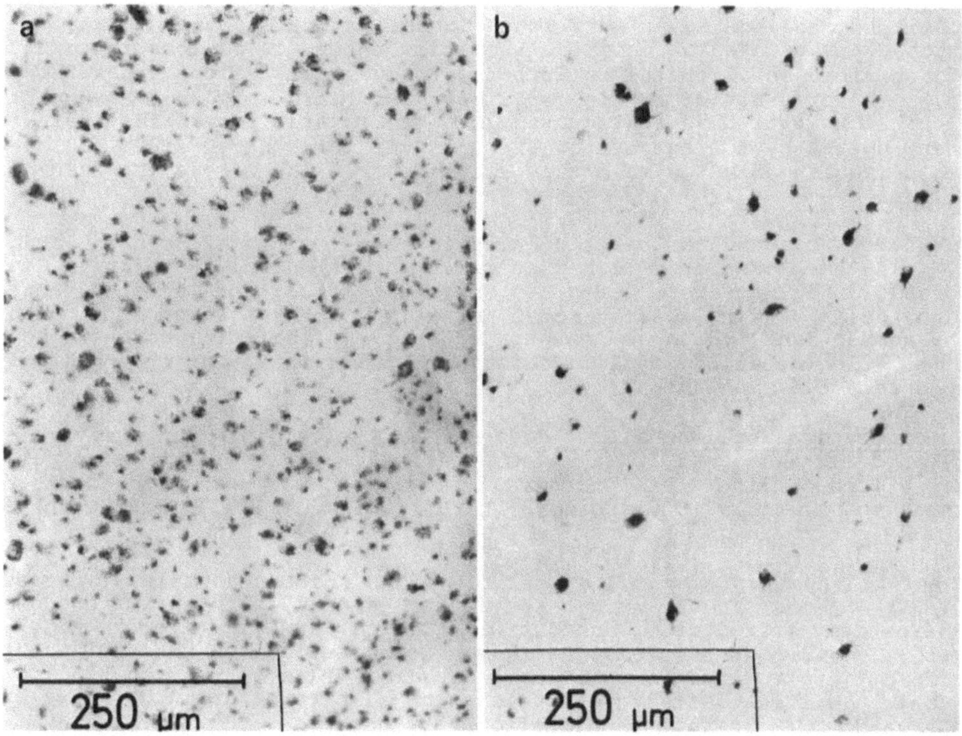

Fig. 5.(a) Density of HRP-labelled ganglion cells 36 h after injection of the optic nerve in an unoperated control rat; (b) density of HRP-labelled ganglion cells 36 h after injection of a sciatic nerve graft 10 mm from the point of anastomosis

Table 1. Total number of HRP-labelled ganglion cells counted in ONA retina

Rats	Total no. HRP-labelled ganglion cells
1	1611
2	1856
3	789
4	1558
5	513
6	915
Mean	1207
SE	220

No filled ganglion cells were found in the retina of control rats, in which HRP was injected into the sciatic nerve, 10 mm from the point of anastomosis with the optic nerve, immediately after operation. Thus, the presence of HRP in ganglion cells represents true transport along axons that have regenerated into the sciatic nerve graft rather than uptake by nonregenerated axons in the optic nerve stump following diffusion of HRP within the sciatic nerve and across the anastomosis.

The number of ganglion cells counted in each retina is shown in Table 1. The small number of labelled cells in ONA animals, relative to normal, is illustrated in Fig.5, where the density of labelled ganglion cells, after sciatic nerve injection of HRP, is compared with the number labelled in an unoperated rat, after optic nerve injection. No labelled ganglion cells were found in any of the control eyes from the same animal.

Frequency of Ganglion Cells. Figure 6 shows the distribution of ganglion cells in normal, ONS, and ONA retinae. Both experimental procedures reduce the total number of ganglion cells and also disturb their normal radial palisading. Large cells survive in ONA, but not in ONS retinae.

The total numbers of ganglion cells in ONS and ONA animals, together with the number in the normal eye of each animal, were estimated and the percentage reduction computed. The results are given in Fig.7. It is clear that anastomosing the optic nerve to the sciatic nerve affords protection for 10% more ganglion cells than with section alone. This 10% difference between the numbers of ganglion cells in ONS and ONA retinae is significant by unpaired 't' testing (P = 0.01-0.001).

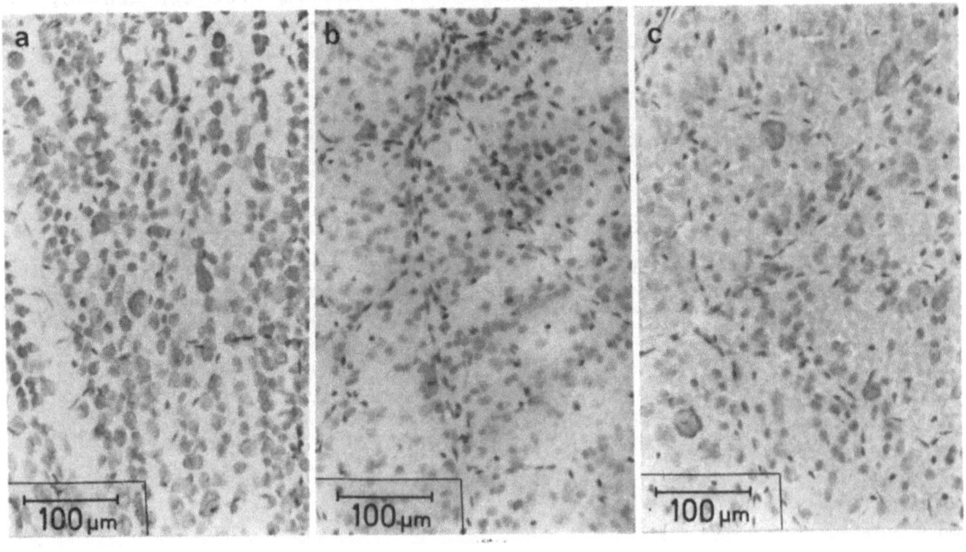

Fig. 6. Ganglion cells in the retinae of normal (a), ONS (b), and ONA (c) rats. Note the reduced numbers of ganglion cells, the absence of palisading in (b) and (c) and the differential survival of large cells in (c), relative to (b) (Nissl-stained retinal whole mount preparations)

The relative frequency distribution of normal ganglion cells of different sizes is unimodal. Most ganglion cells are smaller than 150 μm^2 in diameter but there is a long tail to the distribution extending up to 450 μm^2 (Fig.8). In both ONS and ONA retinae, there is a shift in the distribution of retinal ganglion cells in favour of those of smaller size (Fig.8). However, the distribution of the ONS group is characterised by the following parameters: mean = 61(SD$\pm$35), median = 55, mode = 56; and the ONA group by: mean = 93(SD$\pm$51), median = 80, mode = 70. Thus, peripheral nerve anastomosis probably protects against cell shrinkage so that a complete spectrum of cell sizes survives in the ONA group.

The change in size distribution of ganglion cells in the ONA and ONS groups was further studied by computing the percentage cumulative frequency of incrementing sizes of ganglion cells in all groups

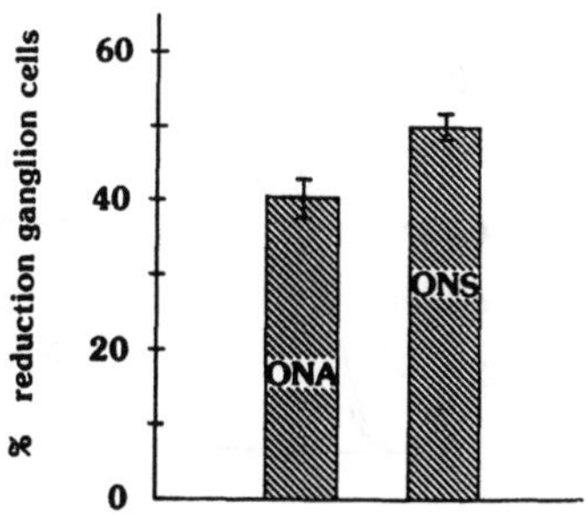

Fig.7. Percentage reduction in the number of ganglion cells in ONA and ONS retinae - see text for further details

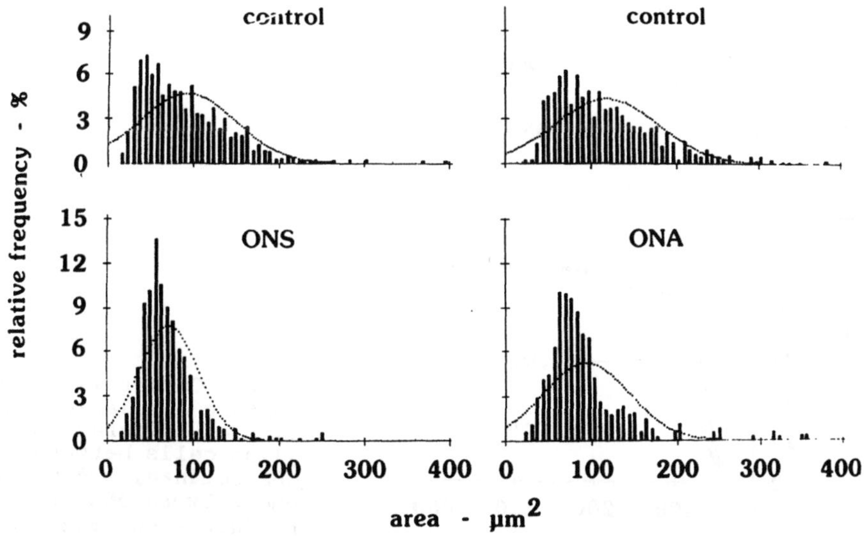

Fig. 8. Relative percentage distributions of different sized ganglion cells in the retinae of ONS and ONA treated rats and their controls. Note that ONS rats show a greater shift in the distribution toward smaller sized cells than the ONA group. In each case the Gaussian curve has been fitted to the distribution and confirms this observation - see text

(Fig.9). Anastomosing the sciatic nerve to the optic nerve prevents shrinkage of all sizes of ganglion cells equally. The percentage cumulative values in the ONA group are always less than normal because of a reduction in the number of all size classes of ganglion cells (see Fig. 5, 7, 8).

Retinal Lesions

Lesions Alone. At 30 dpl, retinal axons regenerate and grow for a long distance in the inner plexiform layer of the retina in random directions (Fig. 10a). HRP applied to a crushed segment of the optic nerve, 4 mm from the optic nerve head of such animals, demonstrates that very few regenerating fibres grow into the optic nerve (Fig. 10b).

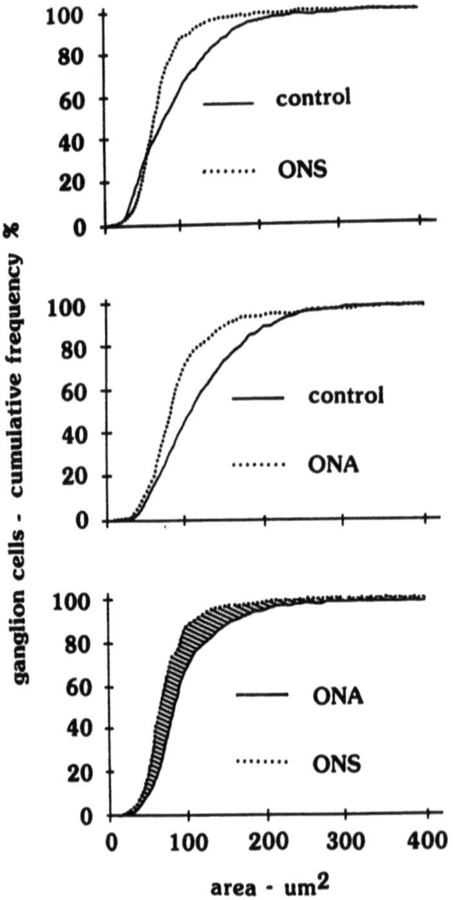

Fig.9. Comparison of the percentage cumulative distribution of different sized ganglion cells between normal, ONS retinae, and ONA retinae, and between ONS and ONA retinae - see text for further details

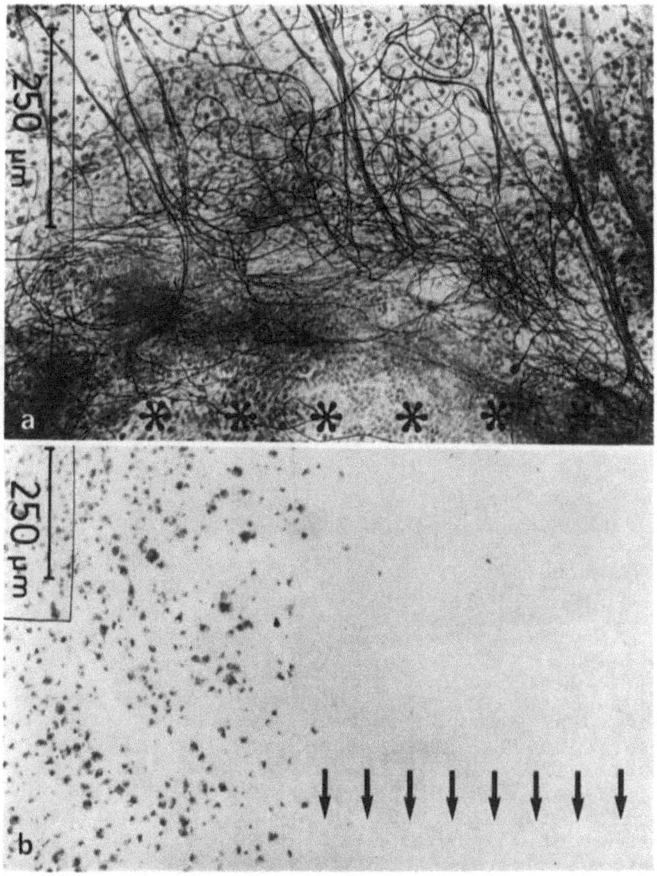

Fig. 10.(a) Randomly growing axons about a retinal lesion (*) 30 dpl; (b) HRP-filled retinal ganglion cells juxtaposed to a lesioned area (arrows) of the retina - note that the area of the retina distal to the lesion contains no filled cells

Implanting a segment of sciatic nerve into a retinal lesion has a dramatic effect on regenerating ganglion cell axons in that, although a few fibres continue to grow randomly, most grow directly into the sciatic nerve transplant (Fig.11). However, the number of ganglion cell axons surviving after retinal lesions does not appear to be greatly influenced by the presence of a sciatic nerve graft in the lesion.

Discussion

The numbers of fibres in the optic nerve are usually estimated to be less than the total number of ganglion cells in the retina (Hughes and Wassle 1976; Vaney and Hughes 1976; Tiao and Blakemore 1976; Hughes 1977), whilst retrograde HRP tracing labels only 50% of the neurons in the ganglion cell layer of normal rats (Cowey and Perry 1979). The discrepancy is probably explained by an inability to dis-

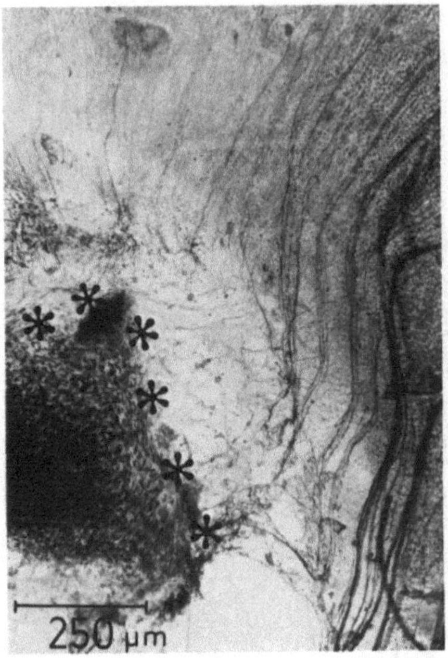

Fig. 11. Representative example of retinal axons invading a sciatic nerve graft (*) implanted into a retinal lesion - note that ganglion cell axons do not grow randomly about the retinal lesion, as in Fig. 10, but directly enter the graft, and that the number of ganglion cell axons surviving is similar to the number surviving in retinal lesions without sciatic nerve implant (Fig.10)

tinguish amacrine cells, displaced into the ganglion cell layer (Perry 1979, 1981), from ganglion cells per se (Stone 1965, 1978; Hughes 1975, 1977). Assuming that the most recent counts by Hughes (1977) of both retinal ganglion cells (190 000) and fibres in the optic nerve (120 000) of agoute rats are the most accurate; the total number of displaced amacrine cells in the rat retina is 70 000. Since we have recorded a reduction in the ganglion cell population of 50% and 40% in ONS and ONA rats respectively, the total number of 'ganglion cells' in each group would be 95 000 and 114 000 respectively.

If the amacrine cell population is unaffected by optic nerve section (since the axons of these cells do not leave the retina) (Allcutt et al. 1984a,b; Grafstein and Ingoglia 1982; Miller and Oberdorfer 1981; Misantone et al. 1984), the actual number of ganglion cells in ONS and ONA retinae is 25 000 and 44 000 respectively. Thus, 19 000 (16%) more cells remain viable in the ONA group, when compared to the number present in the ONS group. Of the 44 000 ganglion cells surviving in the ONA group, on average only 1207 (3%) of their axons regenerate the 10 mm along the sciatic nerve graft to the site of injection of HRP, by 30 dpl. It is difficult to explain this discrepancy between the trophic and tropic effects of the peripheral nerve graft. The operative techniques achieved good approximation of the sciatic nerve graft to the optic nerve stump without the intervention of scar tissue. Thus, access to the peripheral nerve by optic nerve fibres was unimpeded. Ide et al. (1983) have shown that the presence of viable Schwann cells is unnecessary for the growth of severed peripheral axons into a distal stump where the inner surface of the basal lamina of the Schwann cells alone appears sufficient to support and guide growth. Thus, the tropic influences offered to CNS axons by a peripheral nerve graft might be mediated by nondiffusable molecules bound to the inner surface of the basement membrane tubes of Schwann cells. The small number of optic nerve fibres regenerating into the peripheral nerve graft (3%) might indicate that access to the nerve and thus attachment to the basal lamina is a chance event and not dependent on a diffusable tropic factor. On the other hand, trophic influences are clearly operational in ONA animals increasing ganglion cell survival and maintaining ganglion cell size. Such diffusable factors might emanate from Schwann cells. Thus, at the site of anastomosis, in ONA rats, a complex environment probably obtains in which tropic and trophic factors, together with growth inhibitory factors (Berry 1982, 1983, 1984), are all present. The net result of these interactions is an increased viability of ganglion cells of some 16%, but only 3% of their fibres regenerate for any distance into the sciatic nerve graft.

The cell shrinkage observed in this study in both ONS and ONA groups appears to reflect a regressive metabolic change which follows axonal transection (Grafstein and Ingoglia 1982; Miller and Oberdorfer 1981; Misatone et al. 1984). The maintenance of ganglion cell size in the ONA group is probably related to an increase in protein synthetic activity initiated by trophic factors retrogradely transported by regenerating axons from the sciatic graft.

In the case of retinal lesions, it has been suggested that an absence of growth of inhibitory factors in the damaged retina may explain the continued growth of lesioned axons (McConnell and Berry 1982). The random growth of these fibres is presumably due to a lack of guidance. Thus, in contrast to the injury response of ganglion cell axons in the optic nerve, where regeneration does not occur, injury in the retina results in prolonged, undirected growth. Our HRP tracing study shows that, by 30 dpl, few of the randomly growing axons penetrate any distance into the optic nerve. The behaviour of the axons is dramatically changed by placing a peripheral nerve into the retinal lesion, when most severed ganglion cell axons regrow directly into the graft. Presumably, the initial growth towards the centre of the wound, observed by McConnell and Berry (1982), is sufficient to bring most axons under the tropic influence of the sciatic nerve, before they loop back way from the lesion. Although we have the impression that ganglion cell survival is probably not affected by the presence of the graft, it is difficult to come to definitive conclusions without quantitation or further HRP tracer studies.

In conclusion, optic nerve fibres regenerate into a segment of sciatic nerve for at least 10 mm by 30 dpl. Since all axons were severed in the optic nerve before anastomosis, the presence of optic nerve fibres in the sciatic graft is an unequivocal demonstration of regeneration of retinal ganglion cell axons. The numbers of fibres regenerating is probably a very small proportion of the total number of axons surviving, indicating that the tropic influence of the peripheral nerve is not mediated by diffusable molecules and/or is largely overridden by a growth inhibitory factor produced by the damaged optic nerve. By contrast, the trophic influence of the peripheral nerve graft is substantial, resulting in the survival of 16% more ganglion cells than in optic nerve section alone, and largely preventing shrinkage of the ganglion cells which normally follows optic nerve transection. Implantation of a peripheral nerve into the retina prevents the randomly directed growth of regenerating retinal fibres so that regrowing severed fibres directly enter the peripheral nerve graft.

Acknowledgments. We thank Kevin Fitzpatrick and Sarah Smith for the photography, Tony Finch and Mass N'Jie for the histology and Margaret Collins for preparing the manuscript. The work was supported by the M.R.C., U.K.

References

Aguayo AJ, Bray GM, Perkins CS, Duncan ID (1979) Axon-sheath cell interactions in peripheral and central nervous system transplants. Soc Neurosci Symp 4:361-383

Aguayo AJ, David S, Richardson P, Bray GM (1982) Axonal elongation in peripheral and central nervous sustem transplants. In: Fedoroff S, Hertz L (eds) Advances in cellular neurobiology. Academic, New York, pp 215-234

Allcutt D, Berry M, Sievers J (1984a) A quantitative comparison of the reaction of retinal ganglion cells to optic nerve crush in neonatal and adult mice. Brain Res 16:219-230

Allcutt D, Berry M, Sievers J (1984b) A qualitative comparison of the reaction of retinal ganglion cells to optic nerve crush in neonatal and adult mice. Brain Res (to be published)

Benes V (1968) Spinal cord injury. Bailliere, Tindall, Tindall and Cassell, London

Benfey M, Aguayo AJ (1982) Extensive elongation of axons from rat brain into peripheral nerve grafts. Nature 296:150-152

Berry M (1982) Post-injury myelin-breakdown products inhibit axonal growth: A hypothesis to explain the failure of axonal regeneration in the mammalian central nervous sustem. Bibl Anat 23:1-11

Berry M (1983) Regeneration of axons in the central nervous system. In: Navaratnam V, Harris RJ (eds) Progress in anatomy, vol 3, Cambridge University Press, Cambridge, pp 213-233

Berry M (1985) Regeeration and plasticity in the CNS. In: Swash E, Kennard C (eds) Scientific basis of clinical neurology. Chap 47, pp 658-679 Livingstone, Edinburgh

Brown JO, McCouch GP (1949) Abortive regeneration of the transected spinal cord. J Comp Neurol 87:131-137

Chi NN, Bignami A, Bich NT, Dahl D (1980) Autologous sciatic nerve grafts to the rat spinal cord: immunofluorescence studies with neurofilament and gliofilament (GFA) antisera. Exp Neurol 69: 568-580

Chizuka I, Knojiro T, Reiko Y, Tohru N, Satoru O (1983) Schwann cell basal lamina and nerve regeneration. Brain Res 288:61-75

Clark WE le Gos (1943) The problem of neuronal regeneration in the central nervous system: II. The insertion of peripheral nerve stumps into the brain. J Anat 77:251-259

Cowey A, Perry VH (1979) The projection of the temporal retina in rats, studied by retrograde transport of horseradish peroxidase. Exp Brain Res 35:457-464

David S, Aguayo AJ (1981) Axonal elongation into peripheral nervous system bridges after central nervous injury in adult rats. Science 214:931-933

Feigin I, Geller EH, Wolf A (1951) Absence of regeneration in the spinal cord of the young rat. J Neuropathol Exp Neurol 10:420-415

Gamble HJ (1976) Spinal and cranial nerve roots. In: Landon D (ed) The peripheral nerve. Chapman and Hall, London, p 330

Goldberg S, Frank B (1979) The guidance of optic axons in the developing and adult mouse retina. Anat Rec 193:763-774

Grafstein B, Ingoglia NA (1982) Intracranial transection of the optic nerve in adult mice: preliminary observations. Exp Neurol 76: 318-330

Hughes A (1975) A quantative analysis of the cat retinal ganglion cell topology. J Comp Neurol 163:107-128

Hughes A (1977) The pigmented rat optic nerve: fibre count and fibre diameter spectrum. J Comp Neurol 176:2631-268

Hughes A, Wassle H (1976) The cat optic nerve: fibre total count and diameter spectrum. J Comp Neurol 169:171-184

Ide C, Tohyama K, Yokoto R, Nitatori T, Onodera S (1983) Schwann cell basal lamina and nerve regeneration. Brain Res 288:61-75

Kao CC (1974) Comparison of healing process in transected spinal cords grafted with autogenous brain tissue sciatic nerve and nodose ganglion. Exp Neurol 44:424-439

Kao CC, Shimizer Y, Perkins LC, Freeman LW (1970) Experimental use of cultured cerebellar cortical tissue to inhibit the collagenous scar following spinal cord transection. J Neurosurg 33: 127-130

Kao CC, Chang LW, Bloodworth JMB (1977a) Elecdtron microscopic observations of the mechanisms of terminal club formation in transected spinal cord axons. J Neuropathol Exp Neurol 36:140-156

Kao CC, Chang LW, Bloodworth JMB (1977b) Axonal regeneration across transected mammalian spinal cord. An electron microscopic study of delayed microsurgical nerve grafting. Exp Neurol 54:591-615

Kiernan JA (1971) Pituicytes and the regenerative properties of neurosecretory and other axons in the rat. J Anat 109:97-114

Kiernan JA (1981) Histological and Histochemical Methods: Theory and Practice. Pergamon, Oxford, pp 272-273

Kimmel DL, Moyer EK (1947) Dorsal roots following anastomosis of the central stumps. J Comp Neurol 87:289-319

Knowles JF, Berry M (1978) Effects of deoxycorticosterone acetate on regeneration of axons in the mammalian central nervous system. Exp Neurol 62:1-15

Lugaro E (1906) Sulla presentarigenerazione autogena delle radici posteriori. Riv Patol Nerv Ment 11:337-340

McConnell P, Berry M (1982) Regeneration of ganglion cell axons in the adult mouse retina. Brain Res 241:362-365

McCouch GP (1955) Comments on regeneration of functional connections. In: Windle WF (ed) Regeneration in the central nervous system. Thomas, Springfield, pp 171-175

Mesulam M-M (1982) Principles of horseradish peroxidase neurohistochemistry and their applications for tracing neural pathways - axonal transport, enzyme histochemistry and light microscopical analysis. In: Mesulam M-M (ed) Tracing neural connections with horseradish peroxidase. Wiley, Chichester, pp 1-152

Miller NM, Oberdorfer M (1981) Neuronal and neuroglial responses following retinal lesions in the neonatal rats. J Comp Neurol 202: 493-504

Misantone LJ, Gershenbaum M, Murray M (1984) Viability of ganglion cells after optic nerve crush in rats. Brain Res (to be published)

Moyer EK, Kimmel DL, Winborn LW (1953) Regeneration of sensory nerve roots in young and in senile rats. J Comp Neurol 198:283-308

Nonidez JF (1939) Studies on the innervation of the heart. I Distribution of the cardiac nerves, with special reference to the identification of the sympathetic and parasympathetic postganglionics. Am J Anat 65:361-413

Paskind HA (1936) Regeneration of posterior root fibres in cat. Arch Neurol Psychiatry 36:1077-1084

Perry VH (1979) The ganglion cell layer of the retina of the rat: Golgi study. Proc R Soc 204:363

Perry VH (1981) Evidence for an amacrine cell system in the ganglion cell layer of the rat retina. Neuroscience 6:931-944

Richardson PM, McGuiness VM, Aguayo AJ (1980) Axons from CNS neurones regenerate into PNS grafts. Nature 284:264-265

Richardson PM, Aguayo AJ, McGuinness VM (1981) Role of sheath cells in axonal regeneration. In: Kao C, Bunge RP, Reier P (eds) Proceedings of the 1st annual symposium on spinal cord reconstruction. Raven New York, pp 293-304

Sager O, Marcovic G (1972) Regeneration in the central nervous system. Rev Roumanian Neurol 9:23-29

Stensaas LJ, Burgess PR, Horch KW (1979) Regenerating dorsal root axons are blocked by spinal cord astrocytes. Soc Neurosci Abst 5:684

Stone J (1965) A quantitative analysis of the distribution of ganglion cells in the cat's retina. J Comp Neurol 180:753-772

Stone J (1978) The number and distribution of ganglion cells in the cat's retina. J Comp Neurol 180:753-772

Sugar O, Gerard RW (1940) Spinal cord regeneration in the rat. J Neurophysiol 3:1-19

Ten Cate J (1932) Befunde nach der experimentellen Isolierung eines Rueckenmarksabschnittes. Arch neer physiol de l'homme et des animaux 17:149-238

Tiao YC, Blakemore C (1976) Regional specialisation in the golden hamster's retina. J Comp Neurol 168:439-458

Tower SS (1931) A search for trophic influecnes of posterior spinal roots on skeletal muscle, with a note on the nerve fibres found in the proximal stumps of the roots after excision of the root ganglion. Brain 54: 99-110

Vaney DI, Hughes A (1976) The rabbit optic nerve: fibre spectum, fibre count and comparison with a retinal ganglion cell count. J Comp Neurol 170:241-252

Westbrook WHL, Tower SS (1940) An analysis of the problem of emergent fibers in the posterior spinal root, dealing with the rate of growth of extraneous fibers into the roots after ganglionectomy. J Comp Neurol 72:383-397

Wise G, Dollery C, Henkin P (1971) The retinal circulation. Harper and Row, New York

Axonal and Glial Proteins Associated with Development and Response to Injury in the Rat and Goldfish Optic Nerve

J. A. Freeman, S. Bock, M. Deaton, B. McGuire, J. J. Norden, and G. J. Snipes
Vanderbilt University, School of Medicine, Department of Cell Biology, Nashville, TN 37232, USA

Introduction

An understanding of neural growth and regeneration at the molecular level requires a knowledge of the proteins that mediate these events, and of the gene regulation of these proteins. Modern recombinant DNA technology provides the necessary tools to yield some understanding of the gene regulation of nerve growth, so that the identification of different proteins associated with neurogenesis is of considerable interest. An increasing number of investigations of different regenerating neural systems, making use of electrophoretic techniques, have found that the synthesis of a small, specific set of proteins (termed "growth-associated proteins", or "GAPs" by Skene and Willard 1981a-d) is selectively induced following nerve injury. In the present study, we have examined rapidly transported proteins that are selectively expressed by retinal ganglion cells during the development of the rat visual system, and during regeneration of the goldfish optic nerve, using a newly developed computerized method to identify and quantitate proteins, and to correct for local protein synthesis by glial cells. We have found a small subset of GAPs that appear to be homologous in both species, and that are expressed during both developmental as well as regenerative growth. We have also examined the proteins released into their extracellular environment by glial cells during developmental growth of the rat optic and sciatic nerves, and during regenerative growth of the rat sciatic nerve and goldfish optic nerve following injury. We have found that glial cells release a specific subset of proteins (glial-released proteins, or "GRPs") whose expression is developmentally regulated, and which appear to be intimately associated with the functional status of the neurons with which they are in contact. Immunocytochemical studies using a specific antiserum raised against one of these GRPs support the hypothesis that regulation of the expression of GRPs and of axonally transported growth-associated proteins is functionally linked, and likely to play an important role in the response of a neuron to injury.

Methods

Fast-Transported Axonal Proteins

Fast axonally transported proteins synthesized by retinal ganglion cells were identified following intraocular injection of 35S-methionine (35S-met) into the vitreous chamber of the eye in normal goldfish and rats, and in goldfish 2-10 weeks following optic tract

Experimental Brain Research, Suppl. 13

crush. After 3 h in the rat or 4 h in the goldfish, the animals were killed and the optic nerve removed and processed for 2D polyacrylamide gel elctrophoresis (2D-PAGE) after the method of O'Farrell (1975). First dimension isoelectric focussing (IEF) gels contained 4% acrylamide, 9 M urea, 3% pH 3.5-10 ampholytes, and 3% pH 4-6 ampholytes (LKB) and were run at 350 V for 15 h for GRPs and at 500 V for 20 h for axonally transported proteins. Second dimension slab gels utilized the buffer system of Laemmli (1970) and were processed for fluorography as described by Jen and Thatch (1982). Nonequilibrium pH gradient gels (NEPHGE) containing 6% pH 3.5-10 ampholytes were run at 400 V for 3 h as described by O'Farrell et al. (1977). Most of the slowly transported proteins will not have reached the distal part of the optic nerves used for analysis, assuming a maximum rate of slow transport of 9.8 mm/day (Grafstein and Forman 1980). All of the axonally transported proteins to be described are transported to the target tectum within 3-4 h after labeling, and all of them sedimented with a particulate fraction following high speed centrifugation, suggesting that they are membrane-associated proteins (Deaton 1983).

Control experiments in goldfish showed that precursor injected into one eye resulted in little if any labeling of protein in the opposite eye (Benowitz et al. 1981). By contrast, in the adult and especially in the neonatal rat optic nerve, local synthesis of protein by glial cells labeled by diffusion or vascular leakage of precursor presented a significant problem. To correct for local synthesis, we employed a computerized 2D gel analysis system in which fluorograms of 2D gels from intraocularly labeled proteins (containing glial protein contaminants) were compared with fluorograms of locally labeled glial proteins (labeled in vitro as described below), and the glial proteins were then removed from the composite gel. This involved scanning each pair of gels, resolving individual polypeptides, computing the integrated volume density, and storing the resulting peak lists, then bringing the two gels into registration (using common proteins as markers), and computing matching peaks.

Glial-Released Proteins

To visualize glial-released proteins, sciatic and optic nerves from neonatal and adult rats and optic nerves from adult goldfish were excised from their neuron cell bodies at various stages of development or following injury, and incubated in media containing 35S-met (Skene and Shooter 1983; Deaton 1983). The proteins in the supernatant from these short-term cultures were collected, TCA-precipitated, and analyzed by 2D-PAGE. Control gels were run on representative labeled whole nerve segments to insure that the GRPs were released and did not appear in the supernate as a result of cellular lysis (Deaton 1983). These same samples were also representative of locally synthesized glial proteins. Purification of the 37K protein released by sheath cells of the crushed rat sciatic nerve was achieved using preparative 2D-PAGE: the Coomassie-stained spot corresponding to the 37K protein released 2 weeks postcrush was excised, homogenized in incomplete Freund's adjuvant, and injected into rabbits at days 0, 14, 21, and 28, after which they were bled for the 37K antibody. In order to visualize the localization of the 37K protein immunocytochemically, 10 µm frozen sections were cut from brains of adult rats which had received unilateral lesions of either the optic nerve or motor cortex. The sections were fixed in chloroform/methanol, reacted in hydrogen peroxide/methanol to block endogenous peroxidase activity, postfixed in paraformaldehyde and the 37K antigen localized using a standard peroxidase-antiperoxidase technique, with a primary antiserum dilution of 1:500.

Results

Fast-Transported Proteins in the Goldfish Optic Nerve

Following optic tract crush, the great majority of rapidly transported proteins underwent enhanced synthesis, as previously described (Grafstein and Forman 1980; Giulian et al. 1980; McQuarrie and Grafstein 1982; Heacock and Agranoff 1982). Three of these, however, were consistently found to have a much greater increase in synthesis than could be accounted for by the overall increase. Figure 1 shows a comparison of fast axonally transported proteins in the normal goldfish optic nerve (Fig. 1A) and 2 weeks following optic tract crush (Fig. 1B). Three rapidly transported proteins with molecular weights of 21K, 24K, and 46K (pIs of 6.9-7.3, >7.3, and <5.5 respectively) specifically undergo increased synthesis in the regenerating nerve. The rate of synthesis for all of these proteins returned to normal within 10 weeks following optic tract crush. These proteins had a minimum transport velocity of 50 mm/day in the 2 week regenerating animals, as calculated from the time and distance for the transported proteins to reach the optic tectum. These three proteins appear in the tectum within 4 h postinjection: the degree of labeling in the tectum relative to the nerve suggests that they do not accumulate in the nerve, and that they are not quickly degraded in the axon terminals. Removal of the target tectum also induced the synthesis of these three proteins, equivalent to optic tract crush. Evidently the signal for induction does not originate exclusively in the target tissue.

Fast-Transported Proteins in the Rat Optic Nerve

A developmental study was undertaken to identify fast-transported proteins associated with neurogenesis in the optic nerves of developing rats ranging in age from 4 days prenatal to 3 months postnatal. Intraocular injection of 35S-met resulted in a significant artifact due to the amount of label that subsequently became incorporated into protein throughout the brain due to vascular leakage, as well as local synthesis of proteins in the optic nerve due to diffusion of the label and incorporation by glial cells. Figure 2 shows a comparison of the pattern of transported vs locally synthesized proteins following in vitro labeling. Figure 3 shows a computer-analyzed plot of these fluorograms used to correct for local protein synthesis. The expression of one class of proteins which includes a 21K and a 22.5K protein remains constant during development from 2 days prenatal to adulthood. The rate of synthesis is increased 50- to 100-fold as postnatal development proceeds for a second class of proteins which includes a group of basic proteins with MWs of 21.5-23K and a group of acidic 29K proteins, as shown in Fig. 4. A 43K protein is selectively expressed both during development and also following optic nerve crush in the adult rat as shown in Fig. 5. The 43K protein is maximally expressed at postnatal days 8-10 and is barely detectable in the normal adult rat. The expression of this protein correlates well with the time of synaptogenesis in the developing rat visual system, which occurs largely during the first postnatal month.

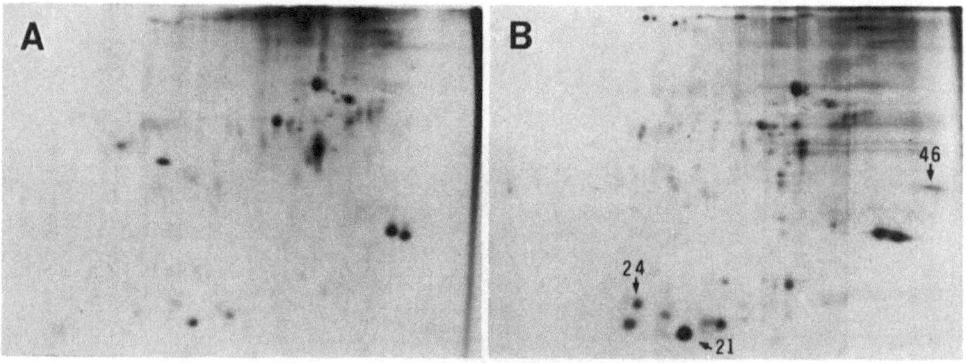

Fig. 1. Fast axonally transported proteins in the goldfish optic nerve of (A) normal fish and (B) 2 weeks following optic tract crush. For this and all subsequent gels, proteins of interest are identified by their molecular weights in kilodaltons. Unless otherwise stated, the 2D-PAGE gels, visualized by fluorography, utilized NEPHGE in the first dimension; proteins migrating to the basic end of the gel are to the left. Two weeks following optic tract crush in goldfish, three proteins (21K, 24K, and 46K) undergo increased synthesis in the optic nerve

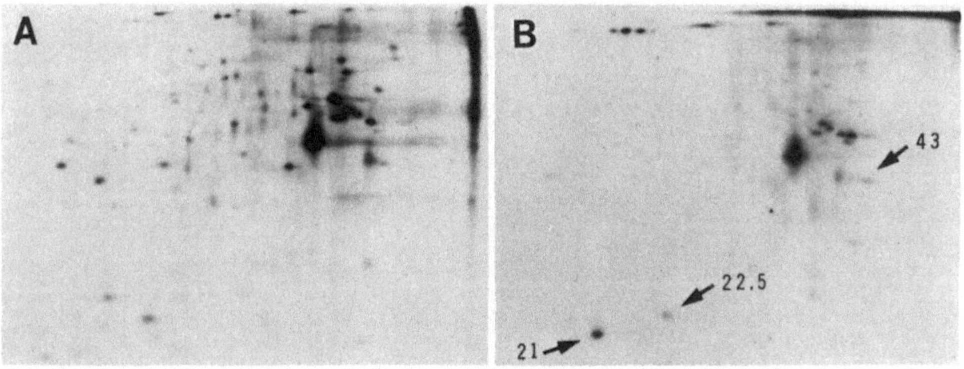

Fig. 2. A comparison of (A) locally synthesized proteins following in vitro labeling and (B) fast axonally transported proteins in rat optic nerve at 19 days of gestation. Arrows indicate three fast axonally transported proteins which are not present following in vitro labeling of the nerve

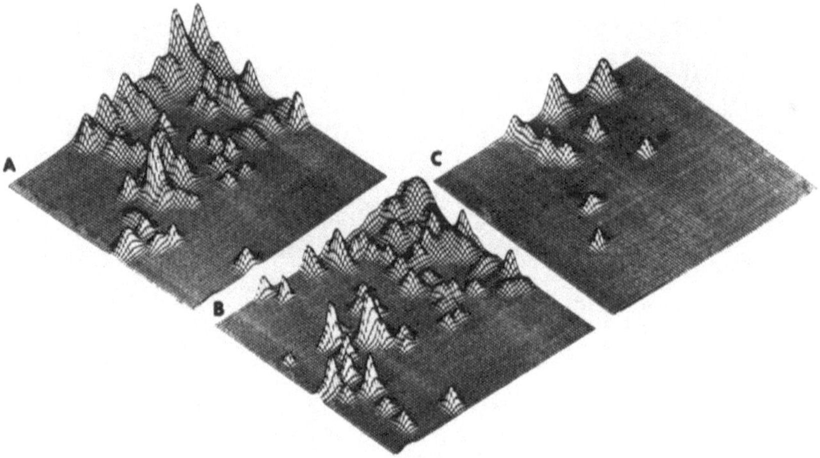

Fig. 3. Computer analysis of proteins expressed in developing rat optic nerve. (A) Proteins labeled following intraocular injection of 35S-met, containing fast axonally transported proteins plus locally synthesized glial proteins. (B) Glial-released proteins following in vitro labeling. (C) Axonally transported proteins on gel A with glial-released proteins removed

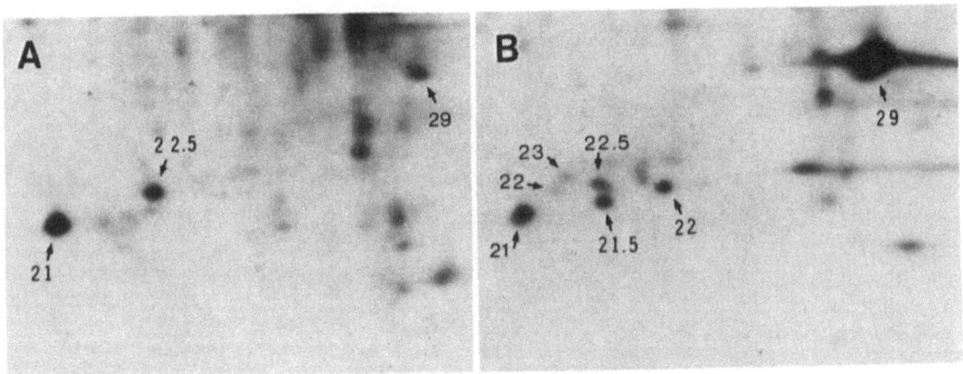

Fig. 4. Fast axonally transported proteins in optic nerve of (A) 2 day prenatal and (B) adult rats. A class of basic proteins with MWs ranging from 21.5-23K and 29K acidic proteins undergo a selective increase in their rate of synthesis during development

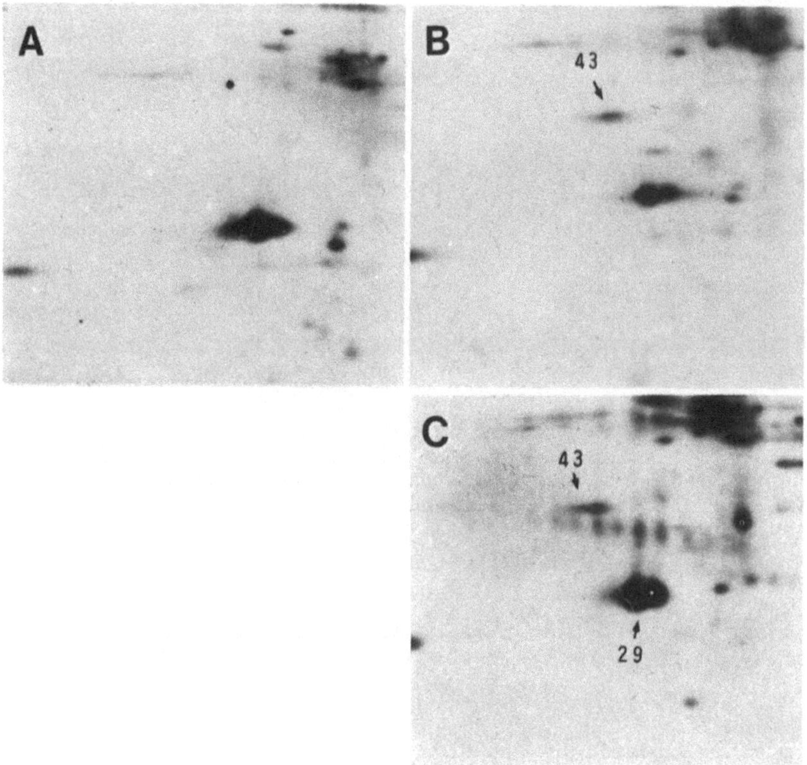

Fig. 5. Expression of a fast axonally transported 43K protein in the rat optic nerve in (A) a normal adult; (B) an 8-day-old neonate; and (C) an adult 12 days following optic nerve crush. The 43K protein is selectively expressed during development and following optic nerve crush but is barely detectable in the normal adult. In these fluorographs of 2D gels, isoelectric focussing was used as the first dimension; the acid end of the gel is towards the left

Proteins Released by Glial Cells of the Goldfish Optic Nerve

Three acidic proteins (MW 16K, 30K, and 42K) are specifically released by optic nerve sheath cells (glial and endothelial cells) after nerve injury in vivo, as shown in Fig. 6. These proteins are released distal to the site of injury and are detected as early as 3 h following nerve lesion. The peak of synthesis occurs at 3 weeks postcrush, and returns to an extremely low level by 6 weeks.

Evidence that the 16K and 42K GRPs are endocytosed and retrogradely transported to the retina was obtained by injecting 35S-met into the optic tract of fish which had received an optic nerve crush 10-14 days previously, and whose nerves were regenerating. Proteins having the same electrophoretic mobilities as the 16K and the 42K proteins appear in the retina within 16 h following injection. We have obtained preliminary evidence (Deaton 1983) that the 30K GRP binds to a receptor localized on the membranes of chronically denervated optic nerves, which contain predominantly glial cells. These obser-

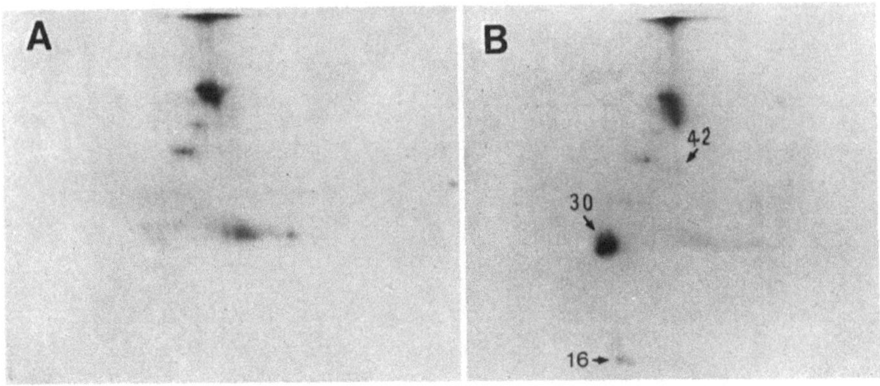

Fig. 6. Proteins released by optic nerve glial cells of goldfish (A) in normal fish and (B) in the distal part of the nerve 10 days following optic nerve crush. Three acidic proteins (MWs 16K, 30K, and 42K) are specifically released after nerve injury. Isoelectric focussing was used as the first dimension

vations suggest that these glial-released proteins participate directly in regenerative mechanisms.

Proteins Released by Rat Sciatic Nerves

We examined alterations in newly synthesized proteins secreted by glial cells of rat sciatic nerves during development. A 35K and a 37K protein are preferentially released during development relative to adult nerve sheath cells, as shown in Fig. 7. In addition, a 51K and a 54K protein, which are undetectable in early postnatal development, rise to relatively high levels by adulthood. The 37K, 51K and 54K proteins appear to be identical to those found by Skene and Shooter (1983) to be regulated following injury to the adult sciatic nerve.

Proteins Released by Glial Cells of Rat Optic Nerves during Development

Proteins released by rat optic nerves during the first 3 weeks of postnatal development were also examined by 2D-PAGE. The expression of a 37K and a 26K protein, initially high, was found to decrease with increasing age. However, even in adult animals, both of these proteins were still present in significant amounts, as shown in Fig. 7. The release of a 35K protein increased during the first postnatal week but was undetectable at 3 weeks postpartum. The 37K protein produced by optic nerve glial cells appears to be homologous to the 37K protein produced by sciatic nerve sheath cells during development and during regeneration and is itself induced following optic nerve injury as reported by Skene and Shooter (1983) and shown in Fig. 7. In general, however, proteins released by glial cells in the developing optic nerve were similar but not identical to those released by the developing sciatic nerve sheath cells.

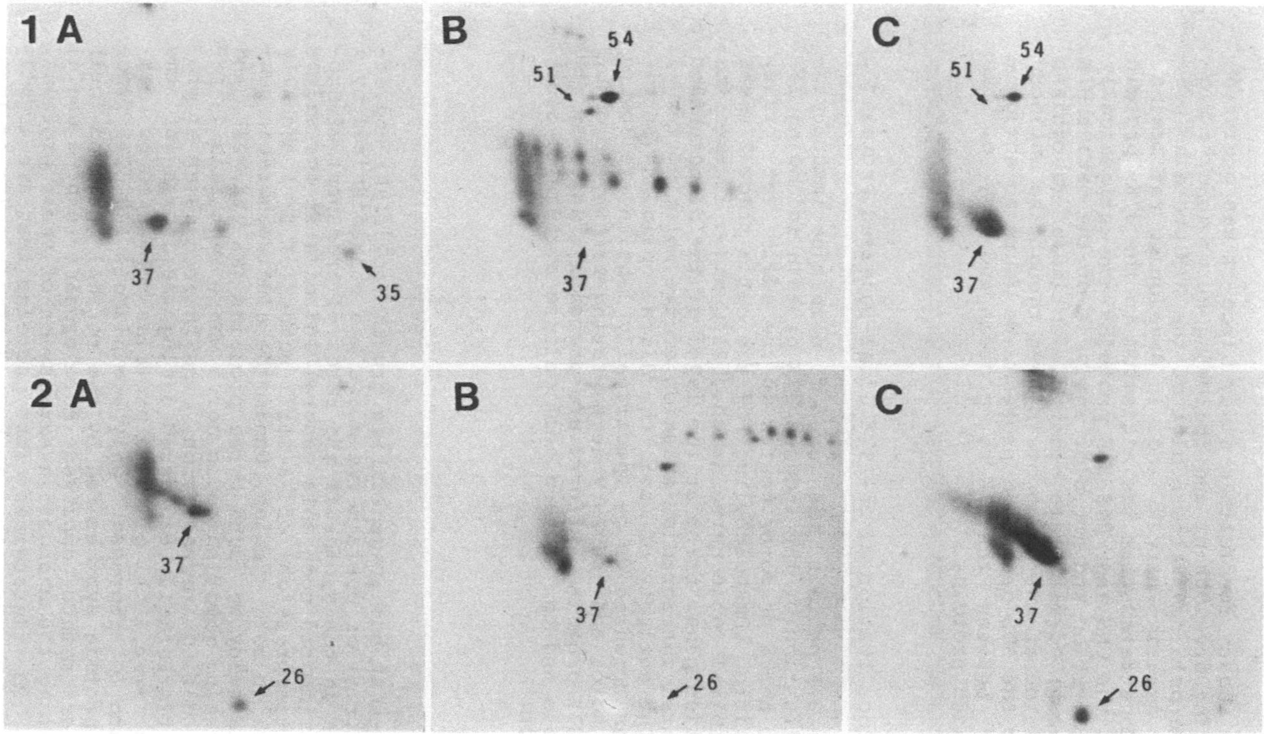

Fig. 7. Newly synthesized acidic proteins secreted by rat sciatic nerve sheath cells (1A-C) and rat optic nerve glial cells (2A-C) in (A) 0.5-day neonates; (B) normal adult; and (C) adult 18 days following nerve crush. In the sciatic nerve, two proteins (MWs 35K and 37K are preferentially released during development relative to release in the adult. Two proteins (51K and 54K), undetectable in early postnatal development, rise to high levels in the adult. In the optic nerve, the expression of two proteins (MWs 37K and 26K) decreases during development, although significant amounts of these proteins are present in the adult. Isoelectric focussing was used as the first dimension

Production of an Antiserum to the 37K Glial-Released Protein

The 37K protein was chosen for further study since it was preferentially expressed during development and following nerve injury. Figure 8 shows an immunostained Western blot of the 37K protein present in regenerating sciatic nerves and in developing rat brains. The Western blot demonstrates that the antiserum to the 37K protein is specific and that the CNS and PNS 37K proteins are antigenically related. Immunoprecipitation and dot blot analysis of the native 37K protein (data not shown) demonstrate that the antiserum recognizes the native conformation of the 37K protein even though the antibody was prepared against the SDS-denatured protein. We found no antigenic cross-reactivity between the 37K rat protein and any of the goldfish glial-released proteins.

Immunocytochemical Localization of the 37K Glial-Released Protein

The antiserum to the 37K protein was used to determine immunocytochemically the localization of the induced 37K protein following lesions of the optic nerve or motor cortex in the adult rat brain. Figure 9 shows specific immunolabeling of the optic tract 2 weeks following an intraorbital optic nerve crush. Also shown is specific labeling of the corticospinal tract at the level of the diencephalon following a unilateral intracranial ablation of the motor cortex 2 weeks previously. The entirety of both the distal optic and corticospinal projections appeared to be specifically labeled following injury. Thus, both motor and sensory degenerating pathways produce injury-associated 37K protein. Efforts are presently underway to purify the 37K protein and to develop cell culture systems to assay its cellular function.

Discussion

This study confirms that both regenerative growth (goldfish) and developmental growth (rat) of optic nerve axons are characterized by a generalized increase in the synthesis and rates of transport of axonal proteins. Superimposed upon this general increase it the selective increase in the expression of a small number of fast axonally transported proteins in both species. Under the electrophoretic conditions used in this study we found three proteins (MW 21K, 24K, and 46K) in the regenerating goldfish optic nerve whose synthesis is selectively enhanced during regeneration, and three classes of proteins that are differentially expressed during the growth of the neonatal rat optic nerve. All of these proteins appear to be membrane associated, and to represent only a miniscule (<0.01%) fraction of the total protein synthesized by the retinal ganglion cell bodies. The computer method employed here appears to be a useful method for quantitating the changes in expression of these proteins, and for correcting 2D fluorograms for proteins that are locally synthesized as opposed to axonally transported. The problem of local synthesis by glial and other cells is unavoidable due to diffusion of label when dealing with very short axons {e.g., Xenopus optic nerves (Szaro et al. 1984)ð or due to leakage into the vascular

Fig. 9. Specific immunocytochemical localization of 37K protein in ► (A) optic tract or (B) corticospinal tract 2 weeks following injury to the optic nerve or motor cortex, respectively

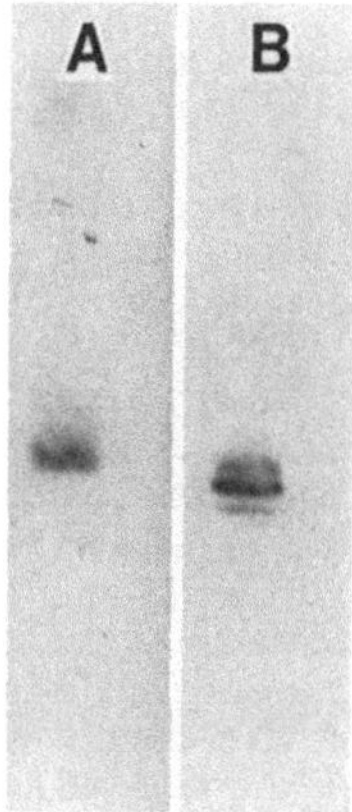

Fig. 8. Western blots of 1D gels of the soluble proteins washed out from minced rat (A) sciatic nerve (adult) 16 days postcrush and (B) brain (5 days postpartum). A 37K protein is present in both regenerating sciatic nerve and developing brain. These blots indicate that the antiserum to the 37K protein is specific and that the sciatic and optic nerve proteins are antigenically related

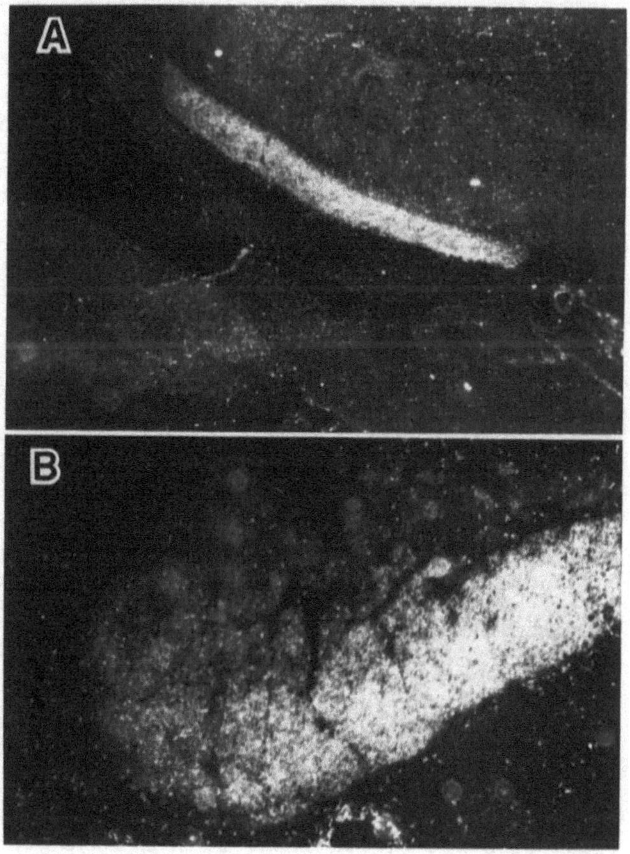

system due to the generally high membrane permeability when dealing with many developing or immature systems.

Previous studies in goldfish using slightyly different 2D-PAGE conditions have revealed similar selective enhancement of the synthesis of a subset of fast axonally transported proteins. Heacock and Agranoff (1982) found an increased synthesis of a 23K, 68K, 75K, and a 45K protein following optic nerve crush, and Benowitz and Lewis (1983) found an increased synthesis of a 24K, 44-49K complex, and a 210K group of proteins up to 30 days postcrush. Benowitz et al. (1983) also found that the expression of several high (>140K) molecular weight proteins, not examined using our electrophoretic conditions, depended on the presence or absence of the target tectum, although the expression of the 44K-49K complex did not. It appears on the basis of similarities in MW, pI, and association with growth, that the 46K protein in goldfish optic nerve and the 43K protein in the rat optic nerve described here, as well as the 43K-49K proteins described in the optic nerves of goldfish (Heacock and Agranoff 1982; Benowitz and Lewis 1983), toad (Skene and Willard 1981a), frog (Szaro et al. 1984), and in the developing optic and crushed hypoglossal nerves of rabbit (Skene and Willard 1981b) are all homologous, and represent an evolutionarily conserved protein that is characteristically expressed during active nerve growth.

It is of considerable interest whether the ability to induce the synthesis of a particular subset of proteins is a necessary prerequisite for a neuron to enter an active growth state, as proposed by Skene and Willard (1981a-d). Two corollaries to this hypothesis are that such proteins should be expressed by growing neurons but not by non-growing ones. The presence of the 24K and 21K proteins in normal goldfish optic nerves appears to contradict the hypothesis, until one considers that the goldfish continues to add new ganglion cells to the retina throughout adult life, such that approximately 1% of the optic nerve fibers are in a state of growth at any given time (Easter et al. 1981). Computer analysis reveals that the 21K and 24K proteins are present in normal nerves at 2-6% of their levels in regenerating nerves, consistent with their being associated with nerve growth. Similarly, the transient elevation we observed in the 43K protein following injury to the adult rat optic nerve might also be related to the recent evidence that following such injury the nerve is capable of at least partial regeneration (Foerster et al. 1983).

Their singular kinetics and association with regeneration and development suggest that a small subset of rapidly transported proteins might play a special role in axonal growth. Delineation of the functional role of these proteins will be greatly aided by development of specific immunological probes, which is presently in progress. The 43K protein appears to be an integral membrane protein (Skene and Willard 1981c) and preliminary evidence suggests that it is enriched in growth cones. One could speculate that this protein might be involved in adhesive interactions between the growth cone and its substrate (and thus might play a role in growth guidance and/or contact recognition); or that it might be associated with the regulation of growth-associated ionic currents in the growth cone (Freeman and Manis 1984). Alternatively, it is possible that these growth-associated proteins are transported as part of a supramolecular subcellular structure, rather than as individual proteins (Tytell et al. 1981).

Our results are in agreement with previous studies which show that glial or nerve sheath cells of both the peripheral and central nervous systems release soluble proteins into their extracellular environment following neuronal injury (Skene and Shooter 1983; Politis et al. 1983). It is not known whether this represents specifically an increase in the rate of synthesis, the rate of release, or the number of cells releasing these proteins, or a combination of effects. The onset of enhanced synthesis of GRPs following nerve injury is more rapid in the goldfish optic nerve than in the rat optic or sciatic nerve. The fact that we found no antigenic cross-reactivity with any of the three goldfish GRPs described here with the 37K protein in rat sciatic nerve may mean that these proteins are functionally and structurally unrelated. Evidently the 37K protein described by Skene and Shooter (1983) in rat sciatic nerve and the 37K and 40K proteins described by Politis et al. (1983) and by Carey and Bunge (1981) are identical. Little is known of the mechanism of induction of these glial-released proteins. Systemic factors are unlikely since glial cells from normal nerves contralateral to crushed goldfish optic nerves do not release them (Deaton 1983): It is more likely that their appearance is triggered by a direct interaction between neurons and glia. This might imply that normal nerves release a locally acting suppressor substance, or that some product of degeneration serves as a trigger for the induction of these proteins. The 16K and 42K GRP in goldfish optic nerve appear to be endocytosed and retrogradely transported to the retinal ganglion cell bodies, whereas the 30K protein appears to bind to a specific receptor on glial cell membranes. It is possible that the 30K protein triggers the synthesis or release of the 16K and 40K proteins, or some other metabolic response of glial cells in goldfish. Alternatively, the trigger mechanism might be related to a drop in the electrical activity of the injured nerve fibers (Skene and Shooter 1983), as suggested by the coincidence in the increase in release of the 37K protein with a drop in the number of Na+ channels in the degenerating rat optic nerve (Politis et al. 1983).

It is instructive to consider possible functions of these glial-released proteins. Clearly, glial cells undergo shifts in their functional states which are influenced by the condition of the neuronal membranes with which they are in contact, and these changes are developmentally regulated. The fact that the 37K protein in rat is found immunocytochemically in association only with those portions of neurons distal to a site of injury (i.e., undergoing degeneration) suggests that this protein might have some function in clearing debris (e.g., a lytic function) or in promoting growth (possibly by laying down an extracellular matrix), or, in later stages, in myelinization. Proteins released by gliomas have been found to promote neurite outgrowth of cultured PC12 cells (Arenander and De Vellis 1981). It is not clear whether the enhanced expression of the 37K protein in the developing rat optic nerve represents a growth-promoting function, or a response to naturally occurring cell death which takes place at this time (Jeffery and Perry 1982). The elucidation of the functions of these glial-released proteins provides an exciting challenge for future research, which should be greatly facilitated by the production of specific antisera to these proteins such as described here for the 37K protein.

Acknowledgments

We would like to thank Dr. Paul Manis for programming assistance and Mr. Michael Blei for help with the neonatal surgery. Supported by NIH Grants NS18103 and EY01117 to JAF and EY03718 to JJN.

References

Arenander AT, De Vellis J (1981) Glial-released proteins: III. influences on neuronal morphological differentiation. Brain Res 224: 117-127

Benowitz LI, Lewis ER (1983) Increased transport of 44,000- to 49,000 dalton acidic proteins during regeneration of the goldfish optic nerve: a two-dimensional gel analysis. J Neurosci 3:2153-2163

Benowitz LI, Shashoua VE, Yoon N MG (1981) Specific changes in rapidly transported proteins during regeneration of the goldfish optic nerve. J Neurosci 1:300-307

Benowitz LI, Yoon MG, Lewis ER (1983) Transported proteins in the regenerating optic nerve: regulation by interactions with the optic tectum. Science 222:185-188

Carey DJ, Bunge RP (1981) Factors influencing the release of proteins by cultured Schwann cells. J Cell Biol 91:666-672

Deaton MA (1983) Alteration in the synthesis of specific proteins underlying regeneration of the goldfish optic nerve. PhD Dissertation, Vanderbilt University, Nashville

Easter SS, Rusoff AC, Kish PE (1981) The growth and organization of the optic nerve and tract in juvenile and adult goldfish. J Neurosci 1:793-811

Foerster A, Holmes M, Diamond J (1983) Optic tract regeneration in the adult rat: physiology. Soc Neurosci Abstr 9:694

Freeman JA, Manis PB (1984) Steady-state ionic currents generated by growth cones. J Neurosci Res (to be published)

Giulian D, Des Ruissaux H, Cowburn D (1980) Biosynthesis and intra-axonal transport of proteins during neuronal regeneration. J Biol Chem 255:6494-6501

Grafstein B, Forman DS (1980) Intracellular transport in neurons. Physiol Rev 60:1167-1283

Heacock AM, Agranoff BW (1982) Protein synthesis and transport in the regenerating goldfish visual system. Neurochem Res 7:771-787

Jeffery G, Perry VH (1982) Evidence for ganglion cell death during development of the ipsilateral retinal projection in the rat. Dev Brain Res 2:176-180

Jen G, Thatch RE (1982) Inhibition of host translation in encephalomyocarditis virus-infected L-cells: a novel mechanism. J Virol 43:250-261

Laemmli UK (1970) Cleavage of structural proteins during the assembly of the head of bacteriophage T4. Nature 227:680-685

McQuarrie IG, Grafstein B (1982) Protein synthesis and axonal transport in goldfish retinal ganglion cells during regeneration accelerated by a conditioning lesion. Brain Res 251:25-37

O'Farrell PZ (1975) High resolution two-dimensional electrophoresis of proteins. J Biol Chem 250:4007-4021

O'Farrell PZ, Goodman HM, O'Farrell PH (1977) High resolution two-dimensional electrophoresis of basic as well as acidic proteins. Cell 12:1133-1141

Politis MJ, Pellegrino RG, Oaklander AL, Ritchie JM (1983) Reactive glial protein synthesis and early disappearance of saxitoxin binding in degenerating rat optic tectum. Brain Res 273:392-395

Skene JHP, Shooter EM (1983) Denervated sheath cells secrete a new protein after nerve injury. Proc Natl Acad Sci USA 80:4169-4173

Skene JHP, Willard M (1981a) Changes in axonally transported proteins during regeneration in toad retinal ganglion cells. J Cell Biol 89:86-95

Skene JHP, Willard M (1981b) Axonally transported proteins associated with axon growth in rabbit central and peripheral nervous systems. J Cell Biol 89:96-103

Skene JHP, Willard M (1981c) Characteristics of growth-associated proteins in regenerating toad retinal ganglion cell axons. J Neurosci 1:419-426

Skene JHP, Willard M (1981d) Electrophoretic analysis of axonally transported proteins in toad retinal ganglion cells. J Neurochem 37: 79-87

Szaro BG, Faulkner LH, Hunt RK, Loh Y (1984) Axonal transport of [35S]methionine labeled proteins in Xenopus optic nerve: phases of transport and the effects of nerve crush on protein patterns. Brain Res 297:337-355

Tytell M, Black MM, Garner JA, Lasek RJ (1981) Axonal transport: each major rate component reflects the movement of distinct macromolecular complexes. Science 214:179-181

Signals of External Origin Can Trigger Injured Adult Rabbit Optic Nerves to Express Regeneration-Like Responses

M. Schwartz[1], M. Belkin[2], M. Hadani[2], A. Harel[1], V. Lavie[1], I. Rachailovich[1], A. Solomon[2], and C. Stein-Izsack[1]

1 Department of Neurology, The Weizmann Institute of Science, Rehovot 76100, Israel
2 Department of Neurosurgery, Maurice and Gabriela Goldschleger Eye Research Institute, Tel-Aviv University Sackler, School of Medicine, Chaim Sheba Medical Center, Tel-Hashomer 52621, Israel

Abstract

The limited regenerative capacity of adult mammalian central nervous system (CNS) neurons is puzzling. This low ability could stem from a deficiency either in the cell body or in the microenvironment of the injured axon. By studying a regenerative model (i.e. the goldfish visual system) we were able to demonstrate that regeneration in this system is accompanied by alteration in type and amount of diffusible and nondiffusible substances derived from the axonal environment. We found that among these changes are those responsible for increased activity of factors which trigger regenerative response in a non-regenerative CNS (i.e. injured optic nerve of adult rabbit). Such a response is induced when a regenerating nerve is grafted or a media conditioned by it is implanted into an injured optic nerve of adult rabbit.

The ability of a nerve to produce diffusible substances which induce a regenerative response is found to correlate with the growth-state of the axon independently of whether the growth is of developing or of regenerating nerves.

Introduction

The process of axonal regeneration is complicated and involves the participation of various cellular and chemical factors. This process can be divided into three major phases: (i) cell body response and formation of new sprouts, (ii) elongation of the new sprouts, and (iii) cessation of axonal elongation and synaptogenesis. Successful regeneration depends on the sequential interplay of the various events that occur in each of the above-mentioned phases. The ability to regenerate may be intrinsically preprogrammed or may depend on the environment. Evidence in favor of both possibilities has been described (Cajal 1928; Aguayo et al. 1978; David and Aguayo 1981; Kao et al. 1977; Kiernan 1979; Longo et al. 1983; Lunberg et al. 1982; Nieto-Sampiedo 1982; Richardson et al. 1980; Varon et al. 1982). However, direct association between the environment and the intrinsic properties of the neuron has not been elucidated.

It is currently believed that for regeneration to occur, a successful growth state of the neuron has to be established. This involved changes in synthesis and transport of specific macromolecules including those specifically geared to reconstruct new axons, collectively named restorative events (Grafstein 1983). Injury induces

Experimental Brain Research, Suppl. 13

changes in the CNS of lower vertebrates and in the peripheral nervous system of mammals (regenerative systems) are characterized by alterations in the synthesis and transport of specific macromolecules (Benowitz et al. 1981, 1983; Bisby 1980; Giulian et al. 1980; Grafstein and Murray 1969; Griffin et al. 1981; Heacock and Agranoff 1976, 1982; Neumann et al. 1983a; Perry and Wilson 1981; Quitschke and Schechter 1983; Skene and Willard 1981a, b) and by sprouting activity (Landreth and Agranoff 1976, 1979; Schwartz and Agranoff 1981), in culture. Such regenerationassociated characteristics have not been observed in mammalian CNS.

Such changes occur throughout the process of regeneration, each change at a distinct time. It is possible that one signal is required to trigger a cascade of events. Another possibility is that each of the sequential changes has its own triggering signal. Signals can be provided either by the axonal injury itself or by both the injury and factors conveyed to the cell body from the environment. Induction of the cell body response by the injury itself would imply that the neurons have the intrinsic capacity to regenerate independently of environmental conditions. Such an intrinsic capacity means that an inhibitory component of the neuron should be removed or modified for the cell body response to occur. This regulatory mechanism may be either neuron derived but target dependent or target derived. Induction of the response by environmental factors would imply that the neuron has the capacity to respond provided that its environment is adequate. Failure in the production of appropriate factors or in response to any of them could account for the failure of mammalian CNS neurons to establish growth state and thus to regenerate.

We studied studied a regenerative CNS of lower vertebrates in order to identify the parameters involved in regeneration and then examine which of them is missing or inadequate in a mammalian nonregenerative system. After establishing that the regeneration of the goldfish optic nerve is accompanied by alterations in the axonal environment, it seemed that the time was ripe to move to a nonregenerative mammalian system. Our approach was to apply a regeneration-supportive environment of a xenogeneic CNS (fish optic nerve) to a nonregenerative system (optic nerve of an adult rabbit) by means of axonal grafting. Such a treatment triggered a regeneration-like response in the corresponding retina of the adult rabbit, and thus indicated that the optic nerve of adult rabbits (representing a nonregenerative CNS) is defective at least with respect to the production of external factors.

Results

Environmental Changes Associated with Regeneration of the Goldfish Optic Nerve

In intact nerves there is a stable relationship between the axon and the nonneuronal cells (Liu 1981). This type of relationship is presumably required for the maintenance and the functional activity of the axon (Gainer et al. 1977), on the one hand, and for the maintenance of the fully differentiated noneuronal cells (Bunge et al. 1980, 1982; Collins 1980; Hanson and Partlow 1980; Hanson et al. 1982), on the other. An injury may disrupt this stable relationship, possibly leading to various changes which may be involved in different events during the process of regeneration. Such changes may include secretion of diffusible substances that might act as signals to the neuronal cell body, and as trophic substances required for growth and elongation of the regenerating fibers, and formation of a guiding substratum needed to direct the fibers toward the specific preexisting target sites.

To demonstrate the direct contribution of environmental elements to the process of regeneration we pursued the following line of research. First, we determined if environmental changes (diffusible and nondiffusible factors) occur as a consequence of an injury in a regenerative model system, and then we examined if and how these changes are relevant to the process of regeneration. This was investigated in both regenerative and nonregenerative systems.

Nonneuronal Cells Involved in the Process of Regeneration. It was shown by several investigators that regeneration is accompanied by proliferation of the surrounding nonneuronal cells (Maxwell and Elam 1980; Stevenson and Yoon 1978, 1981). In order to establish a correlation between proliferating nonneuronal cells and the process of regeneration, we examined the effect of inhibition on nonneuronal cell division on the subsequent regenerating competence. As a means to inhibit cell proliferation we employed X-irradiation (under conditions in which the effect of radiation on other processes was minimal). Since changes in the ability of the neuron to regenerate in vivo are reflected in the ability of the cell body to express characteristics of regeneration in vitro, we used sprouting activity in culture as a parameter of evaluating the degree of regenerative competence (Landreth and Agranoff 1976, 1979; Schwartz et al. 1982a, b).Protein synthesis in the retina and accumulation in the tectum of radiolabeled components, transported from the retina, were used as additional parameters for regeneration competence subsequent to the irradiation. The results indicated that proliferating nonneuronal cells indeed play a role in the process of regeneration. X-irradiation of the visual system at an early stage after the nerve injury (7 days after the crush) had a promoting effect, whereas irradiation at a later stage resulted in a reduced sprouting activity (Table 1). The contradictory consequences of irradiation on regeneration-competence have led us to propose that in response to injury there is an onset of proliferation of at least two types of cells, the kinetic

Table 1. X-irradiation of injured goldish optic nerves affect the sprouting capacity of the retinae

Days postcrush Irradiation	Explanation	$[NGI_X/NGI] \pm SEM$
7	8	4.05 ± 0.9
10	11	0.58 ± 0.6

Optic nerves of goldfish were crushed and exposed to X-irradiation (400 rad) at the indicated days after the lesion. Twenty-four hours after the irradiation the retinae were dissected out and cultured in L-15 medium containing 1% fetal calf serum. Retinae of injured nonirradiated nerves were used as controls. Growth was scored 72 h after explanation and was expressed by NGI. The results are expressed by the ratio between outgrowth from retinae of X-irradiated regenerating nerves versus that from the retinae of the corresponding nonirradiated injured nerves (NGI_X/NGI) $\pm$ SEM of which are probably distinct (Neumann et al. 1983b). The question is how these effects are mediated.

Diffusible Molecules Originating from the Axonal Environment. To screen for specific changes in diffusible substances originating from the nonneuronal cells, the experimental preparations consisted of media conditioned by the various axonal segments. We characterized the released substances (pulsed labeled with [^{35}S] Methionine) by one- and two-dimensional gel electrophoreses. The most prominent observation was the appearance of the 28K polypeptide under conditions that favor regeneration (stimulated by injury with and without X-irradiation, Rachailovich and Schwartz 1984). During the course of this work two other laboratories reported injury-induced changes in substances released from supportive cells of mammalian peripheral nerves (Skene and Shooter 1983) and goldfish visual nerves (Deaton and Freeman 1983). The latter work reported the appearance of a 30K polypeptide.

Nondiffusible Substances. To associate regeneration with any cellular or extracellular responses of the nonneuronal cells, we performed an ultrastructural analysis of the optic nerve of goldfish following its injury. This analysis indicated that regeneration of the goldfish optic nerve is accompanied by the accumulation of fibrous collagen adjacent to astrocytic limitans (Fig. 1). In this respect the nonneuronal cells of the goldfish optic nerve (representing CNS) may resemble Schwann cells in regenerating mammalian peripheral nerve (Bunge et al. 1980, 1982; Nathaniel and Pease 1963; Nathaniel and Nathaniel 1973).

The collagen in the regenerating nerves was found to be limited in amount: however, it was found in significantly higher amounts under conditions which favor regeneration, such as X-irradiation (Neumann et al. 1983b). The apparent correlation between the increased in vitro growth capacity of regenerating retina (Table 1) which was exposed to X-irradiation and the increased amount of accumulated collagen in this preparation may attribute to the observed collagen a permissive role in the process of regeneration.

These ultrastructural observations may provide additional clues as to the nature of components which endow the goldfish visual system with the high ability to regenerate. Other morphological similarities between regenerating fish optic nerves and mammalian peripheral nerves were described by other investigators (Lanners and Grafstein 1980; Murray 1976; Wolberg et al. 1983). However, futher studies are required in order to verify whether the observed collagen does originate from the astrocytes and whether the collagenous environment in regenerating and X-irradiated regenerating nerves is truly supportive of regeneration rather than an epiphenomenon coincident with regeneration or promoted regeneration.

Regeneration-Supportive Environment as a Trigger for Regeneration-like Response

The results described above provide evidence that in a system that has the potential to regenerate, nonneuronal cells surrounding an injured axon demonstrate alterations in the type and amount of substances that they can release or provide. At this stage it is hard to attribute a specific role in the promotion or inhibition of axonal growth (Rachailovich and Schwartz 1984) to any specific proteins whose levels were found to increase or decrease in regeneration.

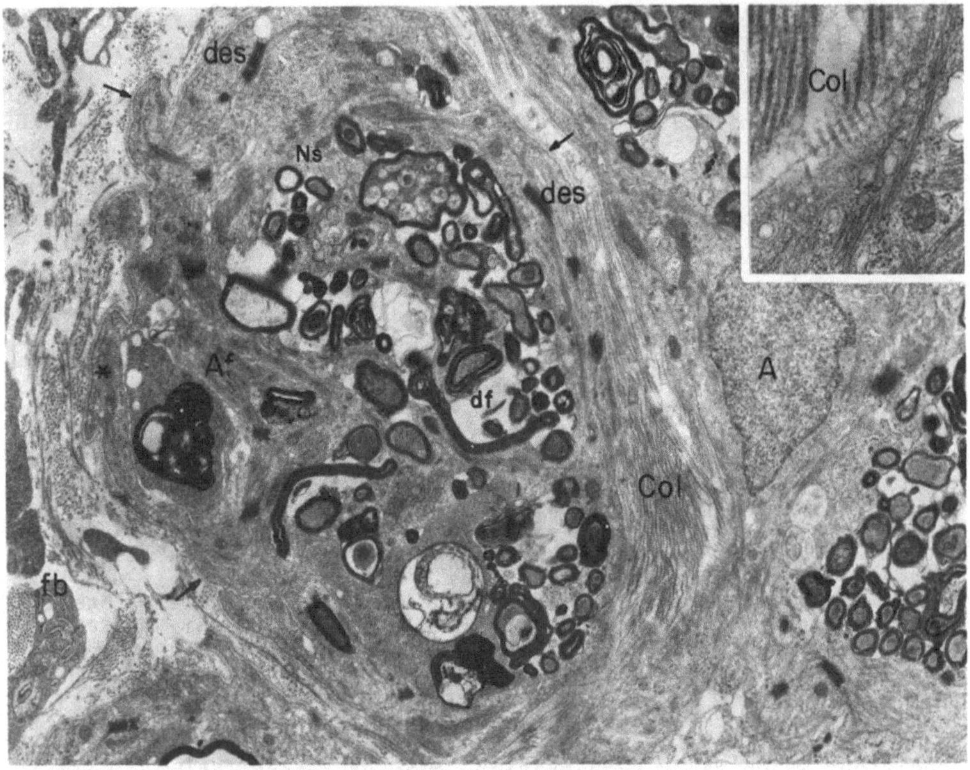

Fig. 1. Accumulation of fibrous collagen at the glial limitans in an optic nerve of goldfish, under conditions which favor regeneration. Cross section through regenerating nerve (stimulated by X-irradiation). The micrograph illustrates the column-like structure formed by astrocytes (A) and astrocytic processes (Ap) connected to each other by desmosomes (des). Note the astrocytic filaments (Af) and the caveolae (*) in these astrocytes. The astrocytes forming the column-like structure enclose degenerating fibers (df) and slender new sprouts (Ns). The astrocytes are surrounded by basal lamina (arrow) facing the perineural spaces. Collagen accumulation at these limitans is observed. Fibrocyte (fb) is also seen in this field. x 10105

Based on the above observations, we wanted to examine whether the environment of a regenerating nerve (presumably regeneration-supportive) can make a neuron from a mammalian CNS (nonregenerative system) express regeneration response. If successful, it would indicate that (i) a nonregenerative systemn is defective in its environmental factors, and (ii) environmental changes in a regenerative system are involved in the production of factors which enable triggering of regeneration response. To carry out such a study, we developed a surgical procedure which enabled the use of an adult rabbit's visual system as our experimental model for a nonregenerative system.

At the onset of this study, we transplanted regenerating carp optic nerve into a transected optic nerve of adult rabbit by suturing these nerves end to end (the use of carp rather than goldfish is for

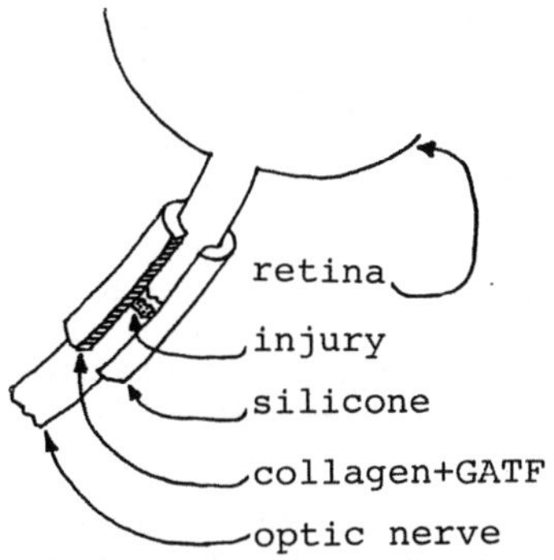

Fig. 2. Schematic presentation of a wrap around implant, which is introduced to an injured optic nerve of adult rabbit. Note: The silicone tube is coated with collagen and is soaked in media conditioned by the appropriate neuronal source, and thus contains the growth-associated triggering factors (GATFs) (Hadani et al. 1984)

practical reasons, i.e., the similarity of nerve diameters). These experiments resulted in regeneration-like response in the rabbit retina. However, immunological rejection was noticed after a few days (Solomon et al 1985). To circumvent the immunological rejection, we replaced the graft with a "wrap around" implant which consisted of a silicone tube coated inside with collagen (Solomon et al. 1985, Fig. 2). Substances from the media conditioned by the regenerating fish optic nerve were allowed to adsorb to the collagen, and the tube was then implanted around a severed optic nerve of an adult rabbit. This experimental manipulation resulted in a general increase in protein synthesis in the retina as was measured by [^{35}S]-methionine incorporation (up to 70% increase relative to the retina of the uninjured, contralateral side). Under the same experimental conditions, nerve injury alone caused a general decrease in protein labeling (about 30% decrease relative to the contralateral side; Schwartz et al. 1985). Such an increased incorporation of [^{35}S] methionine suggests a general increment in metabolic activity as was shown to occur in retina of regenerating fish optic nerve. In addition to the general increase in labeling, induced by substances originating from the regenerating fish optic nerve, there was selective increase in the labeling of several specific polypeptides (Schwartz et al. 1985). The selective increase could be due to either synthesis of new polypeptides or enhanced synthesis of already existing polypeptides. The most prominent changes as analyzed by polyacrylamide gel elctrophoresis (SDS-PAGE) were in polypeptides of apparent molecular weights 130K, 110K, 74K, 64K and 26K (Schwartz et al. 1985). The molecular weights of some of these polypeptides are similar to those of polypeptides previously decribed in regenerating systems (including goldfish and toad optic nerves and mammalian sciatic nerve (Willard and Skene 1982). It is possible that at least a few of these polypeptides may function as specific growth-associated proteins and may be related to the known growth-associated proteins (GAPs) which were defined by Skene and Willard (1981 a, b). However, definitive identification requires further study.

In order to determine if the induced biochemical changes observed in our work are associated with morphological regeneration, we measured the ability of the retina of implanted injured nerve to sprout in culture. Neurons of intact or injured nerves of adult mammalian CNS do not send out neurites in culture. However, neurons of lower vertebrate CNS express a high sprouting ability in culture following nerve injury (Landreth and Agranoff 1976, 1979). Our study demonstrates that retinae of injured rabbit optic nerves which were grafted with regenerating fish optic nerve or were wrapped around with implants containing media conditioned by regenerating fish optic nerves demonstrated ability to sprout in culture (Schwartz et al. 1985).

To find out why a regenerative response does not occur in the retina of an injured optic nerve of adult rabbit under ordinary conditions, we addressed ourselves first to the question of whether the injured optic nerve of adult rabbit can be a source of active conditioned media. If this is not the case, the question arises as to why and when the rabbit lost the ability to provide molecules which can trigger the retinal response.

To answer this question, we further refined our experimental approach using the synthetic wrap around implant. We introduced into the collagen-coated silicone tube media conditioned by optic nerves of neonatal rabbits, adult rabbit optic nerves, and injured optic nerves of adult rabbits. The results of these experiments indicated that implanted substances originating from neonatal nerves could also trigger the retinal response (Hadani et al. 1984). This was demonstrated by the same parameters used above, i.e., a general increase in protein synthesis and by a selective increase in the same specific polypeptides as when transplanted with materials derived from regenerating optic nerve of fish. In addition, we found that optic nerve from intact adult rabbit could also provide the triggering signals (although to a lesser degree), but injured adult optic nerve could not. Such a loss could evolve from the disruption of the bidirectional relationship between the axons and the surrounding nonneuronal cells. These results point out that nonneuronal cells of higher vertebrates are not able to provide regeneration-triggering molecules after injury, in contrast to those of lower vertebrates.

At this stage of our research we cannot rule out the possibility that the activity of the various conditioned media might be mediated by different molecules, although it is likely that this is not the case. It appeared that the implanted substances might represent the signal which triggers the retinal response. Alternatively, the implanted substances may be responsible for providing signals for preceding events, such as the activation of glial cells within the injured adult nerve, and thus induce the production of signals by them. It seems that provision of triggering factors by nonneuronal cells is in correlation with the growth state of the neuron which is responsible for their production. This would justify the consideration of these environment-derived triggering factors as growth-associated triggering factors (GATFs) in analogy with the neuronal intrinsic growth-associated proteins (GAPs). Table 2 summarizes the correlation. It should be noted that these factors are not necessarily the trigger itself; they may be the missing link in an array of events leading to triggering.

Table 2. Correlation between the growth state of an axon and its ability to provide triggering factors

	Rabbit optic nerve (nonregenerative CNS)				Fish optic nerve (regenerative CNS)	
	Adult			Neonatal	Adult	
Neuronal features	Intact	Injured	Injured and Implanted[a]	Intact	Intact	Injured
Type of neuron						
Sprouting from the corresponding retina (in culture)	-	-	+	+	-	+
Alterations in protein synthesis	-	-	+	+	-	+
Activity of media conditioned by these neuronal preparations[b]	+	-	++	+++	+	+++

[a]Implanted with media conditioned by optic nerves of neonatal rabbits
[b]The assigned activities were based on the relative ability of the various conditioned media to cause a general increase in protein synthesis in retinae of injured optic nerves of adult rabbits

The described findings give rise to the question as to whether the triggering of regenerative response is sufficient to ensure the occurrence of subsequent events in regeneration. Our initial studies at the electron microscopic level were performed in order to find out whether the retinal response, including the sprouting in vitro, is accompanied by sprouting in vivo. The results indicated that in the proximal zone (adjacent to the site of injury), the implanted optic nerve exhibited a morphology different than that of a control nerve. In the zone in the control nerves, degeneration fibers and glial scar formed by astrocytic processes were seen. In restricted regions in the implanted nerves, a better maintenance of the tissue and a high frequency of unmyelinated fibers, presumably new sprouts, were observed. In addition, while in the control nerves, growth cones were not detected (seven different nerve preparations were examined), in the implanted nerves we found a few growth cones (six preparations were examined, Fig. 3).

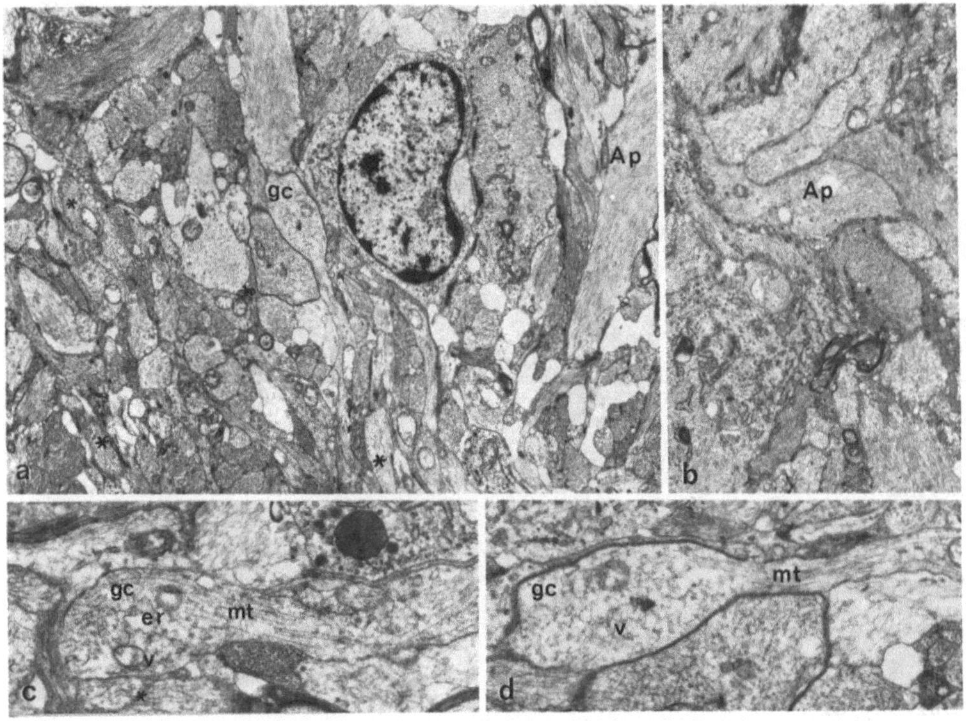

Fig. 3a-d. A wrap around implant of media conditioned by regenerating fish optic nerve induces increased appearance of new sprouts with growth cones. Injured optic nerve of adult rabbit implanted with media conditioned by regenerating fish optic nerve (a,c,d) or as a control with bovine serum albumin (b). These nerves were processed for electron microscopic analysis 2 weeks after the injury and the implantation. Note the appearance of a growth cone with vesicles (v) (a,d) at the tip of an organized fiber with microtubules (mt). The structure in b resembles scar region formed by astrocytic processes (Ap). a,b x 8,480; c,d x 20,930

Our study demonstrated that (i) an injured optic nerve of a nonregenerative system can be induced to express characteristics of regeneration provided that its environment is suitably modified, and (ii) an environmental modification is achieved by diffusible substances derived from a growing nerve independently of whether the growth is by a regenerating or developing nerve.

The use of our approach may provide the means for the identification of the environmental factors which are deficient or are missing in a nonregenerative system, and for the isolation and purification of these factor(s).

Acknowledgments. This work was supported by grants to M.S. from the United States-Israel Binational Foundation and the Israel Ministry of Health, and by the Dr. M. Levine Estate trustees, Sara and James Kaplan, in association with the California chapter of the National Spinal Cord Injury Association.

References

Agranoff BW, Feldman EL, Heacock AM, Schwartz M (1980) The retina as a biochemical model of central nervous system regeneration. Neurochemistry 1:487-500

Aguayo A, Samuel C, Richardson P, Bray G (1978) Axonal elongation in peripheral and central nervous system transplants. Adv Cell Neuobiol 3:215-233

Benowitz LI, Shashoua VE, Yoon MG (1981) Specific changes in rapidly transported proteins during regeneration of the goldfish optic nerve. J Neurosci 1:300-307

Benowitz LI, Yoon MG, Lewis ER (1983) Transported proteins in the regenerating optic nerve: Regulation by interactions with the optic tectum. Science 220:185-187

Bisby MA (1980) Changes in the composition of labeled protein transported in motor axons during their regeneration. J Neurobiol 11:435-445

Bunge MB, Williams AK, Wood PM, Uitto J, Jeffrey JJ (1980) Comparison of nerve cell and nerve cell plus Schwann cell cultures with particular emphasis on basal lamina and collagen formation. J Cell Biol 84:184-202

Bunge MLB, Williams AK, Wood PM (1982) Neuron-Schwann cell interaction in basal lamina formation. Dev Biol 92:449-460

Cajal RYS (1928) In: May RM (ed) Degeneration and regeneration in the nervous system. Oxford University Press, London

Collins F (1980) Neurite outgrowth induced by the substrate associated material from nonneuronal cells. Dev Biol 79:247-252

David S, Aguayo AJ (1981) Axonal elongation into peripheral nervous system "bridges" after central nervous system injury in adult rats. Science 214:931-933

Deaton MA, Freeman JA (1983) Glial cells secrete specific proteins during regeneration of the goldfish optic nerve which may play a role in axon growth. Abstr Neurosci 9:693

Gainer H, Tasaki I, Lasek RJ (1977) Evidence for the glia-neuron protein transfer hypothesis from intracellular perfusion studies of squid giant axons. J Cell Biol 74:524-530

Giulian D, Des Ruisseaux H, Cowburn D (1980) Biosynthesis and intra-axonal transport of proteins during neuronal regeneration. J Biol Chem 255:6494-6499

Grafstein B, Murray M (1969) Transport of protein in goldfish optic nerve during regeneration. Exp Neurol 25:494-508

Griffin JW, Price DL, Drachman DB, Morris J (1981) Incorporation of axonally transported glycoproteins into axolemma during nerve regeneration. J Cell Biol 88:205-214

Hadani M, Harel A, Solomon A, Belkin M, Lavie V, Schwartz M (1984) Substances originating from optic nerve of neonatal rabbit induced regeneration-associated response in injured optic nerve of adult rabbit. Proc Natl Acad Sci USA 81:7965-7969

Hanson GR, Partlow LM (1980) A comparison of two factors affecting the proliferation of non-neuronal (glial) cell in vitro. Brain Res 192:371-381

Hanson GR, Partlow LM, Iverson PL (1982) Neuronal stimulation of nonneuronal (glial) cell proliferation: lack of specificity between different regions of its nervous system. Dev Brain Res 3:547-555

Heacock AM, Agranoff BW (1976) Enhanced labelling of a retinal protein during regeneration of optic nerve in goldfish. Proc Natl Acad Sci USA 73:828-832

Heacock AM, Agranoff BW (1982) Protein synthesis and transport in the regenerating goldfish visual system. Neurochem Res 7:771-778

Kao CC, Chang LW, Bloodworth JMB (1977) Axonal regeneration across transected mammalian spinal cords: an electron microscopic study of delayed microsurgical nerve grafting. Exp Neurol 54:591-615

Kiernan JA (1979) Hypothesis concerned with axonal regeneration in the mammalian nervous system. Biol Rev 54:155-197

Landreth GE, Agranoff BW (1976) Explant culture of adult goldfish retina: effect of prior optic nerve end. Brain Res 118:299-303

Landreth GE, Agranoff BW (1979) Explant culture of adult goldfish retina: a model for the study of CNS regeneration. Brain Res 161:39-53

Lanners HN, Grafstein B (1980) Early stages of axonal regeneration in the goldfish optic tract: An elctron microscopic study. J Neurocytol 9:733-751

Liu HM (1981) Reaction of the central nervous system to injury. In: Liu HM (ed) Biology and pathology of nerve growth, Academic, New York, pp 236-269

Longo FM, Manthorpe M, Skaper SC, Lundborg G, Varon S (1983) Neuronotrophic activities accumulate in vivo within silicone nerve regeneration chambers. Brain Res 261:109-117

Lundborg G, Longo FM, Varon S (1982) Nerve regeneration model and trophic factors in vivo. Brain Res 232:157-161

Maxwell JK, Elam JS (1980) Amino acid incorporation into proteins of degrading and regenerating goldfish optic tracts. Exp Neurol 67:118-130

Murray M (1976) Regeneration of retinal axons into the goldfish optic tectum. J Comp Neurol 168:175-196

Nathaniel EJH, Nathaniel DR (1973) Regeneration of dorsal root fiberes into the adult rat spinal cord. Exp Neurol 40:333-350

Nathaniel EJ, Pease DC (1963) Collagen and basement membrane formation by Schwann cells during nerve regeneration. J Ultrastruct Res 9:550-560

Neumann D, Scherson T, Ginzburg I, Littauer UZ, Schwartz M (1983a) Regulation of mRNA levels for microtubule proteins during nerve regeneration. FEBS Lett 162:270-276

Neumann D, Yerushalmi A, Schwartz M (1983b) Inhibition of non-neuronal cell proliferation in the goldfish visual pathway affects the regenerative capacity of the retina. Brain Res 272:237-245

Nieto-Sampedro M, Lewis ER, Cotman CW, Manthorpe M, Skaper SD, Barbin G, Longo FM (1982) Brain injury causes a time-dependent increase in neurotrophic activity at the lesion site. Science 297:860-861

Perry GW, Wilson DL (1981) Protein synthesis and axonal transport during nerve regeneration. J Neurochem 37:1203-1217

Quitschke N, Schechter N (1983) Specific optic nerve proteins during regeneration of the goldfish retinotectal pathway. Brain Res 258:69-76

Rachailovich I, Schwartz M (1984) Molecular events associated with increased regenerative capacity of the goldfish retinal ganglion cells following X-irradiation: decreased levels of growth inhibitors. Brain Res 306:149-155

Richardson PM, McGuinness UM, Aguayo AJ (1980) Axons from CNS neurons regenerate into PNS grafts. Nature 284:264-265

Schwartz M, Agranoff BW (1981) Multiprocesses emerge from cultured retinal ganglion cells. Brain Res 206:331-336

Schwartz M, Mizrachi Y, Kimhi Y (1982a) Regenerating goldfish retinal explants: induction and maintenance of neurites by conditioned medium from cells originated in the nervous system. Dev Brain Res 3:21-28

Schwartz M, Mizrachi Y, Eshhar N (1982b) Factors from goldfish brain induce neuritic outgrowth from explanted regenerating retinas. Dev Brain Res 3:29-35

Schwartz M, Solomon A, Harel A, Rachailovich I, Lavie V, Hadani M, Belkin M (1985) Regeneration-like response induced in adult rabbit CNS by grafted regenerative fish optic nerve. Science (to be published)

Skene JHP, Shooter EM (1983) Denervated sheath cells secrete a new protein after nerve injury. Proc Natl Acad Sci USA 80:4169-4173

Skene JHP, Willard M (1981a) Changes in axonally transported proteins during axon regeneration in toad retinal ganglion cells. J Cell Biol 89:86-95

Skene JHP, Willard M (1981b) Axonally transported proteins associated with axon growth in rabbit central and peripheral nervous systems. J Cell Biol 89:96-103

Solomon A, Belkin M, Hadani M, Harel A, Rachailovich I, Lavie V, Schwartz M (1985) A new transorbital approach to the rabbit's optic nerve. J Neurosci (to be published)

Stevenson JA, Yoon MG (1978) Regeneration of optic nerve fibers enhances cell proliferation in the goldfish optic tectum. Brain Res 153:345-351

Stevenson JA, Yoon MG (1981) Mitosis of radial cells in the optic tectum of adult goldfish. Neuroscience 1:862-875

Varon S, Longo FM, Manthorpe M (1982) Comparative features of spinal neuronotrophic factors in fluids collected in vitro and in vivo. J Neurosci Res 8:241-250

Willard M, Skene JHP (1982) Molecular events in axonal regeneration. In: Nicholls JG (ed) Repair and regeneration of the nervous system. Springer, Berlin Heidelberg New York, pp 41-56

Wolberg H, Kastner R, Kurz-Isler G (1983) Lack of orthogrowth proticle assemblies and presence of tight functions in astrocytes of goldfish (carassius auratus). Cell Tissue Res 234:389-397

Tissue Damage at the Site of Trauma

Pathophysiology, Consequences and Pharmacological Prevention of Post-Traumatic CNS Ischemia

E. D. Hall, D. L. Wolf, and J. M. Braughler

CNS Diseases Research, The Upjohn Company, Kalamazoo, MI 49001, USA

Introduction

Several laboratories, using a wide variety of anatomical, biochemical, and physiological techniques, have documented the occurrence of a progressive decrease in spinal cord gray and white matter microvascular patency and blood flow (SCBF) within a few hours after severe contusion or compression injury in various species (see review by Means and Anderson 1985). While studied to a lesser extent, pronounced brain ischemia has also been observed as an early event after blunt head trauma in cats (Wagner and Tornheim 1983). In the case of spinal cord injury, the degree of the ischemic response is positively correlated with injury severity, while negatively correlated with the extent of post-traumatic functional recovery (Ducker et al. 1978). This, it is generally accepted that in order to promote recovery after blunt spinal cord injury, an attempt must be made to attenuate the post-traumatic decline in SCBF. Failure to maintain SCBF leads to ischemic hypoxia, ATP depletion, lactic acidosis, and rapid tissue destruction, which is believed to be due, in large part, to oxygen free radical-induced lipid peroxidative reactions (Demopoulos et al. 1980; Hall and Braughler 1982a; Means and Anderson 1984).

Effects of High Dose Methylprednisolone in Acute Spinal Cord Injury

A major focus of research aimed at the acute treatment of spinal cord injury has been on pharmacological agents that would either prevent post-traumatic ischemia or reduce the sensitivity of the affected tissue to the associated hypoxia. In our laboratories, we have conducted a lengthy series of studies on the ability of high dose administration of the synthetic glucocorticoid steroid and methylprednisolone sodium succinate (MPSS) to retard ischemia development and the associated occurrence of oxygen free radical-generated lipid peroxidation after spinal cord contusion in cats. This choice of glucocorticoid has been based largely on the demonstration that high dose MPSS treatment of cats (Means et al. 1981; Demopoulos et al. 1982) and monkeys (Green et al. 1980) can promote neurologic recovery after blunt spinal cord injury. Previous work by us has shown that the i.v. injection of high MPSS doses (15-30 mg/kg) at 30 min following severe spinal contusion can reduce both the accumulation of lactic acid (Braughler and Hall 1983) and the occurrence of lipid peroxidative reactions (Hall and Braughler 1982b) within the injured spinal tissue. While the reduction in lipid peroxidation is probably, in part, secondary to the

Experimental Brain Research, Suppl. 13

apparent support of SCBF, a direct antioxidant effect of the same high dose of MPSS on cat spinal cord tissue has also been shown (Hall and Braughler 1981; Anderson et al. 1983).

Following these earlier investigations, we have recently turned our attention specifically to post-traumatic ischemia in spinal white matter and the possible ability of high doses of MPSS to attenuate its occurrence. The rationale for these studies has been that preservation of white matter represents the key to neurological recovery after spinal cord injury.

Methods

Adult mongrel cats either sex (2.0-3.7 kg) were anesthetized with Na pentobarbital (30 mg/kg i.v.) and supplemental doses given as needed throughout the subsequent experiment. A tracheotomy was performed and a catheter inserted into the right carotid artery for blood pressure monitoring. A dorsal laminectomy exposed the 3rd lumbar spinal cord segment. The animal was then placed in a prone position in a Kopf 1780 spinal unit and rigidly fixed in place by means of hip spikes, a clamp on the 2nd lumbar spinous process and a stereotaxic head holder. Blood pressure, heart rate, and rectal temperature were monitored continuously.

Local SCBF was determined in the dorsolateral funiculus of the 3rd lumbar spinal segment using the hydrogen clearance technique as described previously (Hall et al. 1984), which allows for repeated measurements in the same animal. Following three preinjury determinations at 30 min intervals, the SCBF electrode was removed and spinal cord contusion injury was produced as published earlier (Hall and Braughler 1982b; Braughler and Hall 1983) except that a 500 g-cm injury force was used (50 g weight dropped 10 cm; dura intact). Following the contusion, the electrode was reinserted in the dorsolateral funiculus in the middle of the injury site and SCBF measured at 10 and 30 min, 1, 2, 3 and 4 h.

At 30 min after injury, each animal received a single i.v. (cephalic vein) bolus injection of 15, 30, or 60 mg/kg MPSS prepared from SoluMedrol Sterile Powder or the corresponding vehicle.

Results and Discussion

Table 1 displays the mean white matter SCBF value for each of the experimental groups at 2, 3, and 4 h after the 500 g-cm contusion. In the vehicle-treated group, there was a progressive fall in SCBF over the 4.5-h experimental period to a mean of 7.7 ml/100 g of tissue/minute, or 62.7 % of the mean preinjury flow value. It should be considered that the critical ischemic threshold for spinal white matter is about 10 ml/100g/min since "normal" values between 10 and 15 ml/100g/min have repeatedly been published using the H_2 clearance method (Senter and Venes 1978; Young and Flamm 1982; Young et al. 1982).

Treatment at 30 min after injury with a single 30 mg/kg i.v. dose of MPSS prevented the injury-induced white matter ischemia and SCBF remained well within the normal range. The difference between the vehicle and 30 mg/kg MPSS treated groups was signifiant ($p<0.05$) at both 3 and 4 h after injury. The fact that this is a high dose MPSS

Table 1. Effect of methylprednisolone sodium succinate (MPSS) on post-traumatic spinal white matter ischemia development when given at 30 min after a 500 g-cm cord contusion[a]

MPSS dose (mg/kg)	No. of cats	Combined mean preinjury blood flow[b] (ml/100 g/min)	Blood flow after injury 2×h	3×h	4×h
Vehicle	8	12.3 ± 3.5	8.6 ± 1.1	8.4 ± 1.0	7.4 ± 1.0
15.0	5		11.5 ± 2.6	11.7 ± 2.7	10.4 ± 2.3
30.0	4		12.9 ± 2.2	13.8 ± 2.3	13.6 ± 2.7
60.0	4		10.1 ± 1.3	9.4 ± 1.3	8.9 ± 1.2

[a]Values represent mean ± standard error
[b]Mean preinjury blood flow between groups were not significantly different and thus were combined
[c] $p<0.05$ by t-test (two-tailed) versus vehicle group at same time point

effect is suggested by the lesser protection observed with the 15 mg/kg i.v. dose. Interestingly, by increasing the dose to 60 mg/kg, the support of SCBF is lost. This biphasic dose-response pattern has been reported earlier for both the MPSS-induced attenuation of post-injury lactic acid accumulation (Braughler and Hall 1983) and lipid peroxidation (Hall and Braughler 1982b) and probably represents the possible membrane disruptive effects of excessively high tissue steroid concentrations (Lewis et al. 1970). Young and Flamm (1982) have reported a similar maintenance of white matter SCBF after spinal contusion in cats by 15 mg/kg and more so 30 mg/kg MPSS given at 45 min after injury. However,these investigators did not look at the effect of higher doses.

As a biochemical correlate of this maintenance of SCBF, we have also recently documented a significant preservation of neurofilament protein within the injured cat spinal cord at 4.5 h after injury as an effect of the early (30 min) administration of the 30 mg/kg MPSS dose (Braughler and Hall 1984).

Concerning the cardiovascular response to lumbar spinal injury, there was an immediate but transient increase in mean arterial pressure (MAP). This was followed within a few minutes by a prolonged hypotension of 20 to 25 mmHg. However, the MAP was fairly constant and never fell below 100 mmHg and thus the hypotension was not a contributing factor in the decline in SCBF.

There were significant cardiovascular effects of MPSS. Whereas the 15 mg/kg dose did not affect the post-traumatic hypotension noted above, the 30 and 60 mg/kg doses significantly intensified it at certain time points. Only in the 30 mg/kg treated group at 1 h did the average MAP fall slightly below 80 mmHg. However, the maintenance of white matter SCBF at a normal preinjury level, even in the presence of a significantly greater postinjury hypotension, is consistent with the recently reported finding that when sympathetic function is intact, white matter SCBF is unaffected by changes in MAP over at least an 80-160 mmHg range (Young et al. 1982).

Possible Mediators of Post-traumatic CNS Ischemia

The development of post-traumatic CNS ischemia is most likely due to a number of factors. In blunt spinal cord injury models, some immediate vascular disruption by the injury force generally takes place. This is especially true in gray matter areas which have the highest microvascular density and is clearly evidenced by the occurrence of gray matter petechial hemorrhage (Means and Anderson 1984). However, within the surrounding white matter, blood flow is initially maintained within normal limits and ischemia progressively develops over the ensuing few hours.

Recent work has shown that there is an injury-induced fall in spinal extracellular calcium levels which occurs as a very early event after contusion injury in both cats (Young and Flamm 1982) and dogs (Stokes et al. 1983). Other studies have shown that intra-axonal calcium deposits appear as early as 30 min after spinal cord contusion in rats (Balentine and Spector 1977) and the total spinal calcium levels are increased (Happel et al. 1981) suggesting that injury causes a massive cellular influx and prolonged accumulation of calcium within the injured spinal segment. An increase in the intracellular levels of free calcium would tend to increase the calcium-dependent production, or release, of a number of potential mediators of ischemia (i.e. biochemical factors which promote vasoconstriction and/or platelet aggregation). For example, prompt and dramatic increases in spinal tissue levels of serotonin (5-HT) (Nemecek et al. 1977), prostaglandin F_{2a} (PGF_{2a}) (Jonsson and Daniell 1976) and thromboxane B_2 (TXB_2), the stable metabolite of TXA_2 (Hsu et al. 1983) have been reported to take place after blunt cord injury. Also, the possible involvement of vasoactive leukotrienes has not been excluded. One would predict, though, that a general calcium-induced activation of the arachidonic acid cascade might lead to a concomitant elevation of the vasodilator and platelet aggregation inhibitor prostacyclin (PGI_2) produced by the vascular endothelium which would counter the actions of other prostanoids (e.g. PGF_{2a} and TXA_2). However, tissue acidosis and peroxidative damage to the endothelium may limit PGI_2 production. Consistent with this view, very recent work by Anderson and colleagues (personal communication) has disclosed that compressive spinal injury in cats causes rapid increases in tissue PGF_{2a} and TXB_2 which far outweigh a small increase in PGI_2 production.

Very recent studies in our laboratories, also in cats, have been directed at a pharmacological dissection of the relative contribution of certain of the putative mediators mentioned above to the production of the post-traumatic decline in white matter SCBF. The purpose has been twofold: (1) to rationally identify other possible acute therapies for spinal cord injury and CNS trauma in general, and (2) to provide clues as to the mechanism(s) by which MPSS may be acting to antagonize ischemia development. To date, three sets of experiments have been carried out in this regard and are described below. The first has examined the effects of pretreatment 30 min before a 500 g-cm spinal contusion in cats with compounds that selectively interrupt certain pathways in the arachidonic acid cascade on the development of post-traumatic spinal cord ischemia (i.e. cyclooxygenase, TXA_2 synthetase, and lipoxygenase inhibitors). The second set has investigated the theoretical possibility that a constant i.v. infusion of a stable PGI_2 analog alone, or in combination with a TXA_2 synthetase inhibitor, may serve to support SCBF after blunt injury. The third group of experiments has attempted to ascertain the possible contribution of free radical-induced lipid peroxidative damage to the white matter microvasculature by testing the protective effects of intensive pretreatment with the anti-oxidants alpha tocopherol (vitamin E) and selenium.

Drug Treatments

All test drugs were given i.v. into the cephalic vein either as a bolus and/or a constant infusion.

Cyclooxygenase inhibitors: Two different cyclooxygenase inhibitors were studied. Ibuprofen, Na salt (IBU) was given 30 min before injury as a 10 mg/kg bolus (10 mg/ml in 0.9% saline) in one series of experiments, while in a second series the 10 mg/kg bolus was followed by a second 5 mg/kg bolus at 1.5 h after injury. Meclofenamate, Na salt (MEC) was given as a 2 mg/kg bolus (2 mg/ml in saline) 30 min before injury.

Thromboxane synthetase inhibitor: The selective thromboxane synthetase inhibitor sodium 5-(3'-pyridinylmethyl)-2-benzofuran carboxylate, monohydrate (U-63557A) (Gorman et al. 1983) was injected in a 10 mg/kg bolus (10 mg/ml in saline) 30 min before injury.

Leukotriene synthesis inhibitor: The selective lipoxygenase inhibitor 6,9-deepoxy-6,9-(phenylimino) ^{6,8}-prostaglandin I_1, (piriprost; Johnson et al. 1983) was injected in a 10mg/kg bolus (10 mg/ml solubilized by the addition of an equal weight of tris (hydroxymethyl)-aminomethane) at 30 min before, followed by a second 5 mg/kg bolus at 1.5 h after injury.

Stable prostacyclin: The stable analog of PGI_2 5-(hexahydro-5-hydroxy-6-(3-hydroxy-1-octenyl)-3a-methyl-2(IH)-pentalenylidene (ciprostene calcium, Aristoff et al. 1983) was given as a constant infusion in saline of either 1 or 3 µg/kg/min beginning at 30 min before injury and continuing throughout the experiment. The total infusion volume in both series was 17 ml. In a third series, a 10 mg/kg bolus of the thromboxane synthetase inhibitor U-63557A was given at 30 min before injury followed immediately by a 1 µg/kg/min constant infusion of ciprostene calcium.

Alpha tocopherol and selenium: Pretreatment of animals with a combination of the antioxidants alpha tocopherol (vitamin E) and selenium was carried out as described by Anderson and Means (1984). Cats were orally dosed once daily for the 5 days prior to the experiment with 1000 IU of d,l alpha tocopherol and 50 µg selenium.

Results and Discussion

The influence of pretreatment with the various test drugs on injury-induced white matter ischemia development is indicated in Table 2. Both of the cyclooxygenase inhibitors, IBU and MEC, given 30 min before injury attenuated subsequent ischemia development. While in the IBU (single dose) group the SCBF did fall slightly below 10 ml/100 g/min at 4 h, this probably reflects a short elimination half-life for IBU since in the double-dose IBU group the mean

Table 2. Effects of various drug pretreatment on post-traumatic spinal white matter ischemia development after a 500 g-cm cord contusion[a]

Drug treatment	No. of cats	Blood flow after injury (ml/100 g/min) 10 min	2 h	4 h
Vehicle	8	12.5 ± 1.6	8.6 ± 1.1[b]	7.7 ± 1.0[b]
Ibuprofen (cyclooxygenase inhibitor)	4	15.7 ± 2.4	11.0 ± 1.1	9.2 ± 1.3
Ibuprofen (2 dose)	4	12.5 ± 0.8	12.7 ± 1.5	10.3 ± 1.8
Meclofenamate (cyclooxygenase inhibitor)	4	14.1 ± 2.0	13.5 ± 1.9[c]	12.4 ± 1.3[c]
U-63557A (thromboxane synthesis inhibitor)	4	11.4 ± 0.6	8.0 ± 1.9	7.0 ± 2.0
Piriprost (lipoxygenase inhibitor)	4	12.6 ± 1.8	9.0 ± 0.7	8.2 ± 1.5[b]
Ciprostene calcium (stable PGI_2) (1 μg/kg/min)	4	13.1 ± 0.3	9.0 ± 1.3	9.7 ± 2.0
Ciprostene calcium (3 μg/kg/min)	4	11.6 ± 1.6	8.2 ± 1.9[b]	8.9 ± 2.6
Ciprostene calcium (1 μg/kg/min) + U-63557A	5	12.7 ± 1.1	11.8 ± 0.6	10.9 ± 0.4[c]
Vitamin E + selenium	4	16.1 ± 2.0	13.5 ± 2.3[c]	14.9 ± 2.0[c]

[a]Values represent mean + standard error
[b]$p<0.05$ by paired t-test versus 10-min value
[c]$p<0.05$ by unpaired t-test versus vehicle group at same time point

SCBF did stay up throughout the experiment. Moreover, the similar effect of MEC confirms that cyclooxygenase inhibition is associated with a preservation of postinjury SCBF.

In contrast, neither treatment with the TXA_2 synthesis inhibitor U-63557A nor the lipoxygenase inhibitor piriprost had any effect, thereby implying that TXA_2 and leukotriene do not figure prominently in ischemia development. Concerning TXA_2, this is somewhat surprising since a recent study in rabbits has shown that there is a greater than 4-fold elevation in spinal tissue levels of the stable metabolite TXB_2 by 1 hour after contusion injury. An explanation may be provided by studies showing that cat pial arterioles are unresponsive to topical TXA_2 (Kontos et al. 1981). Furthermore, the TXA_2 synthesis inhibitor OKY-1581 has been reported to have no effect on platelet aggregation induced by mouse pial vessel injury (Rosenblum and El Sabban 1983).

As an alternative to TXA_2, the efficacy of cyclooxygenase inhibition in contrast to the lack of effect of thromboxane synthesis inhibition may indicate that another prostanoid is an important contributor to the production of post-traumatic ischemia. In this regard, a large body of experimental evidence supports PGF_{2a} as a candidate. Much of this evidence had been recently reviewed by White and Hagen (1982) and includes the following: (1) CNS microvessels normally synthesize and contain a large concentration of PGF_{2a}; (2) PGF_{2a} is a consistent cerebrovascular vasoconstrictor; (3) intracarotid infusion of PGF_{2a} dramatically decreases CNS blood flow as measured by xenon clearance, and (4) abnormally elevated cerebrospinal fluid levels of PGF_{2a} have been reported in patients after subarachnoid hemorrhage, cerebral hemorrhage, spinal injury and various neurosurgical procedures. An additional potential pathophysiological feature of PGF_{2a} is that this prostaglandin can activate brain phospholipase A_2 in what might be described as a positive feedback process (Moskowitz et al. 1984).

Direct support for a critical role of PGF_{2a} in the generation of post-traumatic CNS ischemia has been provided by the elegant work of Pappius and Wolfe (1983). They have found that freezing injury of rat brain causes an almost 40-fold elevation in brain tissue levels of PGF_{2a} together with significant depression in brain functional activity as measured by glucose utilization. Furthermore, the postinjury rise in PGF_{2a} and the decrease in glucose utilization are prevented by pretreatment with either ibuprofen or indomethacin.

The current study, in addition to demonstrating a beneficial action of cyclooxygenase inhibition, has provided some data to suggest a limited effect of PGI_2 to antagonize post-traumatic ischemia production. For instance, a constant 1 µg/kg/min i.v. infusion of the stable PGI_2 analog ciprostene calcium appears to maintain SCBF better than no treatment based on the higher mean values, but not as well as cyclooxygenase inhibition. Raising the infusion rate to 3 µg/kg/min does not result in any greater effect and, in fact, the mean SCBF values in this group of cats are generally less than in the group that received the lower ciprostene calcium infusion.

On the other hand, prior administration of the TXA_2 synthesis inhibitor U-63557A is clearly seen in Table 2 to have augmented the support of SCBF by the stable PGI_2 analog. This suggests that TXA_2 may contribute to ischemia development after all. While the injury-induced increase in TXA_2 (Hsu et al. 1983) may be, by itself, relatively inconsequential, it may nevertheless have a sufficient action to limit the efficacy of exogenous PGI_2 (i.e. ciprostene calcium). Thus, when TXA_2 formation is inhibited, the vasodilator and platelet aggregation inhibiting properties of ciprostene calcium are enhanced.

An additional mediator of ischemia may be the immediate precursors for both PGF_{2a} and TXA_2, the cyclic endoperoxides, the formation of which would also be prevented by cyclooxygenase inhibition. The endoperoxides, if produced in sufficient quantity, may directly cause vasospasm due to peroxidative damage to the endothelium and consequent compromise of PGI_2 formation. For example, Sasaki et al. (1981) have observed that intracisternal injection of a hydroperoxide metabolite of arachidonic acid into dogs can directly induce progressive cerebral vasospasm.

The last series of experiments in Table 2 demonstrates that intensive pretreatment with the antioxidants vitamin E and selenium can

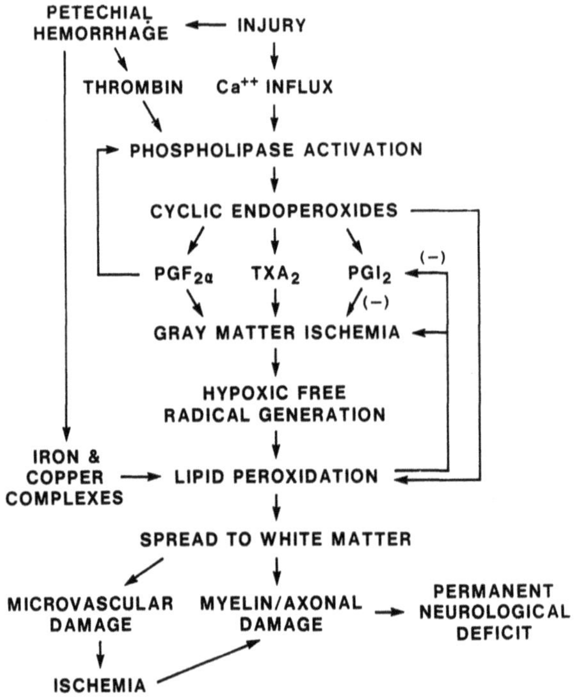

Fig. 1. Pathogenesis of post-traumatic spinal cord ischemia

completely prevent the post-traumatic fall in white matter SCBF. These data support a critical role for hypoxic free radical generation and lipid peroxidative damage to the spinal microvasculature in ischemia development. This conclusion is also supported by biochemical evidence for spinal tissue lipid peroxidation occurring within an hour after contusion injury (Clendenon et al. 1978; Hall and Braughler 1982a; Demopoulos et al. 1982; Pietronigro et al. 1983). It should also be considered that vascular damage by lipid peroxidation would be largely irreversible.

Ischemic hypoxia and the associated oxygen-derived free radical generation, which lead to lipid peroxidation, probably begin in the central gray matter. Gray matter blood flow is decreased very rapidly after spinal injury (Means and Anderson 1985) due to a number of potential factors discussed above, including mechanical damage to the more dense microvasculature, excessive generation of PGF_{2a} and TXA_2, peroxidative impairment of endothelial PGI_2 production, and perhaps the release of vasoactive neurotransmitters (e.g., 5-HT) as well. The peroxidation of membrane lipids is catalyzed by iron and copper complexes provided by gray matter petechial hemorrhages and is a self-perpetuating process that would gradually spread to the circumferential white matter tracts. There, lipid peroxidation would irreversibly damage myelin and axons directly as well as secondarily to ischemia caused by peroxidative microvascular destruction. This sequence of events is depicted in Fig. 1.

Finally, the identical dose-response relationship for the ability of high dose MPSS to attenuate both post-traumatic lipid peroxidation (Hall and Braughler 1982a) and white matter ischemia (Hall et al. 1984) also suggests an association between these two effects. Furthermore, Anderson et al. (1982) have recently demonstrated that early high dose MPSS administration after cat spinal cord compression injury acts to improve cord energy metabolism together with an anatomical preservation of both the gray and white matter microvasculature. The efficacy of MPSS in this regard may be due, in large part, to its antioxidant properties (Hall and Braughler 1981; Anderson et al.1983; Braughler 1984) as well as a direct vasodilator action, a decrease in the vascular responsiveness to vasoconstrictor neurotransmitters, and a reduction in the formation of vasoactive prostanoids (see review by Hall and Braughler 1982a).

References

Anderson DK, Means ED (1984) Free radical induced lipid peroxidation in spinal cord: $FeCl_2$ induction and protection with antioxidants. Neurochem Pathol (to be published)

Anderson DK, Means ED, Waters TR, Green ES (1982) Microvascular perfusion and metabolism in injured spinal cord after methylprednisolone treatment. J Neurosurg 56:106-113

Anderson DK, Means ED, Green ES, Waters TR (1983) Enhancement of Na^+-K^+-ATPase activity and prevention of Fe^{++} initiated free radical-induced lipid peroxidation in spinal cord by methylprednisolone. Neuroscience Abstracts 9:351

Aristoff PA, Johnson PD, Harrison AW (1983) Synthesis of 9-substituted carbacyclin analogs. J Org Chem 48:5341-5348

Balentine JD, Spector M (1977) Calcification of axons in experimental spinal cord trauma. Ann Neurol 2:520-523

Braughler JM (1984) Inhibition of synaptosomal GABA uptake by lipid peroxidation and Ca^{++}: antioxidant effects of glucocorticoids. Neuroscience Abstracts 10:942

Braughler JM, Hall ED (1983) Lactate and pyruvate metabolism in injured cat spinal cord before and after a single large intravenous dose of methylprednisolone. J Neurosurg 59:256-261

Braughler JM, Hall ED (1984) Effects of multi-dose methylprednisolone sodium succinate administration on injured cat spinal cord neurofilament degradation and energy metabolism. J Neurosurg 61: (to be published)

Clendenon NR, Allen N, Gordon WA, Bingham WG (1978) Inhibition of Na^+-K^+-activated ATPase activity following spinal cord trauma. J Neurosurg 491:563-568

Demopoulos HB, Flamm ES, Pietronigro DD, Seligman ML (1980) The free radical pathology and the microcirculation in the major central nervous system disorders. Acta Physiol Scand [Suppl] 492:91-120

Demopoulos HB, Flamm ES, Seligman ML, Pietronigro DD, Tomasula J, DeCrescito V (1982) Further studies on free radical pathology in the major central nervous system disorders: effects of very high doses of methylprednisolone on the functional outcome, morphology and chemistry of experimental spinal cord injury. Can J Physiol Pharmacol 60:1415-1424

Ducker TB, Salcman M, Lucas JT, Garrison WB, Perot PL (1978)Experimental spinal cord trauma: II.Blood flow, tissue oxygen evoked potentials in both paretic and plegic monkeys. Surg Neurol 10:66-72

Gorman RR, Johnson RA, Spilman CH, Aiken JW (1983) Inhibition of platelet thromboxane A_2 synthase activity by sodium 5-(3'-pyridinyl-methyl) benzofuran-2-carboxylate. Prostaglandins 26:325-342

Green BA, Kahn T, Klose KJ (1980) A comparative study of steroid therapy in acute experimental spinal cord injury. Surg Neurol 13: 91-97

Hall ED, Braughler JM (1981) Acute effects of intravenous glucocorticoid pretreatment on the in vitro peroxidation of cat spinal cord tissue. Exp Neurol 73:321-324

Hall ED, Braughler JM (1982a) Glucocorticoid mechanisms in acute spinal cord injury: A review and therapeutic rationale. Surg Neurol 18:320-327

Hall ED, Braughler JM (1982b) Effects of intravenous methylprednisolone on spinal cord lipid peroxidation and (Na^+ + K^+)-ATPase activity: dose-response analysis during the first hour after contusion injury in the cat. J Neurosurg 57:247-253

Hall ED, Wolf DL, Braughler JM (1984) Effects of a single large dose of methylprednisolone sodium succinate on post-traumatic ischemia: dose-response and time-action analysis. J Neurosurg 61: (to be published)

Happel RD, Smith KP, Banik NL, Powers JM, Hogan EL, Balentine JD (1981) Ca^{2+}-accumulation in experimental spinal cord trauma. Brain Res 211:476-479

Hsu CY, Halushka PV, Hogan EL, Banik NL, Lee WA, Perot PL (1983) Altered synthesis of thromboxane and prostacyclin in spinal cord contusion. Neurology 33 [Suppl 2]:146

Johnson HG, McNee ML, Bach MK, Smith HW (1983) The activity of a new, novel inhibition of leukotriene synthesis in Rhesus monkey Ascaris reactors. Int Arch Allergy Appl Immunol 70:169-173

Jonsson HT, Daniell HB (1976) Altered levels of PGF in cat spinal cord tissue following traumatic injury. Prostaglandins 11:51-61

Kontos HA, Wei EP, Ellis EF, Dietrich WD, Povlishock JT (1981) Prostaglandins in physiological and in certain pathological responses of the cerebral circulation. Fed Proc 40:2326-2330

Lewis DA, Symons AM, Ancill RJ (1970) The stabilization-lysis action of anti-inflammatory steroids on lysosomes. J Pharm Pharmacol 22:902-908

Means ED, Anderson DK (1985) The pathophysiology of acute spinal cord injury. In: Davidoff RA (ed) Handbook of the spinal cord, vol 2. Dekker, New York (to be published)

Means ED, Anderson DK, Waters TR, Kalaf L (1981) Effect of methylprednisolone in compression trauma to the feline spinal cord. J Neurosurg 55:200-208

Moskowitz N, Schook W, Puszkin S (1984) Regulation of endogenous calcium-dependent synaptic membrane phospholipase A_2. Brain Res 290:273-280

Nemecek S, Suba P, Cerman J (1977) Serotonin in contused spinal cord. Acta Neurochir 39:53-58

Pappius HM, Wolfe LS (1983) Effects of indomethacin and ibuprofen on cerebral metabolism and blood flow in traumatized brain. J Cereb Blood Flow Metabol 3:448-459

Pietronigro DD, Hovsepian M, Demopoulos HB, Flamm ES (1983) Loss of ascorbic acid from injured feline spinal cord. J Neurochem 41:1072-1076

Rosenblum WI, El-Sabban F (1983) Effect of OKY-1581, a thromboxane synthetase inhibitor on the platelet aggregation and vasodilation induced by injury of mouse pial arterioles. J Cereb Blood Flow Metabol 3:122-126

Sasaki T, Wakai S, Asano T, Watanabe T, Kirino T, Sano K (1981) The effect of a lipid hydroperoxide of arachidonic acid on the canine basilar artery. J Neurosurg 54:357-365

Senter HJ, Venes JL (1978) Altered blood flow and secondary injury in experimental spinal cord trauma. J Neurosurg 49:569-578

Stokes BT, Fox P, Hollinden G (1983) Extracellular calcium activity in the injured spinal cord. Exp Neurol 80:561-572

Wagner KR, Tornheim PA (1983) Acute changes in regional cerebral metabolite concentrations following experimental blunt head trauma. Neuroscience Abstracts 9:599

White RP, Hagen AA (1982) Cerebrovascular actions of prostaglandins. Pharmacol Ther 18:313-331

Young W, Flamm ES (1982) Effect of high-dose corticosteroid therapy on blood flow, evoked potentials, and extracellular calcium in experimental spinal injury. J Neurosurg 57:667-673

Young W, DeCrescito V, Tomasula JJ (1982) Effect of sympathectomy on spinal cord blood flow autoregulation and posttraumatic ischemia. J Neurosurg 56:706-710

Inflammation in Experimental Spinal Cord Injury

E. D. Means, D. K. Anderson, P. Demediuk, R. D. Saunders, and L. A. Horrocks

1 Veterans Administration Medical Center, Cincinnati, OH, USA
2 Department of Physiological Chemistry, Ohio State University, College of Medicine, Columbus, OH, USA

Introduction

We recently reported the presence of acute inflammation and neuronophagia by polymorphonuclear leukocytes (PMN) in feline spinal cord gray matter following experimental spinal cord compression injury (Means and Anderson 1983a). Review of our histological material showed maximal inflammation in gray matter between 8 and 24 h, whereas the peak response was delayed in white matter until 3 days. It is not surprising to observe acute inflammation in traumatic injury to the central nervous system, but the presence of neuronophagia by PMN was unexpected. Additionally, the temporal sequence of the appearance of acute inflammatory cells first in gray matter, then in white matter, may relate to certain pathophysiological mechanisms following blunt trauma.

It has been our thesis (Means and Anderson 1984) that although ischemia plays a role in cellular injury following blunt trauma to the spinal cord, a primary injury to neuronal, glial, and vascular membranes is an important contributing factor. Evidence from our laboratory suggests that free radical induced lipid peroxidation and the release of arachidonic acid presumably by the activation of membrane lipases are early posttraumatic membrane events. Whether free radical induced lipid peroxidation per se can lead to the liberation of arachidonate from neuronal membranes is currently under investigation in our laboratory. The release of arachidonic acid makes this polyunsaturated fatty acid (PUFA) available to the enzymes cyclooxygenase and lipoxygenase, which convert arachidonate to a number of active metabolites including prostaglandins, prostacyclin, thromboxanes, and leukotrienes. The consequences of the formation of these substances when present in significant concentrations are tissue edema, inflammation, and ischemia.

The purpose of this chapter is to report some selected morphological and biochemical studies of experimental spinal cord injury and discuss the relationship of these data to the acute inflammatory response observed in spinal gray and white matter.

Methods and Materials

Animal Model

Adult mongrel cats were anesthetized with intraperitoneal pentobarbital sodium (30 mg/kg), the vertebral column exposed in the upper

Experimental Brain Research, Suppl. 13

lumbar region, and laminectomies performed as previously described (Anderson et al. 1976; Anderson et al. 1978). The spinal cord was compressed by placing a 170 g weight extradurally for either 1, 3, or 5 min.

Light Microscopy

Anesthetized cats were killed by intra-aortic perfusion fixation using a Harvard peristaltic pump as previously described (Means and Anderson 1983a). Animals were killed at 1/2 h, 1 h, 2 h, 4 h, 8 h, 24 h, 3 days, 7 days, and 14 days after injury. Laminectomized control animals were also killed in the same manner at each time period. The tissue was processed in paraffin and 10-μm sections stained with hematoxylin and eosin (H&E), thionin, Masson trichrome, phosphotungstic acid hematoxylin (PTAH), and Holtzer methods.

Cultures

The spinal cords from 12-14 day mouse embryos were removed and placed in Hepes buffer. Trypsin (0.25%) was added to the medium free cords and incubated for 30 min at 35°C in 10% CO_2. Trypsinized cords were added to 3 ml of minimum essential medium [(MEM) - Gibco - #82-0234] to which L-glutamine was added. The trypsin was removed and the tissue triturated. A cell suspension (2-3 ml of MEM/cord - approximately 10^6 cells/ml) was added to each 35-mm glass culture dish containing a Vitrogen coated 22-mm cover slip or 22-mm Vitrogen-coated Aklar plate. Cultures were left in 10% horse serum (HS)/calf serum (CS) in MEM for 12-24 h, then changed to 5% HS/MEM/N_3. Cultures remained undisturbed until pulsed (2-6 days) with fluorodeoxyuridine [(FUDR), Hoffmann-LaRoche]. Subsequently the cultures were fed 2-3 times/week with 5% HS/MEM/N_3, using 1/2 change of medium.

Three-week-old dissociated murine spinal cord cultures were divided into three groups: (1) Culture medium was removed from 35 mm culture dishes containing 22-mm Aklar cover slips and Hepes buffer containing 50 μm ferrous chloride ($FeCl_2$) was added for 1/2 h. Buffer containing $FeCl_2$ was removed and MEM was replaced. Sixteen hours after the addition of $FECl_2$, the Aklar cover slips with neuronal monolayers were air dried and stained with the Wright-Giemsa method. (2) A second group of cultures were treated in an identical fashion as described above but a solution containing 10^4 /ml of PMN was added to the culture dishes following removal of the $FeCl_2$ -buffer. (3) A third group of cultures were treated similarly but PMN were added without prior peroxidation with $FeCl_2$.

Lipids and Eicosanoids

All spinal cord samples were frozen in situ with liquid nitrogen (Anderson et al. 1980) and were stored at -70°C prior to analysis.

Experimental animals were categorized into three groups. Group one consisted of animals for which the vertebral column was exposed and the spinal cord frozen without laminectomy (i.e., through an intact vertebral column). In group two, laminectomies were performed and recovery periods of 10, 60, 90, and 120 min were allowed prior to freezing the spinal cord. Group three animals were laminectomized, allowed to recover for 90 min, and then subjected to compression trauma for 1 min, 3 min, or 5 min followed by immediate freezing; or for 5 min, followed by a 5-min, 15-min, or 30-min recovery period prior to freezing.

Each dura-free frozen sample of spinal cord was prepared for lipid extraction as follows. All work was done in a -20°C stand-up freezer. First, the samples were thoroughly cleaned of meninges and blood by scraping with a scalpel. The samples were next cross-sectioned into three segments. Each segment was then cored with a flat-ended 14 gauge punch into separate portions of either predominantly gray matter (central core) or predominantly white matter (outer area).

Total lipids were extracted either with hexane-isopropanol according to Hara and Radin (1978) or with a two-phase chloroform-methanol-water system (Rehncrona et al. 1980). Lipids extracted with hexane-isopropanol were placed on a small Unisil silicic acid column and eluted with chloroform (α-tocopherol, neutral lipids), methyl formate (prostaglandins and other oxidized metabolites of arachidonic acid, cerebrosides), acetone (remainder of cerebrosides, other glycolipids), and methanol (phospholipids). Chloroform-methanol extracts were subjected to TLC on silica gel G-60 with hexane-diethyl ether-acetic acid (110/90/4) to separate free fatty acids and diglycerides. Lipid bands were visualized with TNS under ultraviolet light (Jones et al. 1982). The free fatty acid and diglyceride bands from TLC were scraped into screw-cap test tubes and esterified in 3.0 ml 0.38 M H_2SO_4 in methanol:toluene (1:1) at 65°C for 4 h. The fatty acid methyl esters were extracted with hexane. Fatty acid methyl esters were separated by gas-liquid chromatography using a Shimadzu GC-8A gas chromatograph. Alltech CS-10 (10% on 100/120 mesh Chrom WAW) was used as the stationary phase in a 5.4 m x 2.5 mm ID glass column with N_2 as the carrier gas and an operating temperature of 195°C. Detector linearity was checked using commercial standards. Chromatograms were recorded and peak areas were calculated using a Nelson Analytical 760 intelligent interface, a Hewlett-Packard 95 desktop computer, and Nelson Analytical software.

Lipid phosphorus (P) was assayed by the method of Rouser et al. (1969). Phospholipids were separated by two dimensional TLC (Horrocks and Sun 1972). Tissue cholesterol was analyzed by the method of Bowman and Wolf (1962). Radioimmunoassay (RIA) was performed directly on methyl formate fraction from silicic acid columns. At least 95% of the prostaglandins are recovered in this fraction and the antibodies cross-react less than 1% (Fertel et al. 1981). Antibodies for radioimmunoassay of prostaglandins E_2, $F_2\alpha$, and 6-keto $F_1\alpha$ and thromboxane B_2 were supplied by Dr. Richard Fertel. Radioimmunoassays were done directly with aliquots of the fractions.

Results

Light Microscopy

Previously reported was the presence of acute inflammation and neuronophagia by PMN in spinal gray matter following spinal cord compression injury in cats (Means and Anderson 1983a). The acute inflammatory response was maximal in gray matter between 8 and 24 h following compression. Only rarely were PMN observed in white matter

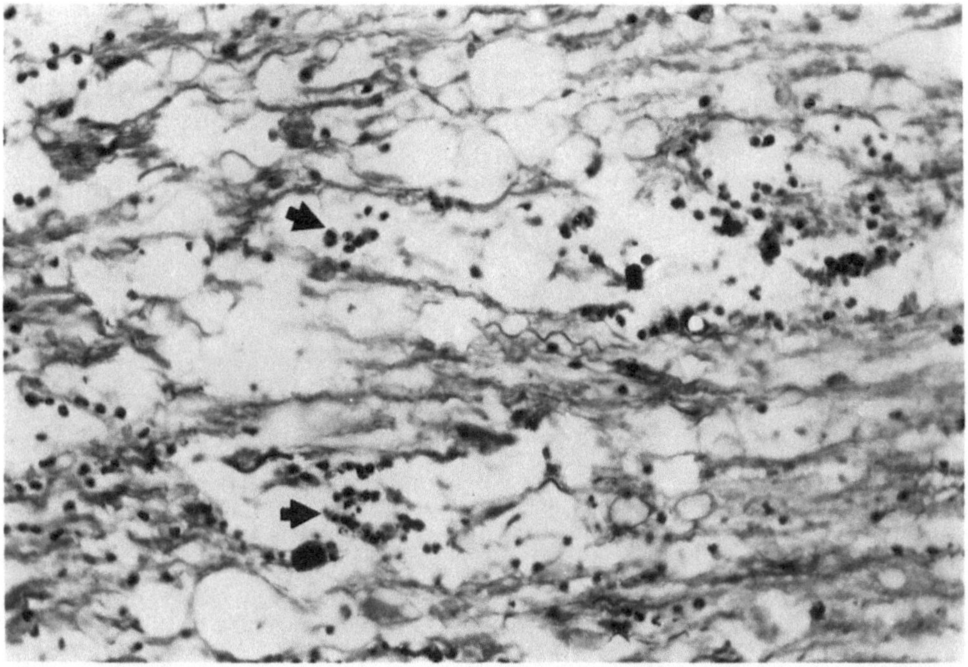

Fig. 1. A saggital section of the feline spinal cord which shows PMN scattered throughout vacuolated white matter. Clusters of PMN (arrows) are present in vacuoles. The animal was killed at 3 days. Light micrograph, PTAH stain, x 1085

at this time. Review of our material showed acute inflammation in white matter which was maximum at 3 days. Polymorphonuclear leukocytes persisted in degeneration white matter through 7 days postinjury. Clusters of PMN which occurred around necrotic blood vessels and enlarged periaxonal spaces were most marked in 3-day material (Fig. 1). Occasional PMN were noted immediately around and within axonal swellings. Whether these PMN were phagocytic could not be determined with light microscopy. Acute inflammatory cells were scattered among necrotic myelinated axons, red blood cells, and macrophages. Few PMN were observed in adjacent necrotic gray matter.

Cultures

Ferrous chloride (50 µM) alone produced only minor changes in neurons in culture at 16 h. Occasionally observed were neurons with granular cytoplasm and fragmentation of neurites. Cultures not exposed to $FeCl_2$ but with PMN added revealed only minor alterations in neurons. Leukocytes tended to become aligned on neuronal processes and on somal surfaces (Fig. 2). Rarely, neurons with adjacent PMN showed necrosis. Cultures treated with $FECl_2$ with added PMN showed marked changes at 8 and 16 h; aggregates of degenerating PMN and neurons were common at 8 h but neurons were rarely observed at 16 h. Granulocytes appeared as inclusions in some degenerating neurons (Fig. 3). Occasionally, degenerating neurons were noted among

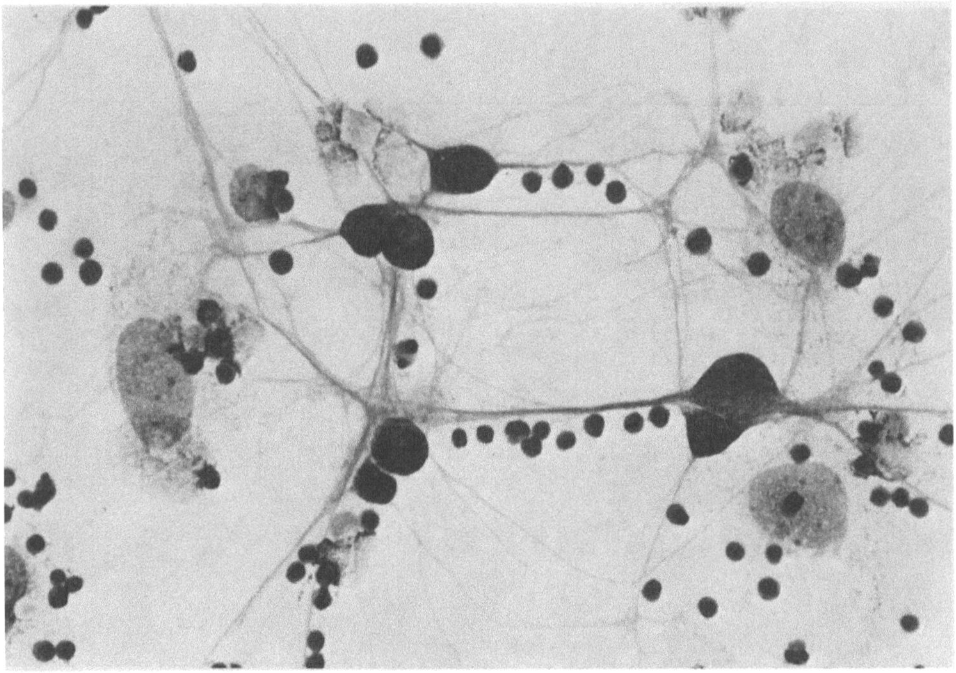

Fig. 2. A murine spinal cord culture showing several neurons. White blood cells, predominantly PMN, are aligned on neuronal processes and scattered throughout culture. White blood cells were added to the culture medium and the tissue was fixed 16 h later. Light micrograph, Wright-Giemsa stain, x 1763

amorphous, granular debris (Fig. 4). The astrocytic monolayer showed some shrinkage of individual cells but cells remained essentially intact.

Lipids and Eicosanoids

Controls for these studies were cats that underwent laminectomy and were then allowed to recover for 90 min before the spinal cord was traumatized. This procedure was necessary because laminectomy caused a 5-fold increase in total free fatty acid (FFA) levels of spinal cord gray matter within 10 min of laminectomy that did not normalize until 60-90 min postlaminectomy. There was a significant fourfold increase in total FFA levels in the gray matter of cat spinal cord within 1 min of compression trauma (Fig. 5). By 3 min the FFA levels were eightfold above control and were still elevated 30 min after the injury had been terminated (Fig. 5). Total FFA concentration in white matter doubled within 1 min after the initiation of compression trauma and remained essentially at this level through 5 min of compression plus 5 min after the weight was removed from the spinal cord. However, by 15 min after the termination of compression, total FFA had risen to levels fourfold above control (Fig. 6). All of the PUFAs analyzed (20:4, 22:5, and 22:6) were elevated in traumatized spinal cord gray matter within 5 min after initiation of compression

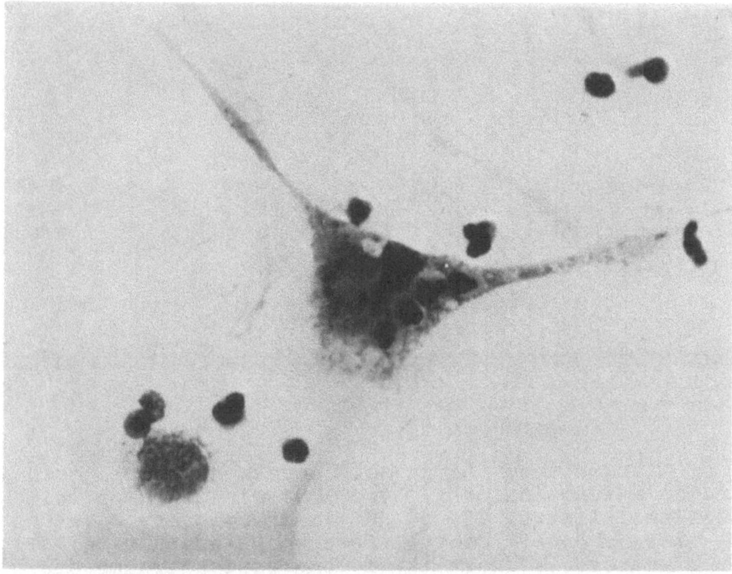

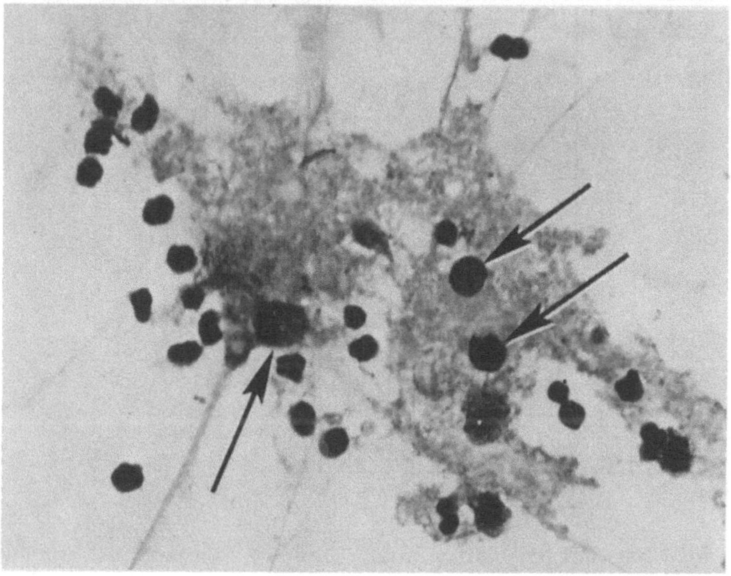

Fig. 3 (above). A murine spinal cord culture showing a degenerating pyramidal shaped cell, probably a neuron. Four PMN are present in vacuolated cytoplasm. The culture was exposed to $FeCl_2$ for 1/2 h, then the medium was replaced and PMN 10^4 /ml added. The tissue was fixed 8 h after $FeCl_2$ exposure. Light micrograph, Wright-Giemsa stain, x 2170

Fig. 4 (below). A murine spinal cord culture showing granular debris, probably neuronal. Leukocytes, predominantly PMN, marginate and are scattered among the debris. Neuronal cuclei (arrows) are apparent. The culture was exposed to $FeCl_2$ for 1/2 h then the medium replaced and PMN added. The tissue was fixed 16 h after $FeCl_2$ exposure. Light micrograph, Wright-Giemsa stain, x 1763

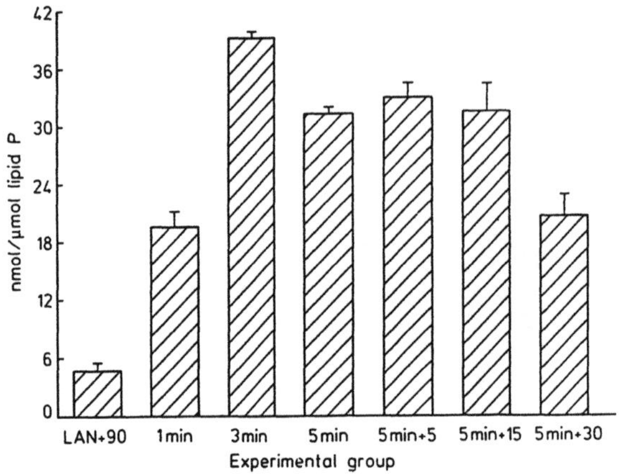

Fig. 5. Total free fatty acid levels in spinal cord gray matter plotted as a function of 1, 3, or 5 min of compression or 5 min of compression plus 5, 15, or 30 min of recovery. Values are means (± SD) of four cats. Controls are cats undergoing laminectomy plus 90 min of recovery. Details of compression trauma are given in the text

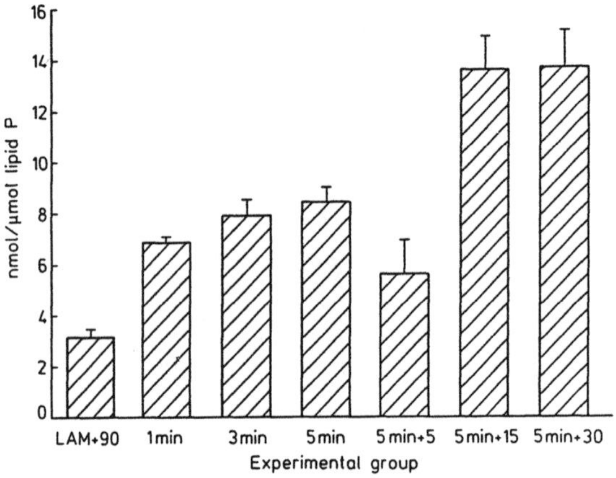

Fig. 6. Total free fatty acid levels in spinal cord white matter after either 1, 3, or 5 min of compression trauma or after either 5, 15, or 30 min of recovery from termination of 5 min of compression. Other details can be found in the legend to Fig. 1 and the text

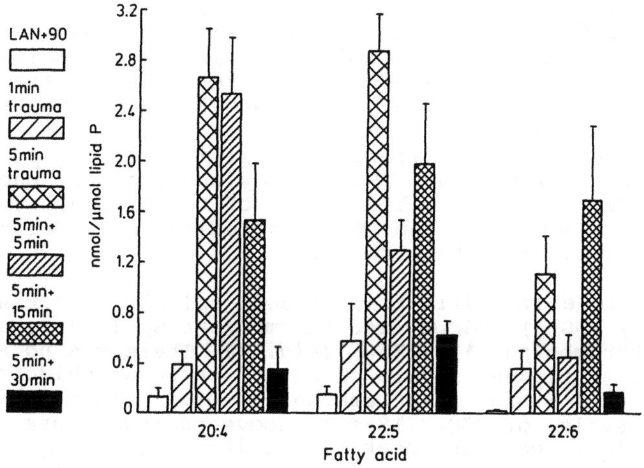

Fig. 7. Major polyunsaturated free fatty acids in feline spinal cord gray matter following either compression trauma or short term recovery from compression trauma to the spinal cord (times noted at left). See Fig. 5 and text for other details

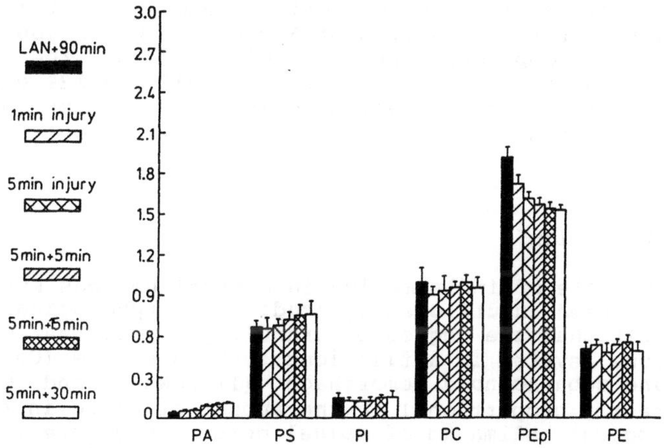

Fig. 8. Phospholipid composition of cat spinal cord after either 1, 3, or 5 min of compression trauma or after 5, 15, or 30 min of recovery from 5 min of compression trauma. Abbreviations: phosphatidic acid (PA), phosphatidylserine (PS), phosphatidylinositol (PI), phosphatidylcholine (PC), ethanolamine plasmalogens (PEplasm), and phosphatidylethanolamine (PE)

(Fig. 7). Arachidonic acid (20:4) levels were maximally elevated (20-fold) by 5 min of compression trauma (Fig. 7). Thereafter, arachidonate levels declined and had normalized by 30 min after the compression trauma had been terminated. The source of the released fatty acids is not known. However, there was a substantial posttraumatic decline in the ethanolamine plasmalogens (PEpl) that was not seen in other glycerophospholipids (i.e. phosphatidylserine, phosphatidylinositol, phosphatidylcholine, or phosphatidylethanolamine) (Fig. 8). Thus, hydrolysis of the plasmalogens must represent a major source of the PUFA (including arachidonate) that accumulate in the tissue after trauma.

Also elevated in traumatized spinal cord tissue were prostaglandin E_2 (PGE_2), $PGF_2\alpha$, and thromboxane B_2 (TXB_2), a stable metabolite of thromboxane A_2 (TXA_2) (Fig. 9). There was no significant increase in the tissue levels of any of these arachidonate oxidation products until 5 min after termination of 5 min of compression trauma. Peak levels of PGE_2 (18-fold above control) were seen at this time period, whereas peak tissue levels of $PGF_{2\alpha}$ (25-fold above control) were not achieved until 30 min after termination of the compression injury. Tissue concentrations of TXB_2 were tenfold above control by 5 min after injury and remained at this level for at least 30 min. Tissue levels of prostacyclin (as measured by its stable metabolite 6-keto $PGF_1\alpha$) were not altered significantly by compression trauma.

Within 1 min following compression trauma, the apparent tissue cholesterol level had declined 11% (Fig. 10). By 5 min, tissue cholesterol was 15% below control and was still at this level 30 min after the termination of 5 min of compression trauma. Since cholesterol comprises approximately 45% of the total lipid of plasma membranes (Houslay and Stanley 1982), a 12% decline would suggest that there is substantial damage to the membranes within 1 min after the initiation of trauma.

Discussion

The role of inflammation in central nervous system trauma and stroke has been insufficiently studied. Morphological investigations suggest that the degree of granulocyte infiltration of the central nervous system in afflictions such as stroke (Garcia and Kamijyo 1974) and subarachnoid hemorrhage (Iwanowski and Olszewski 1960) is directly proportional to the amount of blood in the tissue. The histological hallmarks of spinal cord injury are progressive hemorrhagic necrosis of gray and white matter and infiltration of the tissue with PMN. In addition to acute inflammation, we found neuronophagia by PMN in feline spinal cord gray matter following compression injury. This observation suggests that a chemoattractant is formed by the neuron following blunt trauma to the spinal cord which leads to directional migration of PMN toward neurons. We have proposed that the progressive necrosis (autodestruction) of the spinal cord which occurs in the first 24 h following trauma is due to a series of pathogenetic factors (Means and Anderson 1984). One such factor is blood which extravasates into spinal cord parenchyma following trauma and consequently is directly apposed to neuronal and glial membranes. Blood contains several substances which may affect the neuronal membrane, including elemental iron or iron complexed to a variety of organic molecules such as hemoglobin. Both iron and iron complexes are potent free radical generators which can lead to peroxidation of membrane PUFAs. Moreover, iron "decompartmentalized" from intracellular stores at the time of impact (i.e. from mito-

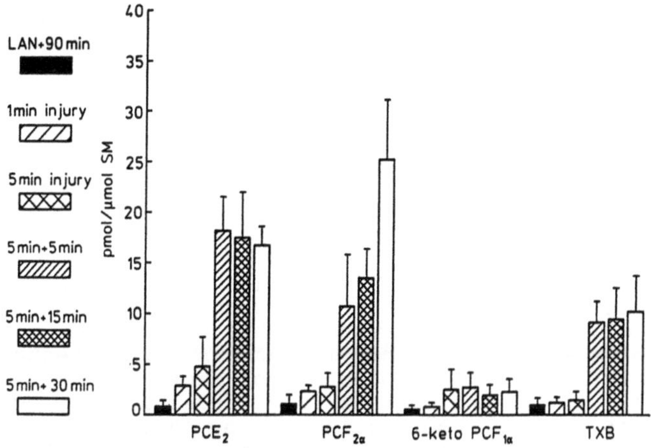

Fig. 9. Prostaglandin, thromboxane, and prostacyclin levels of feline spinal cord following compression trauma and termination of compression trauma (times on abscissa). Abbreviations: prostaglandin E_2 (PEG_2), prostaglandin $F_{2\alpha}$ ($PGF_{2\alpha}$), 6-keto-prostaglandin $F_{1\alpha}$ (6-keto $PGF_{1\alpha}$) and thromboxane B_2 (TXB_2)

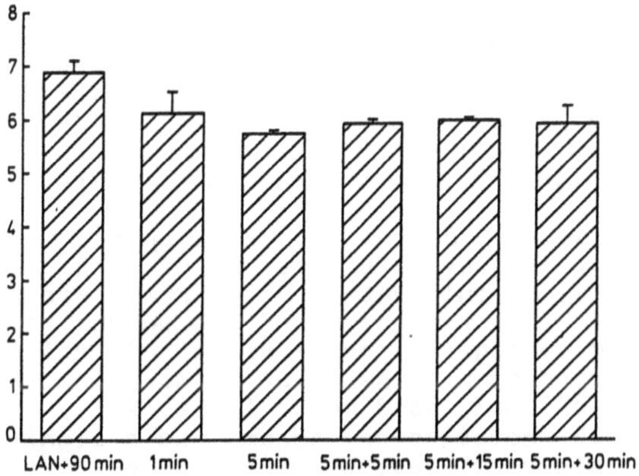

Fig. 10. Spinal cord cholesterol levels following the lengths of compression trauma or times after termination of 5 min of compression trauma indicated on the abscissa. Sphingomyelin was used as the reference base for the cholesterol measurements

chondria) might contribute to peroxidation of susceptible PUFA molecules at an early stage. Blood also contains serine proteases, for example thrombin, which recently has been reported to bind to neurons (Means and Anderson 1983b). Thrombin is a potent stimulus for arachidonic acid release in platelets and could potentially produce a similar effect in neurons and glia following spinal cord injury. It is, therefore, possible that there are a variety of substances contained in blood which perturb the neuronal membrane. Some of these may lead to the release of arachidonate, which subsequently is metabolized to molecules which are chemotactic for PMN.

We have shown that PMN produce marked neuronal necrosis in murine spinal cord cultures following exposure of the culture to a free radical generating system. This observation adds additional evidence to recent reports that in certain experimental systems, free radicals lead to the production of factors chemotactic for PMN. Perez et al. (1979) showed that when arachidonic acid was acted upon by a free radical generating system that produced superoxide anion, a lipid product was formed which possessed potent chemotactic activity for PMN. Petrone and colleagues (1980) described the inhibition of leukocyte infiltration into a challenged site by the intravenous administration of superoxide dismutase. This was used as evidence to suggest that neutrophil-generated superoxide reacts with an extracellular precursor to generate a substance chemotactic for neutrophils. The chemotactic factor described by the above investigators has to our knowledge not been identified. It is possible that free radicals lead to the liberation of arachidonic acid from neuronal membranes (Del Maestro et al. 1980; Chan et al. 1982) which subsequently is oxidized by lipoxygenase, forming leukotrienes, some of which are chemotactic for PMN. Whether arachidonate is nonenzymatically oxidized by free radicals forming chemotactic molecules (Means and Anderson 1983a; Perez et al. 1979) or they liberate arachidonate acid by activating membrane lipases is problematic. We are currently testing this hypothesis in our laboratory.

These data demonstrate that spinal cord trauma causes a rapid elevation in gray matter levels of total FFAs, including the polyunsaturate, arachidonate. Total FFAs were also elevated in white matter but this rise occurred subsequent to the increase in gray matter, indicating the radial spread of this injury. Upon termination of compression, there was a sharp increase in the biologically active arachidonic acid metabolites, PGE_2, $PGF_{2\alpha}$, and TXB_2. Also associated with compression trauma was a significant decline in spinal cord cholesterol levels within the first minute after initiation of injury.

The posttraumatic elevation in spinal cord FFAs indicates that membrane lipases were activated as a result of trauma. The stimulus for this early activation is not known although there are several possibilities, including the posttraumatic production of free radicals (or free radical induced peroxidation of acylated PUFAs) and/or the traumatically induced deformation of membranes. The increase in tissue concentrations of prostaglandins and thromboxanes indicates the released arachidonic acid was rapidly oxidized by cyclooxygenase. Free arachidonate can also be oxidized by lipoxygenase, resulting in the formation of leukotrienes.

The contribution of arachidonate and its cyclooxygenase metabolites, the prostaglandins and thromboxanes, to inflammation is probably limited to alterations in vascular permeability and tissue edema. Chan et al. (1983b) reported that arachidonic acid played an important role in inducing cytotoxic edema using an in vitro system of rat cortical brain slices. These authors (Chan et al. 1983a) were also able to induce vasogenic and cytotoxic brain edema in rat cerebral cortex after the intracerebral injection of arachidonate. Prostacyclin as well as PGE_2 and $PGF_{2\alpha}$ may contribute to edema by potentiating in the edema caused by certain vasoactive agents such as histamine and bradykinin (Williams and Peck 1977). Arachidonate and the stable prostaglandins PGE_2 and $PGF_{2\alpha}$ contribute to vascular permeability and tissue edema but these substances apparently have no direct effect on PMN chemotaxis (Goetzl and Gorman 1978). Certain of the leukotrienes may also enhance capillary permeability and potentiate edema (Dahlen et al. 1981). Although we did not measure leukotriene levels in spinal cord tissue after compression injury, it is likely that this unique class of compounds were elevated since arachidonate was available as a substrate and the cyclooxygenase pathway was activated. Leukotriene B_4 (LTB_4) has little effect on vascular permeability, but is a powerful stimulant of PMNs, causing these cells to migrate, aggregate, and degranulate (Fountain et al. 1980; Hafstrom et al. 1981). It is therefore apparent that arachidonic acid and its cyclooxygenase and lipoxygenase metabolites play an important role in the acute inflammatory response and may contribute significantly to the autodestructive process extant in the spinal cord following blunt trauma.

O'Flaherty (1982) has clearly and succinctly summarized the course of inflammation following initiation. Using his summary and our data, we propose the following hypothetical scheme for the autodestructive changes in spinal cord tissue following blunt trauma.Blunt trauma to the spinal cord results in an immediate alteration to neuronal membranes causing a decrease in cholesterol and ethanolamine plasmalogens. Arachidonate is released due to activation of membrane lipases, making this substrate available to cyclooxygenase and lipoxygenase. Concurrently, decompartmentalized iron initiates free radical reactions in the neuronal plasma membrane and membranes of subcellular organelles. Prostaglandins, thromboxanes, and leukotrienes are formed which alter blood flow, vascular permeability, and platelet aggregatility, leading to ischemia, edema formation, and the activation, endothelial attachment, extravasation, and directional migration of leukocytes to the target site. It is important to mention that ischemia may play a role in the inflammatory process, as central nervous system ischemia is known to release fatty acids from cellular membranes, further contributing to the pool of free arachidonic acid (Horrocks, to be published). Blood extravasated into spinal cord parenchyma minutes after trauma further compounds the neuronal membrane abnormalities via iron-initiated free radical-induced lipid peroxidation and the liberation of arachidonate by thrombin (and perhaps other substances) activation of membrane lipases. The PMN near the inciting agent, perhaps LTB_4, are maximally activated and phagocytic. They release histolytic enzymes, free radical species, and proinflammatory factors such as PGE_2 and LTB_4 which can diffuse back to capillaries and promote further inflammation. This process is self-generating until the provocator is eliminated and mediators are no longer formed. Thus, trauma to the spinal cord initiates a series of self-propagating events which ultimately lead to tissue necrosis, cavitation, and the clinical accompaniment, paraplegia.

References

Anderson DK, Prockop LD, Means ED, Hartley LE (1976) Cerebrospinal lactate and electrolyte levels following experimental spinal cord injury. J Neurosurg 44:715-722

Anderson DK, Nicolosi GR, Means ED, Hartley LE (1978) Effects of laminectomy on spinal cord blood flow. J Neurosurg 48:232-238

Anderson DK, Means ED, Waters TR (1980) Feline spinal cord energy metabolism, normal and post-laminectomy. J Neurosurg 52:494-496

Bowman RE, Wolf RC (1962) A rapid and specific ultramicro method for total serum cholesterol. Clin Chem 8:302-309

Chan DH, Yurko M, Fishman RA (1982) Phospholipid degradation and cellular edema induced by free radicals in brain cortical slices. J Neurochem 381:525-531

Chan DK, Fishman RA, Caronna J, Schmidley JW, Prioleau C, Lee J (1983a) Induction of brain edema following the intracerebral injection of arachidonic acid. Ann Neurol 13:6251-632

Chan DH, Fishman RA, Chen S, Chew S (1983b) Effects of temperature on arachidonic acid-induced cellular edema and membrane perturbation in rat brain cortical slices. J Neurochem 41: 1550-1557

Dahlen S-E, Bjork J, Hedquist D, Arfors K-E, Hammarstrom S, Lindgrem J-A, Samuelsson B (1981) Leukotrienes promote plasma leakage and leukocyte adhesion in postcapillary venules: in vivo effects with relevance to the acute inflammatory response. Proc Natl Acad Sci USA 78:3887-3892

Del Maestro FR, Thaw HH, Bjork J, Planker M, Arfors K-E (1980) Free radicals as mediators of tissue injury. Acta Physiol Scand [Suppl] 492:43-57

Fertel R, Yetiv JZ, Coleman MA, Schwarz RD, Greenwald JE, Bianchine JR (1981) Formation of antibodies to prostaglandins in the yolk of chicken eggs. Biochem Biophys Res Commun 102:1028-1033

Fountain SW, Martin BA, Musclow CE, Cooper JD (1980) Pulmonary leucostasis and its relationship to pulmonary function in sheep and rabbits. Circ Res 46:175-180

Garcia JH, Kamijyo Y (1974) Evolution of histopathological changes after occlusion of a middle cerebral artery in primates. J Neuropathol Exp Neurol 33:408-421

Goetzl EJ, Gorman RR (1978) Chemotactic and chemokinetic stimulation of human eosinophil and neutrophil leucocytes by 12-L-hydroxy-5, 8, 10-heptadecatrienoic acid (HHT). J Immunol 120:526-531

Hafstrom I, Palmblad J, Malmsten CL, Radmark O, Samuelsson B (1981) Leucotriene B_4-a stereospecific stimulator for the release of lysosomal enzymes from neutrophils. FEBS Lett 130:146-148

Hara A, Radin NS (1978) Lipid extraction of tissues with a low-toxicity solvent. Anal Biochem 90:420

Horrocks LA (to be published) Metabolism and function of free fatty acids in brain. In: Eichberg J (ed) Phospholipids in the nervous system. Wiley, New York

Horrocks LA, Sun GY (1972) In: Marks N, Rodnight R (eds) Research methods in neurochemistry, vol 1. Plenum, New York, pp 223-231

Houslay MD, Stanley KK (1982) Dynamics of biological membranes. Wiley, New York, p 20

Iwanowski L, Olszewski J (1960) The effects of subarachnoid injections of iron-containing substances on the central nervous system. J Neuropathol Exp Neurol 19:433-438

Jones M, Keenan RW, Horowitz P (1982) Use of 6-p-toluidino-2-naphthalene sulfonic acid to quantitate lipids after thin layer chromatography. J Chromatogr 237:522-524

Means ED, Anderson DK (1983a) Neuronophagia by leucocytes in experimental spinal cord injury. J Neuropathol Exp Neurol 42:707-708

Means ED, Anderson DK (1983b) Thrombin binds to cells in murine spinal cord cultures. Neuroscience Abstracts 9:348

Means ED, Anderson DK (1984) In: Davidoff RA (ed) The pathophysiology of acute spinal cord injury. Dekker (to be published) (The spinal cord handbook, vol 5)

O'Flaherty (1982) Lipid mediators of inflammation and allergy. Lab Invest 47:314-329

Perez HD, Weksler B, Goldstein F (1979) A new mechanism for the generation of biologically active products from arachidonic acid. Clin Res 27:464

Petrone WF, English DK, Wong K, McCord JM (1980) Free radicals and inflammation: superoxide-dependent activation of a neutrophil chemotactic factor in plasma. Proc Natl Acad Sci USA 77:1159-1163

Rehncrona S, Smith DS, Abesson B, Westerberg E, Siesjo B (1980) Peroxidative changes in brain cortical fatty acids and phospholipids, as characterized during Fe^{2+} and ascorbic acid-stimulated lipid peroxidation in vitro. J Neurochem 34:1630-1638

Rouser G, Siakotos V, Fleischer S (1969) Quantitative analysis of phospholipids by thin layer chromatography and phosphorus analysis of spots. Lipids 1:85-86

VanRollins M, Ho SHK, Greenwald JE, Alexander M, Dorman NJ, Wong LK, Horrocks LA (1980) Complete separation by high performance liquid chromatography of metabolites of arachidonic acid from incubation with human and rabbit platelets. Prostaglandins 20:571-577

VanRollins M, Aveldano MI, Sprecher HW, Horrocks LA (1982) HPLC and TLC separations of underivatized prostanoids and precursosrs. Methods Enzymol 86:518-530

Williams TJ, Peck MJ (1977) Role of prostaglandin mediated vasodilatation in inflammation. Nature 270:530-532

Histopathology of Neural Transplants and Host Brain

G. D. Das

Department of Biological Sciences, Purdue University, West Lafayette, IN 47907, USA

Introduction

Transplantation of neural tissue in the mammalian brain involves manipulation of embryonic neural tissue and infliction of surgical trauma to the host brain in the process of transplantation. These procedures induce degeneration in the transplants and a pathologic reaction in the host brain, which vary in their severity depending upon the methods used. Thus, neural transplantation starts with pathologic events, and it is their resolution that results in successful or unsuccessful transplantation. It is therefore important to comprehend the nature of these pathologic processes and the course of their resolution if a fuller understanding of neural transplantation in relation to neural trauma is to be gained. Some observations on these aspects of neural transplantation are presented in this paper.

Neural transplantation depends upon many procedures adapted from experimental embryology, neuroanatomy, and microsurgery (Das 1974; Das et al. 1979). The first includes collection and dissection of embryos, and dissection and preparation of embryonic neural tissue for transplantation. The second involves preparation of the host animals for the reception of transplants, whether by a freehand (Das 1974; Das et al. 1979) or a stereotaxic method (Das and Ross 1982) or with the aid of micromanipulators. The precision of these procedures depends upon a sound knowledge of the surface anatomy of the brain, the anatomy of the deeper neural structures, the distribution of major blood vessels, and the location of sinuses. The third involves various aspects of microsurgery that are related to the injection or deposition of the neural transplants inside the host brain. Failure to observe adequate precautions contributes to flaws at any one or all of these steps, and they are likely to result in massive necrosis in the host brain and complete degeneration of the neural transplants. Thus, instead of achieving successful neural transplantation, one may achieve nothing but large cysts in the host brain, filled with fragments of atrophying neural transplants and surrounded by necrotic host brain tissue. However, with appropriate methods and observation of various precautions, it is possible to achieve successful transplantation consistently in a large number of preparations.

Of the various steps involved in neural transplantation, preparation of embryonic neural tissues and techniques of transplanting them in the host brain have a great bearing on the initial survival of the transplants and on pathologic changes in both the transplants and the host brain. Inadequate preparation of embryonic neural tissue contributes to nonviability of the transplants, and following trans-

Experimental Brain Research, Suppl. 13

plantation they undergo rapid necrosis and degeneration. Very little is gained by transplanting such nonviable neural tissue. Further, since all methods of transplantation are invasive by nature, they induce trauma to the host brain. Different methods induce different degrees of trauma. Those methods that cause massive damage to the host brain and produce extensive hemorrhagic conditions at the site of transplantation have adverse effects not only on the host brain but also on the very survival of the transplants. At the other end of the scale are the methods that are conservative, in that they induce minimum trauma to the host brain and cause the least bleeding owing to the injection procedures for neural tissue. These methods yield successful results in a large number of preparations. They involve slow and careful injection of small volumes of neural tissues with the aid of a syringe with a glass needle attached to it. Such injections can be done freehand or stereotaxically (Das et al. 1979; Das and Ross 1982). In this presentation, information is provided on the histopathological changes observed in the transplants and the surrounding host brain parenchyma following transplantation by the methods which induce minimum degeneration in the transplants and the least trauma to the host brain. Conditions that induce extensive damage to the host brain and that do not foster adequate survival of the transplants are then described in a separate section.

Materials and Procedures

Neocortical tissue obtained from 16-day rat embryos was used for transplantation, and it was transplanted in 3- to 4-month-old host animals. The transplants were injected in the forebrain region at the level of the head of the caudate nucleus. The methods of obtaining and preparing neural tissue and transplanting it were similar to those described earlier (Das et al. 1979). Each host animal received one transplant unilaterally, and it was 2.0 - 3.0 mm^3 in volume. The host animals were killed in a developmental sequence, at 4 h, and 1, 2, 3, 4, 5, 7, 10, 12, 18, 30, and 90 days after transplantation. The portions of brain containing transplants were histologically processed, and sections were stained with cresyl violet, H & E, and Bodian stains. The material was evaluated quantitatively as well as qualitatively.

Although the findings presented below are based upon intraparenchymal transplantation of neural tissue, other studies were conducted to evaluate the effects of various other experimental variables on the pathologic reaction in the host brain and survival of the transplants. They were use of nonviable neural tissue for transplantation, extensive bleeding caused inside the host brain at the site of transplantation, extensive damage to the host brain tissue at the site of transplantation, and extensive damage to the host surface brain structures. Observations on these variables are presented separately.

Observations

Some Quantitative Considerations

Although the factors contributing to the histopathologic reaction in the transplants and the host brain are different, there is still some degree of interaction between them. The developmental characteristics of neural transplants related to initial necrosis, recovery, and growth have a strong influence on the end result of the pathologic reaction in the host brain. The nature and degree of pathologic changes induced at the time of transplantation, also influence the growth, normal differentiation, and integration of the transplants.

Comprehension of the interaction between these events requires examination of some quantiative data. Various studies have addressed the quantitative aspects of transplants (Das 1983a, this volume; Das et al. 1980, 1983a,b; Oblinger and Das 1982, 1983), but only a brief overview of these findings is provided in this paper, which will aid in comparative evaluation of the effects of other experimental variables on pathologic changes in transplants and host brains.

Broadly speaking, quantitative analysis of a growing transplant and the host brain can be performed for the following variables: initial volume of the transplant and the volume of host brain tissue displaced; extent of necrosis in the transplant and the host brain; volume of blood in an around the transplant; growth of the transplant in relation to its growth potential; final volume of the transplant and volume of host brain displaced or lost; and changes in angiogenesis in relation to the growing transplant.

Studies devoted to these issues have shown, first, that when neural tissue is injected into the host brain it settles in a cavity. This cavity is created by the pressure of the injection; and the size of the cavity matches the volume of the neural transplant (Fig. 1a). However, if the transplantation technique is faulty the cavity will always be larger than the transplant, and the space between the two will be filled with blood. A cavity larger than the volume of the neural transplant is indicative of an unduly excessive trauma to the host brain.

Secondly, the manipulation of neural tissue prior to their transplantation inevitably results in some parts of the tissue becoming necrotic. Generally, about 50% of neural tissue becomes necrotic within 2 - 3 days following transplantation; the remaining tissue survives and shows subsequent growth (Das et al. 1983a). This means that necrotic reaction in an abnormally large volume of a neural transplant should be taken as indicative of inadequately monitored or excessive manipulation of the transplant or of the introduction of contamination into the host brain along with the transplant. At the same time, in the host brain a band 500 - 1000 µm thick immediately bordering the cavity is seen to show spongy characteristics indicating edema. Within this zone of edema the nerve cells appear shrunken and hyperchromatic and gradually die away. From the quantitative viewpoint, the nerve cells that die initially are confined within this region of edema. However, if extensive damage is caused to the host brain parenchyma, or if extraneous fluids, such as Ringer's solution and tissue culture media, or air bubbles are injected along with the neural tissue, the region of edematous reaction extends deep into many structures of the host brain. This may lead to a loss of large number of nerve cells in many neural structures of the host brain. Under such extensive pathologic conditions collapse of blood vessels leading to ischemic conditions is seen not only near the site of transplantation but also in neural structures far distant.

Thirdly, during the phase of initial necrosis and recovery in the transplant there is an accumulation of blood in and around the transplanted tissue. The volume of blood changes rapidly as the necrotic mass is cleared and the transplant grows. In general, the volume of blood at any time is seen to be equal to that of the space left empty by clearing of the necrotic mass, shrinkage in the host brain parenchyma as the edematous reaction subsides, and shrinkage of the transplanted tissue itself as it recovers. As the transplant grows and the necrotic mass is cleared, the volume of blood gradually declines until it is completely removed. In essence, the presence of blood is found mainly during the phase of initial necrosis and recovery. If there is any evidence of continued presence of blood after this phase, with no signs of subsidence, such a condition reflects extensive hemorrhage inside the host brain, caused in all likelihood by poor transplantation techniques. The adverse effects of continued hemorrhagic conditions are seen in masive necrosis in the transplant and the host brain parenchyma, and an ischemic condition in different neural structures far distant from the transplantation site.

Fourthly, during the phase of growth and differentiation of neural transplants dramatic changes are seen in their volume. The transplants containing neuroepithelial cells, such as those from the neocortex of 15-day embryos, increase markedly in volume at a rapid rate. At the other extreme, neural transplants composed of mostly differentiated neuroblasts, such as those obtained from the neocortical region of 22-day embryos, grow very little and at a slow rate. Other neural transplants, containing variable mixtures of neuroepithelial cells and undifferentiated, partly differentiated, and mostly differentiated neuroblasts, may show variable growth related to their cellular composition. The differential growth chracteristics of neural transplants are seen to have differential roles in the displacement and elimination of debris, filling of the cavity in the host brain parenchyma and prevention of glial scar formation at the regions of apposition between the transplants and the host brain parenchyma. If the neural transplants fail to grow according to their growth potential it is very likely that the cause may be the use of nonviable neural tissues. For instance, if the neocortical tissue from 15-day donor embryos, which is endowed with high growth potential, fails to grow as expected it may be that tissue from nonviable embryos has been used, or tissue that has become nonviable during its dissection and preparation. If nonviability of the tissue is due to the latter, various steps in the tissue preparation should be very closely analyzed. Other factors causing unexpectedly little growth are extensive bleeding at the site of transplantation, severely pathologic condition of the host brain surrounding the transplant, or simply errors in accounting for the tissue that has ooozed out from the site of transplantation and in measuring the initial volume of the transplant. On the other hand, if a neural transplant with low growth potential shows an unexpectedly high increase in its final volume an inaccurate method of determining the initial volume of the transplant is usually responsible.

Fifthly, a neural transplant within the parenchyma of the host brain grows at the cost of loss of host brain tissue. Generally, the volume of host brain tissue lost is commensurate with the final volume of the transplant, and the two are seen to fit together very well. In these conditions the neural transplants are seen to be fully integrated with the host brain. All or most of the surface area of the transplant apposing the host brain tissue is parenchymally continuous with the host brain. There is a viable neural interface between the two. If the techniques of transplantation involve causing extensive damage to the host brain, such as making cavities by surgical means for placement of neural transplants, the mass of host brain tissue lost is generally greater than the final volume of the transplant, integration between the two is never complete, and there is some space left between the transplant and the host brain. The interface, in these conditions, is seen only along restricted regions of apposition between the transplant and the host brain parenchyma, and the rest of the region of apposition is bounded by either glial scar formation or cysts (Das 1983b,c). Thus, loss of a large amount of host brain tissue in comparison with the volume of the transplant is indicative of extensive damage caused to the host brain, necrosis of a relatively large mass of the transplant, or the introduction of other pathogenic agents into the host brain along with the transplant.

Sixthly, at the time of transplantation the neural tissue is left in a cavity without any direct vascular supply from the host brain, although it is surrounded by blood. The damaged blood vessels, in particular the capillaries, in the host brain surrounding the cavity show regeneration and growth in the transplant within 24 h after transplantation. Angiogenesis in the transplant, as evidenced by increase in the density of the capillaries, follows a rapid course for about 8 - 10 days. During this stage the capillaries in the transplant increase not only in number but also in size. The caliber of the newly growing capillaries is large and variable. Following this phase, the density of the capillary network is seen first to remain steady at a high level and then to decline to the final stable level. The reduction in the density of capillary distribution before the final stable level is achieved may be due more to an increase in the volume of the differentiating neural transplant than to an actual retraction of the blood vessels. But during this stage the caliber of the capillaries is seen to decline to a normal level. In a normal, well-integrated neural transplant the size and density of capillaries are similar to those seen in the normal adult brain. However, if the pathologic conditions in the host brain are extensive and persist for quite a long time, or if the transplant is undergoing gradual atrophy and degeneration, as in the case of extraparenchymal transplants (Das 1982), the density of the capillary distribution in the transplant remains high and variable. The capillaries appear irregular in size and highly distended. This abnormal condition of the capillary network persists as long as the pathologic conditions in the transplant and the host brain exist.

Qualitative Findings

From an overall viewpoint, the course of histopathologic changes observed in the host brain is very similar to that described by other workers in various traumatic conditions (Ramon y Cajal 1959; Greenfield and Meyer 1963; Greenfield and Russell 1963; Bielschowsky 1965; Ducker et al. 1971; Zülch 1971; Osterholm 1974; Wakefield and Eidelberg 1975). The major difference between these studies and the present one is that in traumatic conditions the pathologic changes resolve into permanent glia scarring and large cysts at the site of the lesion, whereas in conditions of neural transplantation such residual pathologic effects are normally not seen. Pathologic host tissue is replaced by the apparently normal neural transplant. Further, in the case of neural transplants, pathologic changes follow a similar pattern to that seen in the embryonic brain following its exposure to various teratogenic agents. Under teratologic conditions the damaged embryonic neural tissue shows necrosis, some degree of recovery, and ultimately permanent teratologic characteristics in the form of stunted growth and differentiation. But in the case of neural transplants, following initial necrosis and recovery the growth and differentiation closely resemble those seen in normally developing neural structures. The transplants actively interact and integrate with the host brain. They become neural structures in the host brain, with proper cytological and afferent and efferent characteristics and a normal appearance.

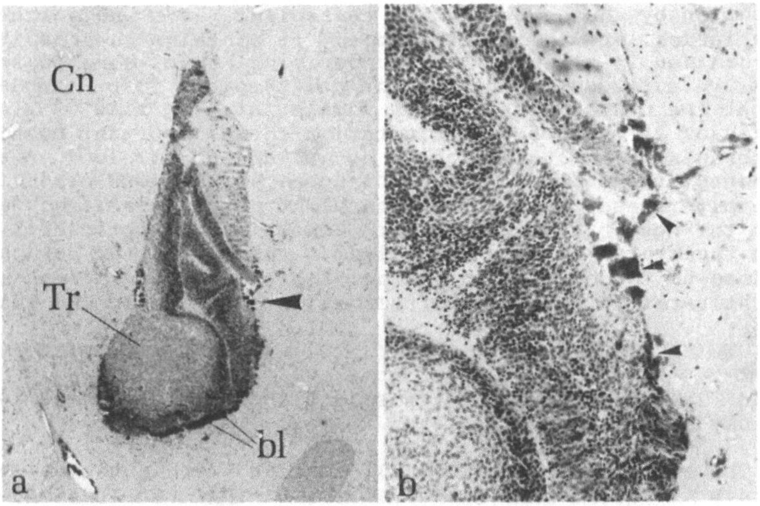

Fig. 1a,b. Neocortical transplant from a 16-day embryo transplanted in the head of the caudate nucleus, 4 h after transplantation.

a The transplant (Tr) is compact, and is located in a cavity in the parenchyma of the caudate nucleus (Cn). The transplant fits snugly in the cavity, and is juxtaposed to the host brain tissue without any meningeal membranes or glial mass or any other foreign material between. Note the layer of neuroepithelial cells (dark staining), and some undifferentiated neuroblasts formed soon after surgery (light staining) in the transplant. There is no evidence of extensive bleeding at the site of transplantation, or extensive damage caused to the host brain tissue. Some bleeding caused by penetration by the needle has contributed to an accumulation of blood (bl) along the basal aspects of the transplant. These facts show that this technique of transplantation helps keep the initial trauma to the host brain and the transplant to a minimum. The region indicated by an arrowhead is shown at a higher magnification in b. H & E; x 10

b Neocortical transplant in close apposition to the host brain tissue. A few small patches of blood (arrowheads) are interposed between the two. The transplant contains viable neuroepithelial cells and neuroblasts. Note that the tissue characteristics of the transplant are still preserved; its constituent cells are held together according to their embryonic cytoarchitecture. This is the result of careful and minimal manipulation of the tissue. H & E; x 100

Histopathologic events taking place in the transplant and the host brain can be viewed together in three different stages. The first stage, which lasts from the time of transplantation to 1 day after surgery, has some interesting early features. With this technique of intraparenchymal transplantation, up to about 4 h after transplantation very little bleeding is seen at the site of injection of the tissue. But the very act of injection seems to have created a cavity in the host brain parenchyma, and the transplant can be seen within the cavity (Fig. 1a). It is completely retained inside the host brain. The transplant is seen closely juxtaposed to the host brain tissue, except for a few patchy areas where some accumulation of blood is noted (Fig. 1b). This is the initial bleeding due to penetration by the needle. The transplant as a whole shows the characteristics of a compact tissue. It contains neuroepithelial cells organized in a layer, and underlying them some neuroblasts formed early after surgery. All the cells appear to be viable, indicating that the neural tissue was viable at the time of transplantation. Within 24 h after transplantation three important changes are seen: necrosis in the transplant, accumulation of blood in and around the transplant, and early signs of edema in the host brain (Fig. 2a-c). Despite these changes the transplant as a whole is closely juxtaposed to the host brain parenchyma. Generally, necrotic cells are located in the center of the transplant, while, at the periphery, which is close to the host brain tissue, the cells seem to remain viable. This feature of differential distribution of necrotic and viable masses of

Fig. 2a-e. Neocortical transplant in the head of the caudate nucleus ▶
1 day after transplantation.

a The transplant (Tr) is still confined within the cavity of the host brain tissue. Accumulation of blood (bl) has increased owing to delayed bleeding. One large and irregular cyst (cy) is seen between the transplant and the host brain. Very probably this cyst, like other small cysts within the transplant, actually contained blood, which was washed away during perfusion of the animal and subsequent histological processing. The region indicated by an arrowhead is shown at a higher magnification in b. Other illustrations are obtained from the same transplant but in other sections. H & E; x 25

b The transplant shows a close apposition to the parenchyma of the caudate nucleus (right). Increased accumulation of blood (bl) is generally seen around the fragments of the transplant. Within the transplant necrosis of cells is evident (left). The necrotic cells have darkly staining nuclei and are loosely scattered, with no tissue cohesiveness holding them together. Necrotic cells are generally located in the center of the transplant. Along the periphery, surviving neuroepithelial cells (ne) have reorganized themselves into loosely structured rosettes. H & E; x 100

c Neuroepithelial cells, organized in pseudocolumnar layer, in the transplant appear viable. Their internal lumen (arrowheads) is apposed to the host brain parenchyma. On the right, scattered cells with darkly staining nuclei are the necrotic neuroblasts.
H & E; x 250

d Mass of necrotic neuroblasts in the transplant shows very little original tissue cohesiveness. These cells containing shrunken and darkly staining nuclei are scattered loosely. Within this mass, a small cluster of viable neuroblasts (arrowheads) characterized by large, light-staining, and granular nuclei is seen. Such clusters of viable and surviving neuroblasts may be found scattered throughout the transplant without any apparent order. H & E; x 250

e Neuropil of the caudate nucleus near the site of transplantation appears spongy, indicating edematous condition. The clusters of small, dark-staining nuclei (arrowheads) are extravasated blood cells. Other scattered cells with large, spherical, and dark-staining nuclei which are surrounded by thin pale-staining cytoplasm are the necrotic neurons of the caudate nucleus. In this zone of edema surrounding the cavity all the shrunken and hyperchromatic cells (both neurons and glia cells) characterized as necrotic cells die.
H & E; x 250

cells in a transplant may vary considerably with different types of neural tissue used as transplants, and with different techniques used for transplantation. From the cytological viewpoint, in a transplant neuroblastic elements tend to become necrotic more readily than neuroepithelial cells (Fig. 2b-d). In other words the bulk of the necrotic mass in a transplant may be composed of neuroblasts rather than neuroepithelial cells. In some regions of necrotic mass, it is possible to find a small number of neuroblasts still viable even when they are surrounded by a large number of necrotic neuroblastic elements (Fig. 2d). It is likely that these neuroblasts survive for a long time and differentiate into neurons. The neuroepithelial cells generally maintain their pseudocolumnar laminar organization (Fig. 2c), but occasionally they may show early patterns of reorganization

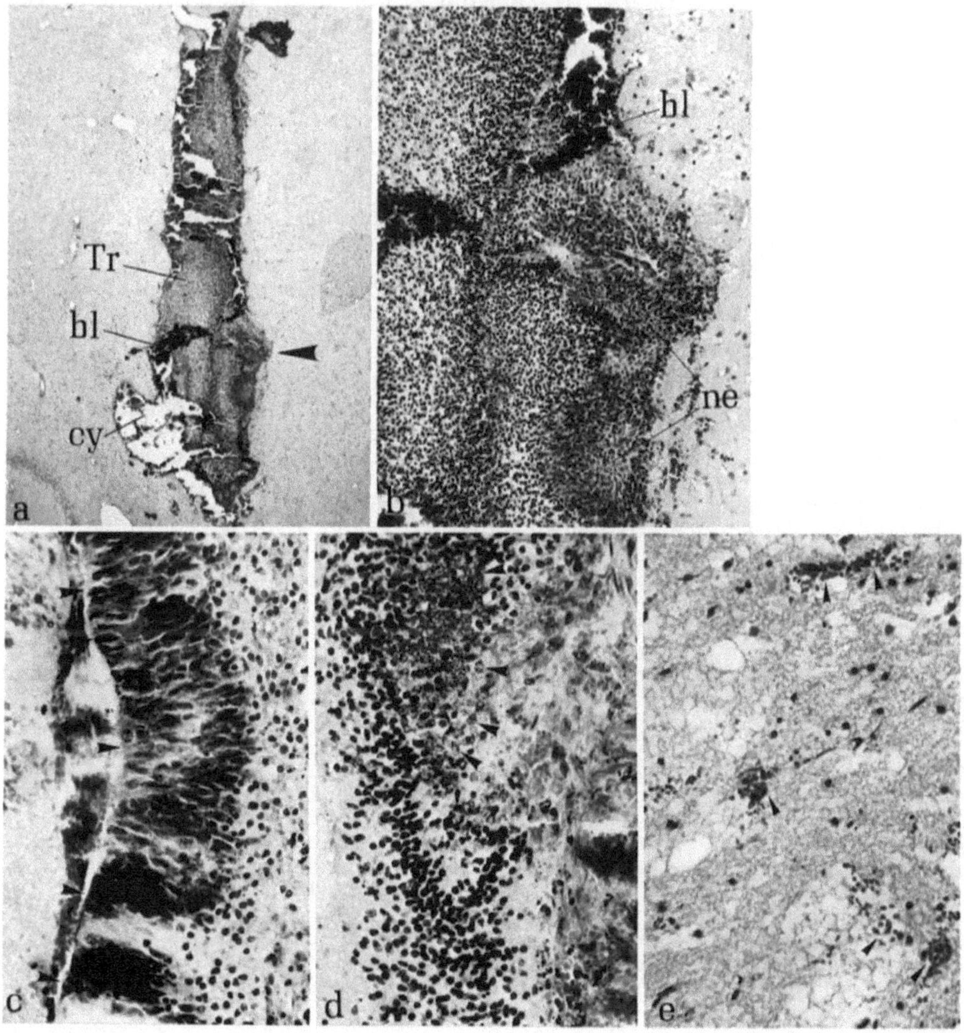

in the form of rosettes (Fig. 2b). These rosettes, in particular, being juxtaposed to the host brain, show early signs of parenchymal attachment with it. From an overall viewpoint, the level of necrotic reaction observed in these preparations is the lowest. With different methods of manipulation of neural tissues and different techniques of transplantation the amount of necrosis in the transplants is seen to be large. The delayed necrotic reaction noted in these preparations is not unique to the situation of transplantation. It is also observed in other experimental conditions, where embryonic nervous tissue is subjected to teratogenic materials or irradiation. If trauma to the neural transplants as a result of excessive manipulation or any other variable were severe, necrosis of cells would be noticeable within hours and would be far more extensive. Therefore, delayed necrosis in transplants can in fact be taken as indicative of careful and less traumatic methods of manipulation and preparation of the tissues. Further, during this period bleeding at the site of transplantation is seen to increase (Fig. 2b). The blood accumulates in pockets between fragments of transplants, and in the space between the transplants and the host brain tissue. Erythrocytes and leukocytes are densely packed in these pockets. This delayed bleeding may be associated with gradual changes in the structural integrity of the blood vessels due to indirect and subthreshold trauma. In addition to this, there may be massive extravasation of blood cells from the intact blood vessels near the vicinity of transplantation. Moreover, at this stage development of cysts in an around the transplants is seen. More often than not this may be an artifact of the fixation procedure or caused by fragmentation of necrotic masses of the transplants. True cyst formation takes place at a later stage, following phagocytosis of debris and shrinkage of the host brain tissue.

The host brain tissue along the borders of the cavity seems to acquire spongy characteristics (Fig. 2e). This is due to edema. The edematous reaction does not extend to deep inside the tissue of the caudate nucleus; it is confined to the borders of the cavity in the form of a ring. In this zone of edema, neurons and glia cells appear necrotic. The loss of these cells is permanent. In addition to this, capillaries are seen to have become dilated, and surrounding them clusters of extravasated blood cells are observed. It is important to comment that these distended capillaries, although showing a pathologic reaction, are the source of new blood vessels that will grow into the transplants. With other techniques of transplantation, which cause extensive damage to the host brain, the region of edema is very large, with a large number of neurons and glia cells showing necrotic reaction. In such cases a massive hemmorhage is generally observed as well as edema.

The second stage lasts 2 - 6 days after transplantation, and is characterized by increased necrosis in the transplant, increased accumulation of blood, phagocytosis of the necrotic mass, emergence of cysts within the transplant, and subsidence of edema in the host brain tissue. During this stage there is an increase in necrosis in the transplant, which is confined in its center (Fig. 3a,b). It reaches its peak by day 3 after transplantation (Das et al. 1983a). Generally, the necrotic mass is found to account for as much as 50% of the total neural tissue transplanted. This is true when the whole transplant is composed of viable neural tissue. If it contains some nonviable portions the volume of necrotic mass may be much higher. Closely associated with this change is the increased blood accumulation, which is seen in the regions of necrosis in the transplant. In small patchy areas, small clusters of erythrocytes and leukocytes may be observed in parts of the transplant. From the viewpoint of histological processing the regions containing the necrotic mass and blood

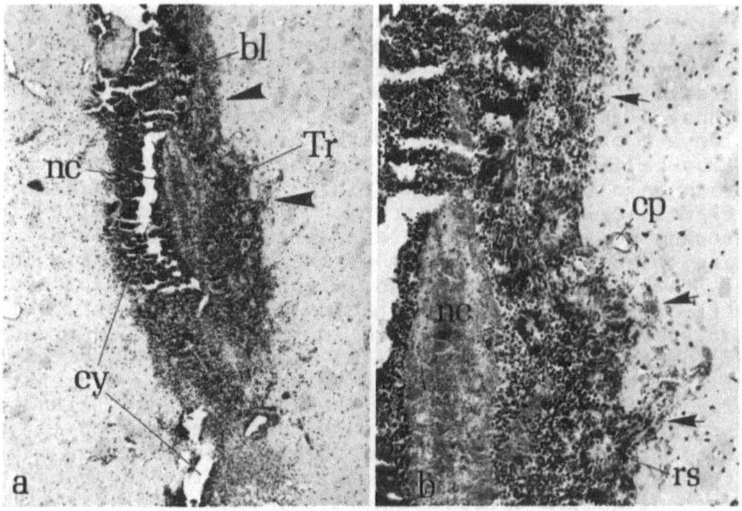

Fig. 3a,b. Neocortical transplant 3 days after transplantation.

a The transplant contains viable tissue (Tr) confined to the periphery, necrotic mass (nc) in the center of the transplant, patchy accumulation of blood (bl) within and around the transplant, and cysts (cy) appearing in the regions of necrosis as it is cleared. Note that the transplant appears to be very close to the host brain tissue along its entire surface. There are no crevices or meningeal membranes between transplant and the neuropil of the caudate nucleus. The region indicated by arrowheads is shown at a higher magnification in b. H & E; x 40

b Viable portion of transplant, along the periphery, shows neuroepithelial cells organized in the form of rosettes (rs). These rosettes, being in different planes of the section, show different forms and shapes. The neuroblasts emerging from them seem to be pushing into the parenchyma of the host brain (arrows). Along its entire surface the transplant appears to be parenchymally attached to the host brain tissue. The necrotic mass is confined to the center of the transplant. Accumulation of blood (black patches) is seen within the center of the transplant and close to the necrotic mass. An enlarged capillary (cp), when traced in serial sections, is seen to run from the caudate nucleus into the transplant, where it shows some branching and a reduced diameter. The caudate nucleus (right) does not show spongy characteristics, indicating that edema has subsided. H & E; x 100

accumulations are slightly harder and are brittle in comparison with the viable portions of the transplant and the host brain tissue. Because of this, during sectioning of the tissue cracks are seen to develop in this region, which may leave an erroneous impression of cysts developing in the transplant. In time, this artifact is eliminated as necrotic mass and blood are cleared. In addition to blood cells, macrophages are seen to have penetrated the region of necrosis. Although most of them may have come through blood circulation, some show evidence of proliferation locally. They are large, and their cytoplasm is highly vacuolated. They are seen to contain phagocytic material, indicating their active role in the phagocytosis of the necrotic mass. In these preparations there is no evidence to suggest that any of these phagocytic elements are derived from the transformed microglia cells of the host brain. It is important to stress that these macrophages are never seen within the viable portions of the transplant or the normal host brain tissue; they are always closely associated with the necrotic tissue. By this time some large-caliber capillaries may be seen to penetrate from the host brain into the transplant (Fig. 3b), where they branch into a network of small-caliber capillaries. They are confined to the viable and growing portions of the transplant. Capillaries are rarely, if ever, seen to grow in the region of the necrotic mass. The viable portion of the transplant as a whole appears densely packed, and contains well-organized rosette formations surrounded by clusters of undifferentiated neuroblasts. There is no apparent order in the distribution of rosettes of neuroepithelial cells. The unsystematic distribution of rosettes is very probably determined by unsystematic fragmentation of viable neuroepithelium immediately after transplantation. Each fragment of neuroepithelium settles wherever it is and reorganizes itself into a rosette. Thus, whether the rosettes are in a line or in a cluster is a reflection of the pattern in which the fragments of neuroepithelium settled. However, the clusters of undifferentiated neuroblasts arising out of those rosettes show early signs of pushing into the host brain parenchyma and establishing the initial parenchymal attachment. An increased volume of the viable portions of the transplant owing to proliferation of neuroepithelial cells in the rosettes, and close apposition of these rosettes and neuroblasts to the apparently normal host brain tissue suggest that in all likelihood the outgrowing clusters of neuroblasts have penetrated through the edematous region of the host brain and reached the apparently normal host brain parenchyma to establish parenchymal attachment. This initial parenchymal attachment is of critical importance for the neural transplant, so that it can establish an interface with the host brain for afferent and efferent connections. Finally during this period, edema in the host brain tissue surrounding the transplant is seen to subside. There is no indication of sponginess of the tissue. It is possible that the clusters of undifferentiated neuroblasts arising out of the neuroepithelial rosettes readily penetrate through the loosely textured edematous tissue, thereby causing its disappearance within a short time. If this is indeed the case, the necrotic neurons and glia cells found earlier in the edematous region may actually be present between the migrating clusters of the neuroblasts; being shrunken and hyperchromatic they may be difficult to distinguish from the necrotic cells of the transplant. In essence, the growing transplant, by virtue of its clusters of neuroblasts penetrating through the edematous region, has not only achieved parenchymal attachment with the normal host brain tissue, but has also aided in the elimination of the edematous tissue and its constituent necrotic cells.

Around day 5 after transplantation, necrosis and accumulation of blood in the transplant are somewhat reduced. The portion of the transplant containing necrotic mass and blood still develops cracks during histological sectioning (Fig. 4a), but at the same time some cyst formation is also seen in that region. The cysts are small, and they are the result of clearing of the necrotic mass and blood. If some flaw in the transplantation technique were to cause bleeding to continue for a long time the whole transplant would be surrounded by blood, most of the transplant would be necrotic, and cyst formation would not be evident. In such excessively pathologic preparations, as soon as some of the necrotic mass is removed the region is filled up by blood, and therefore true cyst formation never takes place. However, in the present preparations the cysts are real, and are very probably filled with cerebrospinal fluid, which is washed away when

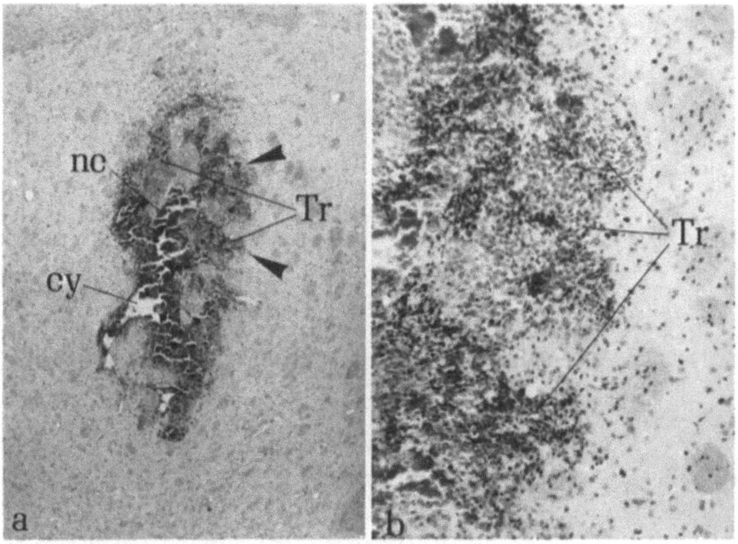

Fig. 4. Neocortical transplant in the caudate nucleus 5 days after transplantation.

a The viable portion of transplant (Tr) has grown on all sides. It is still dense and compact, and appears to be penetrating into the host brain parenchyma. The necrotic mass (nc) is relatively smaller in volume. In small regions where necrotic mass and blood have been cleared away small cyst formations (cy) are seen. The region marked by arrowheads is shown at a higher magnification in b. H & E; x 25

b The viable portion of the transplant (Tr) contains small clusters of neuroepithelial cells (dark staining) and undifferentiated and partially differentiated neuroblasts (light staining). The latter appear to be pushing into the host brain parenchyma. The neuroblasts are somewhat loosely clustered in comparison with the neuroepithelial cells. H & E; x 100

the animals are perfused. In the vicinity of the necrotic mass some small pockets of accumulated blood are still present. In and around this region a large number of macrophages are seen, which indicates a heightened phagocytosis of the debris. In the transplant itself the size and number of rosettes of neuroepithelial cells are considerably reduced (Fig. 4b). Whatever is left of these rosettes looks like small clusters of neuroepithelial cells; these cells are very probably in their final stages of proliferative activity. A large area of the transplant is now occupied by clusters of neuroblasts, containing undifferentiated and partially differentiated elements in their early stages of differentiation. These clusters of neuroblasts are seen to migrate outward in groups and push in the host brain parenchyma. With this mode, at least in the case of the transplants composed of neuroepithelial cells, the interface already established between the transplant and the host brain is being pushed more and more deeply into the normal host brain tissue. This contributes to the expansion of the surface area of the transplant and thus of the interface area. In the host brain there is no indication of edema surrounding the transplant. The increase in the density of glia cells, however, suggests that there is gliosis in the caudate nucleus. This appears to be a transient phenomenon, for in a few days it subsides and glial distribution is normal.

The third stage of the histopathological reaction lasts from day 6 to day 10 after transplantation. During this period necrosis in the transplant and blood are greatly reduced (Fig. 5a). The necrotic mass seen at this stage is not a new entity, being the residual mass that has not yet been phagocytized. It is during this stage that true cyst formation is seen. Cysts of various sizes, generally located in the center of the transplant are a conspicuous feature of the transplant. These cysts are very probably filled with cerebrospinal fluid. Their boundary is lined by the undifferentiated and partially differentiated cells of the transplant. There is no evidence of pia mater or any glial scar formation lining the borders of the cysts. By the end of this stage most of the cysts are closed in by the growing transplant.

Fig. 5a-c. Neocortical transplant in the caudate nucleus of the host ► animal.

a Portion of the transplant 7 days after surgery. The necrotic mass (nc) is very small; many large and small cysts (cy) can be noted. These cysts are clear, and do not contain any necrotic mass or blood. The transplant (bounded by arrowheads) has grown large. The neurons in it appear to be somewhat differentiated, and are dispersed. Note that the transplant has an apparently normal interface with the host brain tissue. H & E; x 40

b Neocortical transplant 12 days after transplantation is seen to have grown large. There is no evidence of any persisting pathologic reaction in the transplant or the host brain. Arrowheads indicate interface between transplant and parenchyma of host brain. H & E; x 25

c Transplant 30 days after surgery. It is now fully differentiated and completely integrated with the host brain tissue. Although it does not show laminar cytoarchitecture of the cerebral cortex it stands out distinctly from the surrounding caudate nucleus. Arrowheads indicate interface between transplant and caudate nucleus. It is significant that with the transplantation technique used in this study there is no evidence of a lasting pathologic reaction in the transplant, in the caudate nucleus, or in the cerebral cortex, through which the needle containing the transplant was inserted at the time of transplantation. A transplant with these normal histological features is an integral part of the host brain. Cresyl violet; x 10

From this viewpoint the cysts in a transplant can be considered to provide space for the neural transplant to grow. However, this is true only in the case of neural transplants with high and moderate growth potential. When neural transplants with low growth potential, such as those composed of brain stem or spinal cord tissue from 15- or 16-day embryos, are used, because of their inability to grow large they fail to fill in the cysts. In such preparations the presence of cysts within and surrounding the transplants is a very common feature. The transplant appears to have increased in size. The main features of its growth during this period are differentiation of neuroblasts and increase in neuropil. These two events are very closely interrelated. The net result of these developmental changes is dispersion of neurons in the transplant. Further, the interface that the transplant has established thus far by pushing out the tissues affected by pathologic changes and usurping their place is seen to be maintained during the entire phase of its outward growth. In other

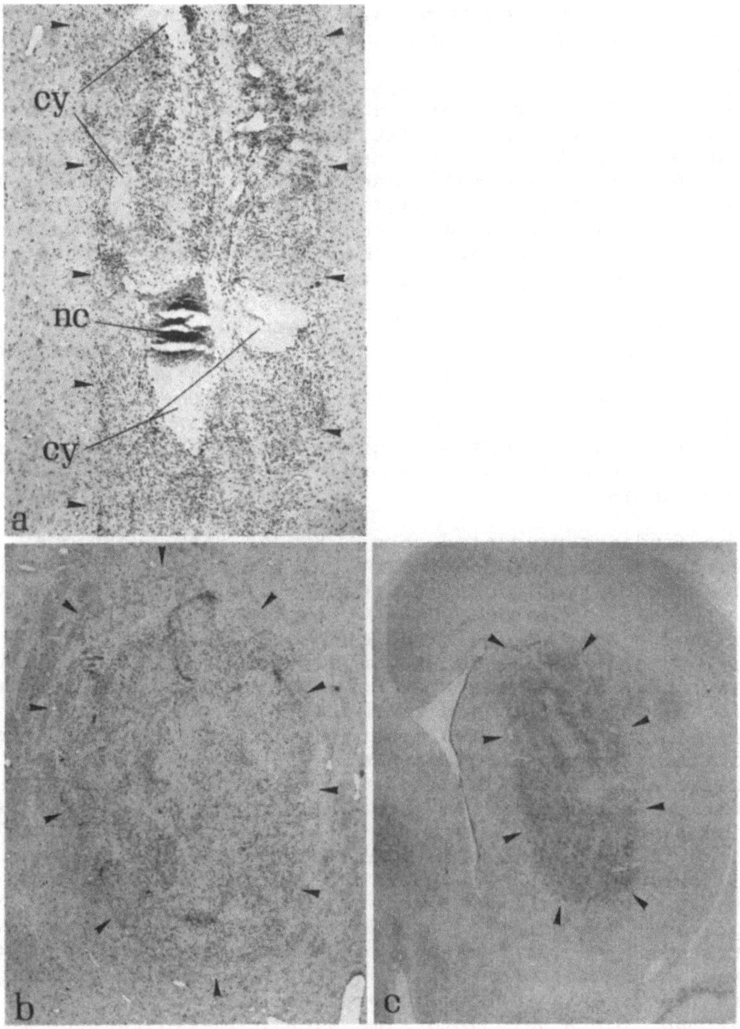

words, once a transplant has established an interface with the host brain it generally continues to maintain it during its period of growth, without any danger of future pathologic incidents.

Following the stages described above, which have a bearing on the histopathological changes in the transplant as well as in the host brain, the transplant simply grows larger. All pathologic changes observed earlier are absent. The transplant appears to be integrated with the host brain from all aspects. The growth of the transplant from this stage onward is determined by cellular differentiation and increase in the neuropil. In the host brain, too, no pathologic changes are observed (Fig. 5b, c). However, the extended growth of the transplant indicates that it takes up progressively more room within the host brain parenchyma until it achieves its full size. This is possible at the cost of either a loss of some amount of host neural tissue, or outward displacement of the host neural structure and other brain structures surrounding it. According to the first hypothesis, as the transplant grows it penetrates actively into the host brain parenchyma, and thereby some amount of host brain tissue is lost. This implies that the growth of the transplant is invasive in nature. As it penetrates into the host brain parenchyma to occupy a territory, the neurons of the host brain in that territory become necrotic and die. In histological preparations no such extended necrosis of neurons around the growing transplant is seen. Alternatively, it is possible that the host brain tissue is not in fact lost but the neural structure, and other structures of the host brain surrounding it, are simply pushed out. This hypothesis implies that the transplants grow by expansion. During their growth they do not penetrate or infiltrate into the normal host brain parenchyma with which they have already established an interface. This growth simply involves physical pushing outward or displacement of the host neural tissues. This hypothesis is supported by the fact that during the last phase of growth of the transplant there is no indication of necrosis of neurons in the host brain tissue immediately surrounding the transplant. Since these two hypotheses are not mutually exclusive, it is possible that both these events, in different measures at different stages of growth, may be taking place during the growth of a transplant. A detailed quantitative analysis of the growth of the transplant will provide a basis for supporting one or both of these hypotheses.

The findings presented above raise many questions, two of which are worthy of further consideration. The growth of neural transplants, at least those containing neuroepithelial cells and showing high growth potential, is largely, if not exclusively, determined by an increase in the number of cells following proliferation of the neuroepithelial cells and active outward migration of clusters of undifferentiated neuroblasts into the host brain. These events have direct interaction with the host brain tissue. It is relevant to ask whether these events themselves also constitute pathologic conditions for the host brain. The two developmental events, viz. increase in the number of cells and penetration of clusters of neuroblasts into the host brain, follow a definite time-course very similar to that observed during neuroembryogenesis in a normally developing embryo. These two events come to a stop at some time, and the neuroblasts settle and start to differentiate into apparently normal neurons. The time-course of these events for a neural transplant is determined primarily by the previous histogenetic history of the neural tissue when obtained from the donor embryo. However, if instead of normal embryonic neural tissue some other non-neural tissue with a high growth potential, such as that from a brain tumor, were transplanted it would grow and expand for a very long time with no definite time-course. It would show all the characteristics of malignancy. It would be a pathologic structure incapable of anatomical integration with the host brain, and it might cause extensive damage to other neural structures by invasive growth. It would very probably seriously interfere with the normal functioning of other neural structures of the host brain. However, a neural transplant does not show these features of malignant growth. As it grows it becomes anatomically integrated with the host brain. In its fully differentiated state it has the characteristics of a normal neural tissue, and is part and parcel of the host nervous system. It is highly probable that rather than interfering with the functioning of the rest of the brain, it becomes an integral part of the whole functioning brain. In view of these considerations a growing neural transplant cannot be considered a pathologic entity or a factor contributing to development of any other ensuing pathologic conditions in the host brain.

Further, neural transplantation is indubitably traumatic to both the transplant and the host brain. The obvious question would be: Does a traumatized neural transplant actually aid in alleviating the pathologic changes in the traumatized host brain? When a lesion is made inside the brain, various pathologic changes, including necrosis of the brain tissue, edema, and hemorrhage, are observed. They linger for a considerable time, finally culminating in cyst formation and the development of glial scar formation. Under these conditions not only is the neural tissue that has been traumatized lost, but there are also permanent residual effects of the pathologic reaction. When a neural tissue is transplanted at the site of the lesion (in the present study the lesion was made by the penetration of the needle and injection of the neural tissue), by virtue of its growth it aids in pushing out the necrotic mass, filling the cysts, preventing development of glial scar formation, and filling the void left behind by the loss of the host brain tissue by a normal neural tissue. In addition to this, it is also possible that a growing neural transplant may aid in reducing necrosis in the host brain tissue, controlling hemorrhagic response following the trauma, and reducing retrograde degenerative changes in the neurons far distant from the site of lesion and transplantation. Of course these beneficial effects of a neural transplant are possible only when it is composed of tissue with high or moderately high growth potential. If a neural transplant fails to meet these requirements of growth potential it will fail to displace the necrotic mass of the host brain, it will not be able to fill the traumatized region completely, and it will not grow fast enough to prevent the development of glial scar formation. Such a transplant will probably serve very little anatomical and physiological purpose.

Conditions Contributing to Extensive Pathological Changes

The course of histopathological changes described above pertains to the experimental conditions, which are highly controlled so as not to cause more than the unavoidable minimum of trauma to the transplant and the host brain. In such preparations the pathologic changes are seen to resolve themselves within a few days after transplantation, and the fully grown transplant looks normal and well integrated with the host brain. However, there are many steps in neural transplantation at which an extensive pathologic reaction and loss of the whole or a large part of the transplant itself can be caused if control is inadequate. Some of the important conditions leading to an extensive pathologic reaction are use of nonviable neural transplants, use of contaminated neural transplants, extensive hemorrhagic reaction at the site of transplantation, and excessive trauma to the superficial and deep structures of the host brain.

Conclusions

The findings presented above can be summarized in the context of trauma to the central nervous system and histopathologic changes in the transplant and the host brain following microsurgery for neural transplantation.

1. Transplantation of embryonic neural tissues into normal and intact host brains yields the most successful results. Pathologic conditions produced in these preparations are minimal, and they follow a course of progression and regression at such a low level that they do not adversely affect the initial survival of the transplants. The growth-related changes in the transplants and the fast subsidence of pathologic conditions provide the best conditions for the transplants to achieve unimpeded growth and maximum integration with the host brain tissue.

2. Transplantation of neural tissues in severe pathologic conditions does not yield successful results. Extensive necrosis in the host brain and hemorrhage at the site of transplantation do not provide a milieu conducive to the initial survival and subsequent growth of the transplants. Even if the transplants are able to survive and grow to some extent, the pathologic conditions overwhelm them and retard their growth to full potential and their achievement of proper anatomical integration with the host brain tissue. Host brain tissue that is itself undergoing necrotic and degenerative changes may not be the most desirable tissue as a substratum for normal initial parenchymal attachment and subsequent integration to the neural transplant.

3. Even these severe pathologic conditions in the host brain, when necrosis of the host brain tissue is somewhat contained and hemorrhage is not protracted, it is nonetheless possible to achieve some successful transplantations when neural tissue with high growth potential, such as neocortical tissue from 15- and 16-day embryos, is used for the transplants. The high growth potential of these types of tissue and the relatively high resistance of neuroepithelial cells to trauma ensure that these transplants will grow through the necrotic mass of the host brain at the site of transplantation and become integrated with the host brain parenchyma. It is possible that a fast-growing transplant, which has displaced the necrotic mass at a fast rate and has become anatomically integrated with the normal parenchyma of the host brain, may have beneficial trophic influences on the host brain, reducing if not completely eliminating the delayed and progressive degeneration developing in different neural structures. However, neural tissues with moderate or low growth potential may not be as beneficial to the host brain in conditions of severe trauma.

References

Bielschowsky M (1965) Histopathology of nerve cells. In: Penfield W (ed) Cytology and cellular pathology of the nervous system, vol I. Hafner, New York, p 147

Das GD (1974) Transplantation of embryonic neural tissue in the mammalian brain. I: Growth and differentiation of neuroblasts from various regions of the embryonic brain in the cerebellum of neonate rats. T.-I.-T.J.Life Sci 4:93-124

Das GD (1982) Extraparenchymal neural transplants: Their cytology and survivability. Brain Res 24:182-186

Das GD (1983a) Neural transplantation in mammalian brain: Some conceptual and technical considerations. In: Wallace RB, Das GD (eds) Neural tissue transplantation research. Springer, Berlin Heidelberg New York, p 1

Das GD (1983b) Neural transplantation in the spinal cord of the adult mammals. In: Kao CC, Bunge RP, Reier PJ (eds) Spinal cord reconstruction. Raven, New York, p 367

Das GD (1983c) Neural transplantation in the spinal cord of adult rats. Conditions, survival, cytology and connectivity of the transplants. J Neurol Sci 62:191-210

Das GD, Ross DT (1982) Stereotaxic technique for transplantation of neural tissues in the brain of adult rats. Experientia 38:848-850

Das GD, Halls B, Das KG (1979) Transplantation of neural tissues in the brains of laboratory mammals: Technical details and comments. Experientia 35: 143-153

Das GD, Halls B, Das KG (1980) Transplantation of brain tissue in the brain of the rat. I: Growth characteristics of transplants from embryos of different ages. Am J Anat 158:135-145

Das GD, Das KG, Brasko J, Aleman-Gomez J (1983a) Neural transplants: Volumetric analysis of their growth and histopathological changes. Neurosci Lett 41:73-79

Das GD, Houle JD, Brasko J, Das KG (1983b) Freezing of neural tissues and their transplantation in the brain of rats: Technical details and histological observations. J Neurosci Methods 8:1-15

Ducker TB, Kindt GW, Kempe LG (1971) Pathological findings in acute experimental spinal cord trauma. J Neurosurg 35:700-708

Greenfield JG, Meyer A (1963) General pathology of the nerve cell and neuroglia. In: Blackwood W, McMenemey WH, Meyer A, Norman RM, Russell DS (eds) Greenfield's Neuropathology. Williams and Wilkins, Baltimore, p 1

Greenfield JG, Russell DS (1963) Traumatic lesions of the central and peripheral nervous system. In: Blackwood W, McMenemey WH, Meyer A, Norman RM and Russell DS (eds) Greenfield's Neuropathology. Williams and Wilkins, Baltimore, p 441

Oblinger MM, Das GD (1982) Connectivity of neural transplants in adult rats: Analysis of afferents and efferents of neocortical transplants in the cerebellar hemisphere. Brain Res 249:31-49

Oblinger MM, Das GD (1983) Connectivity of transplants in the cerebellum: A model of developmental differences in neuroplasticity. In: Wallace RB and Das GD (eds) Neural tissue transplantation research. Springer, Berlin Heidelberg New York, p 105

Osterholm JL (1974) The pathophysiological response to spinal cord injury. J Neurosurg 40:5-33

Ramon Y Cajal S (1959) Degeneration and regeneration of the nervous system, vol II. Hafner, New York

Wakefield CL, Eidelberg E (1975) Electron microscopic observations of the delayed effects of spinal cord compression. Exp Neurol 48: 637-646

Zülch KJ (1971) Hemorrhage, thrombosis, embolism. In: Minckler J (ed) Pathology of the nervous system, vol II. McGraw-Hill, New York, p 1499

Plasticity of Intact Neurons After Trauma

Correlative Morphological and Physiological Studies on Sprouting of Excitatory and Inhibitory Synapses of Red Nucleus Neurons

N. Tsukahara and F. Murakami

Department of Biophysical Engineering, Faculty of Engineering Science, Osaka University, Toyonaka and National Istitute for Physiological Sciences, Okazaki, Japan

Introduction

Although much evidence has been accumulated on the sprouting and formation of new synapses in the central nervous system, less attention has been paid to the correlative morphological and physiological investigation of sprouting (Cotman et al. 1981; Tsukahara 1981 for reviews). It is important to establish sprouting and the formation of new functional connections morphologically as well as physiologically in order to gain insight into its molecular basis and to establish an unequivocal assay system for such molecular approaches.

The mammalian red nucleus (RN) is particularly suitable for such correlative studies, because of its unique synaptic organization which enables us to characterize the location of synapses by electrophysiological means. The segregation of the two major inputs to RN, cerebral cortex and the nucleus interpositus (IP) of the cerebellum, determines the characteristics of postsynaptic potentials recorded in RN cells. By taking advantage of this synaptic organization, it was shown in adult cats that sprouting and the formation of functional synapses occur from the originally dendritic corticorubral input to the proximal portion of the soma-dendritic membrane of RN cells after destruction of the somatic IP input (Tsukahara et al. 1975; Murakami et al. 1982). Furthermore, there is also evidence to suggest that corticorubral sprouting occurs in adult cats after cross-innervation of peripheral flexor and extensor nerves (Tsukahara and Fujito 1976; Tsukahara et al. 1982; Fujito et al. 1982; Murakami et al. 1984).

In kittens, following lesions of the ipsilateral corticorubral fibers which were performed less than 2 months after birth, stimulation of the contralateral sensorimotor cortex or its fibers in the cerebral peduncle (CP) produced in RN cells a slow rising EPSP (Tsukahara and Fujito 1981; Tsukahara et al. 1983). This outcome was interpreted as being a result of sprouting and the formation of new synapses from the contralateral cerebral cortex. The histological investigations by Nah and Leong (1976) in rats and Villablanca et al. (1982) and Kosar et al. (to be published) in cats have indeed demonstrated that crossed corticorubral fibers do exist after lesions of the ipsilateral cerebral cortex.

The most common method of inducing sprouting in the brain is to destroy some of the synaptic inputs of central neurons. However, it is also possible to induce sprouting without destroying the direct synaptic inputs of these neurons by means of such procedures as

Experimental Brain Research, Suppl. 13

cross-innervation of peripheral flexor and extensor nerves. In this article correlative morphological and physiological studies of RN neurons using these two kinds of experimental paradigms will be described.

Lesion-Induced Sprouting of Excitatory Corticorubral Synapses and Properties of Their Synaptic Transmission

Electrophysiological studies have shown that sprouting and the formation of new functional synapses occur from three sources neonatal lesions of the ipsilateral corticorubral fibers in kittens – from contralateral cerebral cortex, from ipsilateral nucleus interpositus, and from contralateral nucleus interpositus (Fig. 1). Fig.1G shows an example of the newly appeared contralateral CP-EPSPs induced by stimulating the contralateral CP in an RN cell in kittens in which ipsilateral corticorubral projections were destroyed 27 days after birth. As reported previously in normal kittens, the corticorubral EPSPs were only produced by stimulating the ipsilateral side, and not from the contralateral CP (Fig. 1B, C) (Tsukahara et al. 1983). The EPSPs had a slow rising time course which was not significantly different from that of the normal corticorubral PSPs (Fig. 1G). This physiological result is corroborated by the morphological result obtained using HRP (Kosar et al., to be published). In animals with lesions of ipsilateral corticorubral input, similar slow rising EPSPs were produced by stimulating the ipsilateral IP and the slow-rising component was superimposed on the fast rising IP-EPSPs. These results were taken to indicate that sprouting and the formation of functional synapses occur after lesions of the ipsilateral corticorubral input.

Since some of the normal corticorubral fibers are collaterals of the pyramidal tract, the question would arise whether the newly appeared crossed corticorubral fibers are also collaterals of the pyramidal tract. By stimulating the pyramidal tract (PT) similar slow rising EPSPs appeared with a latency of about 1.5 ms (Kosar et al., to be published). A collision experiment between contralateral CP-EPSPs and pyramidally induced EPSP (PT-EPSP) was performed, and the specimen records are partly shown in Fig. 2. Contralateral CP-EPSPs were followed at various stimulus intervals by PT-EPSPs. PT-EPSPs were depressed at stimulus intervals of 1.8 ms. With a stimulus interval of 1.6 ms the test PT-EPSPs decreased in amplitude to 50%, and at a stimulus interval of 1.4 ms they were virtually abolished. The time interval of 1.6 ms may be accounted for by the conduction time between contralateral CP and PT (1.1 ms) and the absolute refractory period of the corticorubral fibers (0.5 ms). Therefore, it is likely that contralateral CP-EPSPs are mediated, at least in part, by the collaterals of the pyramidal tract fibers. From latency measurement of these EPSPs and from the time interval of collision, it is possible to estimate the branching point of collaterals, which turned out to be within the cerebral peduncle, very close to the contralateral RN.

The next question concerns the properties of synaptic transmission of the newly formed synapses. We tested the posttetanic effect of the corticorubral stimulation upon the synaptic transmission of the corticorubral synapses. After tetanic stimulation of 100 Hz for 3 min, the EPSPs increased in size within 1 min, attained their peak after about another minute, and lasted for more than 10 min. The recording was made in a preparation where partial removal of the ipsilateral corticorubral projection was performed so that a comparison

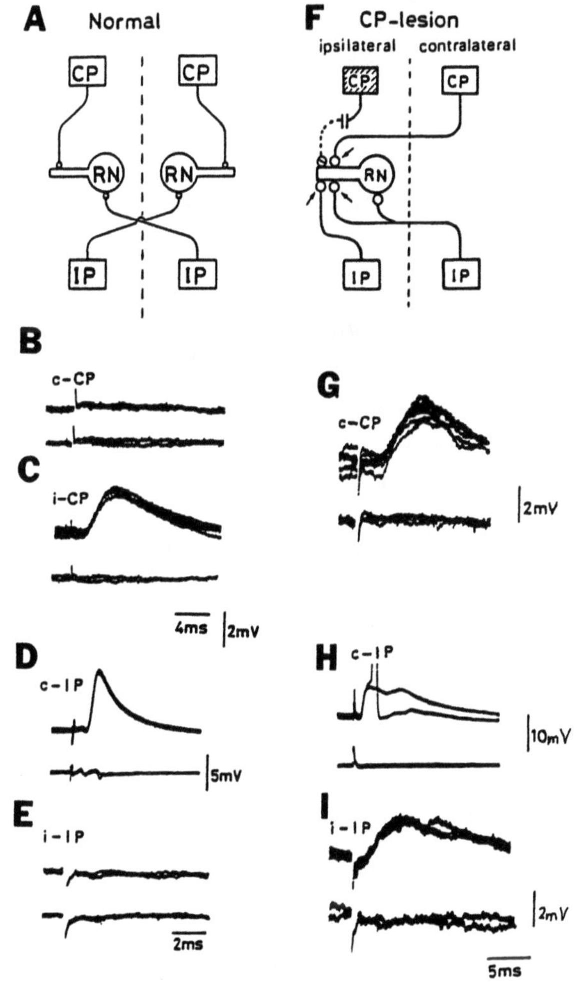

Fig. 1 A-I. Synaptic organization of normal kitten red nucleus and that of sprouted terminals after destruction of the cerebrorubral input. A: Synaptic organization of normal RN. Broken line indicates the midline. CP, cerebral peduncle; RN, red nucleus; IP, nucleus interpositus of the cerebellum. B-E: EPSPs induced in an RN cell by stimulating the ipsilateral CP (i-CP) (C) and the contralateral IP (C-IP) (D). No response was induced by stimulating the contralateral CP (c-CP) (B) or the ipsilateral IP (i-IP) (E). Upper traces, intracellular potentials, lower traces, corresponding extracellular field potentials. F: Synaptic organization in kittens with lesions of the ipsilateral cerebrorubral input. Diagram of three sources of sprouting after removal of the ipsilateral cerebral input. Sprouting sources are indicated by arrows. Midline is expressed by a vertical interrupted line. G-I: Newly appearing EPSPs induced in an RN cell by stimulating the contralateral CP (G), slow rising component superimposed on the fast rising IP-EPSPs by stimulating the contralateral IP (H), and slow rising EPSPs by stimulating the ipsilateral IP (I), in a kitten in which ipsilateral cerebrorubral input was destroyed on the 27th day after birth. (Modified from Tsukahara et al. 1983)

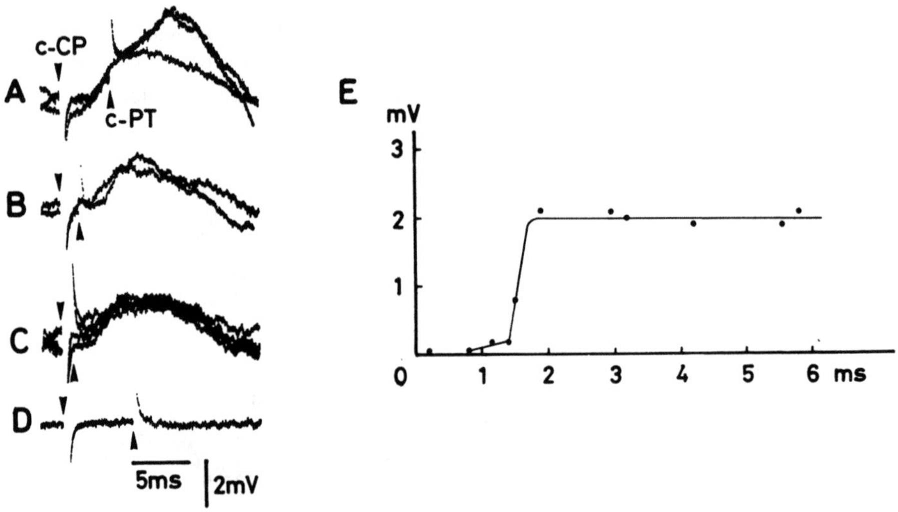

Fig. 2. Collision of the EPSP elicited by stimulation of contralateral cerebral peduncle (c-CP) and contralateral pyramidal tract (c-PT). From A to C, the stimulus interval was gradually reduced. D: Extracellular recording. E: The relation between the amplitude of the c-PT EPSP and the stimulus interval. (Kosar et al., to be published)

could be made in the same cell on the posttetanic potentiation of both ipsilateral and contralateral CP-EPSPs. Potentiation was observed for the ipsilateral CP-EPSPs. There was no increase, and even a transient depression, in the amplitude of the contralateral CP-EPSPs. Therefore, the posttetanic effect of the newly formed connections is less potent as compared to the original connections.

Lesion-Induced Sprouting of Inhibitory Synapses

The RN neurons are disynaptically inhibited by collaterals of the fast-conducting pyramidal tract fibers in addition to being excited monosynaptically by slow conducting pyramidal tract and corticorubral fibers. The transmitter mediating the inhibition seems to be gamma-aminobutyric acid (GABA) for the following reasons. GABA hyperpolarizes the RN cell membrane and is associated with a conductance increase when tested in vivo (Altman et al. 1976) and in vitro in the slice preparation (Sakaguchi et al. 1984). GABA release from RN cells in the in vitro slice preparation has been demonstrated (Kubota et al. 1983). The existence of glutamic acid decarboxylase (GAD) in the RN region has been shown (Nieoullon and Dustricier 1981). Murakami et al. (1983) demonstrated GABAergic synapses on RN cell using immunohistochemistry of GAD combined with the intracellular injection of horseradish peroxidase (HRP). GAD-immunoreactive axon terminals outline the somata and large dendrites of RN cells. Furthermore, Katsumaru et al. (1984) have shown that small GAD-immunoreactive neurons with a major diameter of 16 µm were observed in the RN under both light and electron microscopy. They were uniformly distributed throughout the RN. These neurons were considered to be the inhibitory interneurons, since they receive afferent fibers from

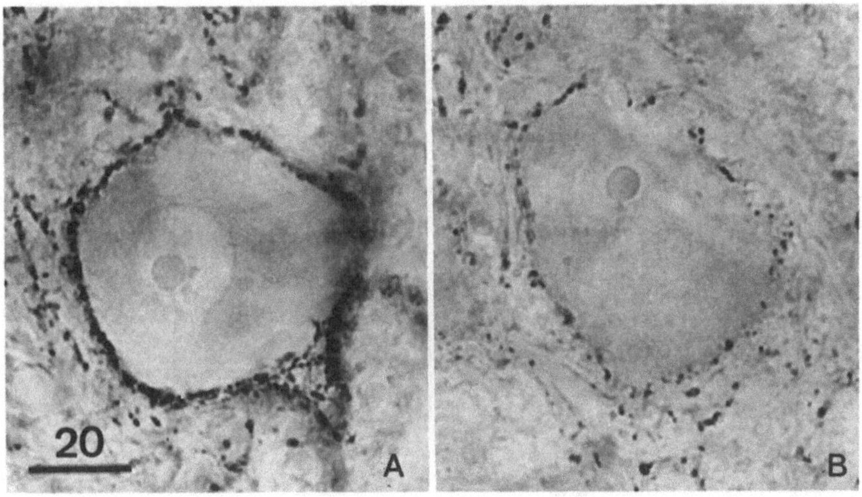

Fig. 3 A,B. GAD positive terminals on the somata of RN cells in the cat with unilateral lesion of the nucleus interpositus of the cerebellum. A: Experimental side with lesions of the interpositorubral input 37 days before acute experiment. B: Control side without lesion of the interpositorubral input. Bar, 20 µm (modified from Katsumaru et al., to be published)

the cerebral sensorimotor cortex. The existence of such interneurons was expected from the presence of disynaptic inhibition in RN neurons by activation of fast conducting pyramidal tract fibers.

Whether or not GABAergic inhibitory axon terminals also undergo modification after lesions of the nucleus interpositus can be answered by observing the number of GAD-positive terminals of RN neurons. A possible change of the GAD-positive terminals was tested in kitten RN after lesions of the major synaptic input, IP, and a marked increase of GAD-positive terminals was found on the somata of the RN neurons (Fig. 3). The area occupied by the GAD-positive terminals increased to 199%. This suggests that sprouting of the inhibitory synapses also takes place, but the evidence is not conclusive.

Sprouting in the Intact Central Neurons

We found that to induce sprouting in the RN neurons, lesions of the direct synaptic inputs are unnecessary. It is possible to induce sprouting by such procedures as cross-innervation of peripheral flexor and extensor nerves. The corticorubral EPSPs recorded in cats in which cross-innervation of peripheral flexor and extensor nerves of the forelimb was performed more than 2 months previously, have a fast rising component superimposed on the original slow rising dendritic EPSPs. This phenomenon is similar to that found after lesions of the interpositorubral input and was interpreted to be due to the formation of new synapses on the proximal portion of the soma dendritic membrane of RN cells. The changing time course of the corticorubral EPSPs was found not only in the composite EPSPs but also in the corticorubral unitary EPSP. Figure 4 shows the relation between the time-to-peak and amplitudes of these corticorubral unitary

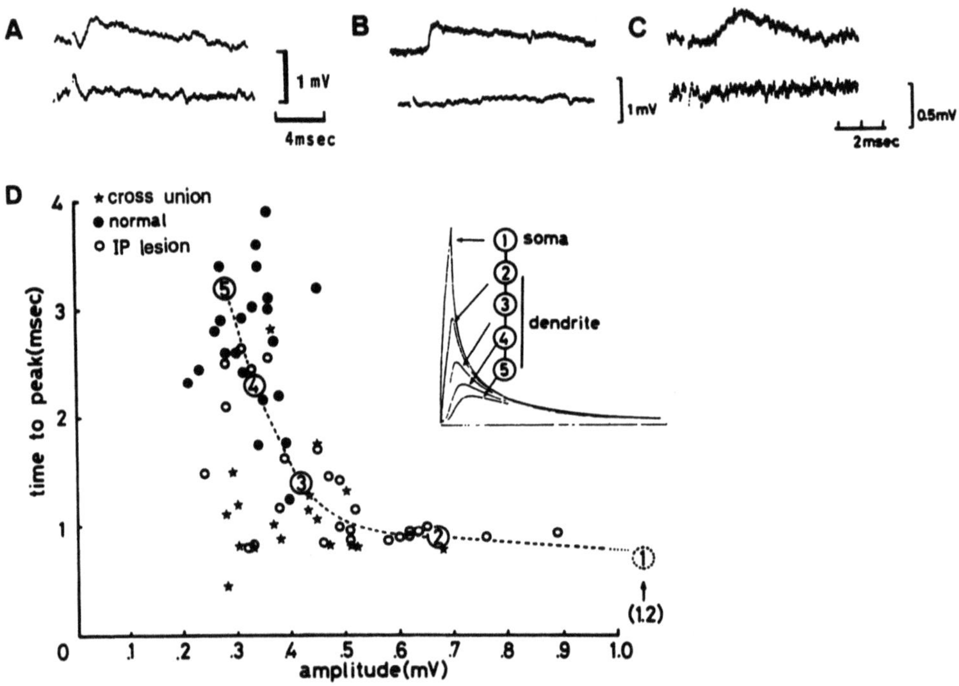

Fig. 4 A-D. Corticorubral unitary EPSPs. A: Intracellular EPSP evoked by stimulation of the CP in an RN cell of a cross-innervated cat. B: Same as A but evoked by stimulation of the sensorimotor cortex (SM) in a cat with a lesion of the nucleus interpositus (IP) 27 days before acute experiment. C: Same as B but in a normal cat. A-C: Upper traces, intracellular potentials; lower traces, extracellular field potential corresponding to the upper traces. D: Relation between time-to-peak and amplitude of the unitary EPSPs. o, unitary EPSPs of IP-lesioned cats; * those of cross-innervated cats more than 2 months before recording; ● , those of normal cats. O, time-to-peak and amplitude of theoretical EPSPs derived from Rall's compartment model initiated at each compartment of a five-compartment chain. The time course of the theoretical EPSPs generated in these compartments is shown in the inset. (Tsukahara 1981)

EPSPs. We noted a tendency for unitary EPSPs with shorter time-to-peak values to have larger amplitudes. This relation is predicted theoretically by Rall's compartment model, as schematically illustrated. These results suggest that sprouting takes place on the proximal portion of the soma-dendritic membrane of RN cells.

In order to correlate these physiological findings to electron microscopic data, we developed a computer system which enabled us to measure the dendritic diameters along the center line of the dendritic tree as a function of the distance from the somatic center. It was found that in normal as well as in cross-innervated cats, dendritic diameters of RN cells decreased monotonically as the distance from the somatic center increased. Based on this measurement it is possible to estimate the location of synapses on an RN cell's dendrites by measuring the smallest diameter of the dendritic cross-section in electron micrographs of HRP-stained RN cells.

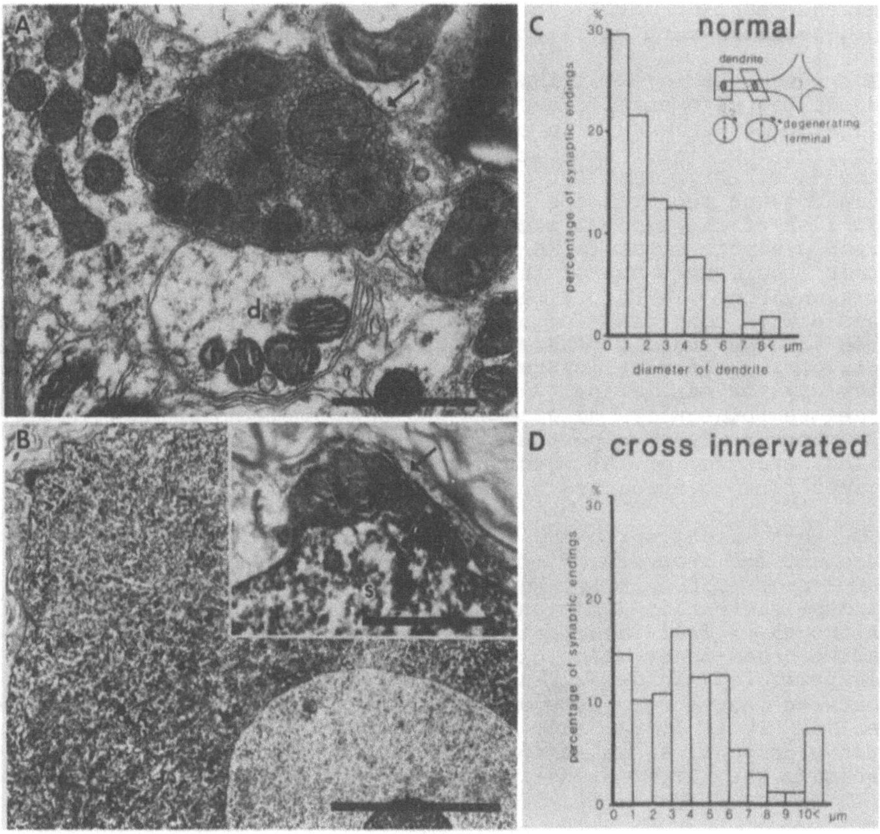

Fig. 5. Corticorubral axon terminals synapsing upon dendrite (A) and soma (B) of RN neurons in a normal cat (A) and in a cat with cross-innervation of peripheral flexor and extensor nerves (B). Degenerating corticorubral terminals (arrowheads) synapsing on HRP-filled RN neurons. Bar in A, 1 µm; bar in B, 10 µm; inset bar, 1 µm. d, dendrite; s, soma; C,D: Histograms illustrating the relation between the number of degenerating corticorubral terminals and the diameter of dendrites. C: The distribution of degenerating terminals in normal cats. Ordinate, number of degenerating synaptic endings as expressed by the percentage of terminals; abscissa, minor diameter of dendrites contacted by degenerating terminals which were measured from electron micrographs. D.: Similar distribution of degenerating terminals found in cats with cross-innervation. (Murakami et al. 1984)

Further corticorubral synapses can be identified by the presence of degenerating terminals after lesions of the sensorimotor cortex.

The degenerating terminals in normal cats are found mainly on the dendrites with small diameters (Fig. 5A, C). On the other hand, degenerating terminals in cats with cross-innervation are found not only on the distal dendrites but also on the proximal dendrites and the soma (Fig. 5B, D). This is consistent with the physiological finding and supports the formation of new synapses on the proximal portion of the RN cell membrane. Since there is no damage to the direct synaptic input to RN neurons after cross-innervation, this result represents one of the few examples of sprouting of intact central neurons. Therefore, lesions of direct synaptic inputs are not a necessary condition for sprouting to be initiated, since sprouting can also be induced after peripheral nerve cross-innervation. In the peripheral nervous system, Brown and Ironton (1977) showed that a few days of muscle inactivity can act as a stimulus to sprouting at the neuromuscular junction. Rotshenker and McMahan (1976) and Rotshenker (1979) reported that in frog with intact innervated muscles, crushing of the motor nerve of the cutaneous pectoris muscles gives rise to sprouting on the contralateral side.

To what extent sprouting and formation of new functional synapses account for recovery of function after brain damage remains to be explored. For example, some motor recovery has been reported after cross-innervation (Brinkan and Porter 1977; Tsukahara et al. 1982). Although a full interpretation of the behavioral adaption observed after cross-innervation is a tremendous task and remains to be investigated in detail, it may be speculated that some correlation between behavioral recovery and the appearance of sprouting may exist. It is likely that sprouting and the formation of new connections account, at least in part, for the behavioral recovery that animals show after cross-innervation and perhaps after brain damage.

References

Altmann H, ten Bruggencate G, Pickelmann P, Steinberg R (1976) Effects of GABA, glycine, picrotoxin and bicuculline methochloride on rubrospinal neurones in cats. Brain Res 111:337-345

Brinkman J, Porter R (1977) Plasticity of motor behavior in monkeys with crossed forelimb nerves. Proc Int Congr Physiol Sci 13:P96

Brown MC, Ironton R (1977) Motor neuron sprouting induced by prolonged tetrodotoxin block of nerve action potentials. Nature 265:459-461

Cotman CW, Nieto-Sampedro M, Harris EW (1981) Synapse replacement in the nervous system of vertebrates. Physiol Rev 61:684-784

Fujito Y, Tsukahara N, Oda Y, Yoshida M (1982) Formation of functional synapses in the adult cat red nucleus from the cerebrum following cross-innervation of forelimb flexor and extensor nerves. II. Analysis of newly-appeared synaptic potentials. Exp Brain Res 45:13-18

Katsumaru H, Murakami F, Wu J-Y, Tsukahara N (1984) GABAergic intrinsic interneurons in the red nucleus of the cat demonstrated with combined immunocytochemistry and anterograde degeneration methods. Neurosci Res 1:35-44

Katsumaru H, Murakami F, Wu J-Y, Tsukahara N (1985) Plasticity of GABAergic synapses in the red nucleus after lesion of the nucleus interpositus of the cat: an immunocytochemical study. (to be published)

Kosar E, Fujito Y, Murakami F, Tsukahara N (1985) Morphological and electrophysiological study of sprouting corticorubral fibers after lesions of the contralateral cerebrum in kitten. (to be published)

Kubota M, Sakaguchi H, Tsukahara N (1983) Release of endogenous GABA from the cat red nucleus slices. Brain Res 270:190-192

Murakami F, Katsumaru H, Saito K, Tsukahara N (1982) A quantitative study of synaptic reorganization in red nucleus neurons after lesion of the nucleus interpositus of the cat: an electron microscopic study involving intracellular injection of horseradish peroxidase. Brain Res 242:41-53

Murakami F, Katsumaru H, Wu J-Y, Matsuda T, Tsukahara N (1983) Immunocytochemical demonstration of GABAergic synapses on identified rubrospinal neurons. Brain Res 267:357-360

Murakami F, Katsumaru H, Maeda J, Tsukahara N (1984) Reorganisation of corticorubral synapses following cross-innervation of flexor and extensor nerves of adult cat: a quantitative electron microscopic study. Brain Res 306:299-306

Nah SH, Leong SK (1976) Bilateral corticofugal projection to the red nucleus after neonatal lesions in the albino rat. Brain Res 107:433-436

Nieoullon A, Dusticier N (1981) Glutamate decarboxylase distribution in discrete motor nuclei in the cat brain. J Neurochem 37:202-209

Rotshenker S (1979) Synapse formation in intact innervated cutaneous pectoris muscles of the frog following denervation of the opposite muscle. J Physiol 292:535-547

Rotshenker S, MacMahan UJ (1976) Altered pattern of innervation in frog muscle after denervation. J Neurocytol 5:719-730

Sakaguchi H, Kubota M, Nakamura M, Tsukahara N (1984) Effects of amino acids on cat red nucleus neurons in vitro. Exp Brain Res 54:150-156

Tsukahara N (1981) Synaptic plasticity in the mammalian central nervous system. Annu Rev Neurosci 4:351-379

Tsukahara N, Fujito Y (1976) Physiological evidence of formation of new synapses from cerebrum in the red nucleus neurons following cross-union of forelimb nerves. Brain Res 106:184-188

Tsukahara N, Fujito Y (1981) Neuronal plasticity in the newborn and adult feline red nucleus. In: Flohr H, Precht W (eds) Lesion-induced neuronal plasticity in sensorimotor systems. Springer, Berlin Heidelberg New York, p 64

Tsukahara N, Hultborn H, Murakami F, Fujito Y (1975) Electrophysiological study of formation of new synapses and collateral sprouting in red nucleus neurons after partial denervation. J Neurophysiol 38:1359-1372

Tsukahara N, Fujito Y, Oda Y, Maeda J (1982) Formation of functional synapses in adult cat red nucleus from the cerebrum following cross-innervation of forelimb flexor and extensor nerves. I. Appearance of new synaptic potentials. Exp Brain Res 45:1-12

Tsukahara N, Fujito Y, Kubota M (1983) Specificity of the newly--formed corticorubral synapses in the kitten red nucleus. Exp Brain Res 51:45-56

Villablanca JR, Olmstead CE, Sonnier BJ, McAllister II JP, Gomez F (1982) Evidence of a crossed corticorubral projection in cats with one cerebral hemisphere removed neonatally. Neurosci Lett 33:241-246

Neurochemical Events Underlying Continued Function Despite Injury to Monoaminergic Systems

M. J. Zigmond, M. K. Stachowiak, T. W. Berger, and E. M. Stricker
Department of Biological Sciences, University of Pittsburgh, Pittsburgh, PA 15260, USA

Introduction

Pharmacological blockade of monoamine receptors usually precipitates severe physiological or behavioral impairments. Yet monoaminergic systems can withstand a considerable amount of damage before such functional deficits emerge. This was noted first by Cannon et al. (1929) in their studies of injury to the sympathetic nervous system, and analogous results have been observed subsequently for noradrenergic, dopaminergic, and serotonergic systems in the central nervous system. Perhaps the most dramatic case in point is that of Parkinson's disease. The severe neurological impairments associated with this disorder appear to be due to primarily to the loss of dopaminergic neurons in the nigrostriatal bundle. However, such impairments usually are not detectable until at least 70% of these neurons have degenerated, and severe impairments do not emerge until more than 90% degeneration (Bernheimer et al. 1973). A similar phenomenon can be demonstrated in an experimental model of the disorder using adult rats. Near-total destruction of the nigrostriatal bundle in these animals leads to a syndrome of akinesia, catalepsy and sensory neglect, yet no such deficits occur if 5-10% of the dopaminergic neurons survive.

In this report we review our recent studies of the possible cellular bases of these phenomena. First, we present some electrophysiological results demonstrating at a cellular level that function can continue despite substantial loss of dopaminergic neurons. Second, we discuss the results of biochemical studies indicating that after injury to the nigrostriatal bundle, the dopaminergic neurons that remain compensate for the lesion by increasing their delivery of dopamine (DA) to target cells. The residual neurons do so, in part, because of an associated increase in their capacity for DA synthesis, and evidence to support this coupling of synthesis to secretion is presented. Finally, we propose a general model whereby the influence of monoaminergic neurons on their target cells can be maintained despite extensive injury.

Spontaneous Firing Rate of Striatal Cells After 6-HDA

In each of the following experiments adult Sprague-Dawley derived rats received the neurotoxin, 6-hydroxydopamine (6-HDA), by way of the lateral cerebroventricle. Desmethylimipramine (25 mg/kg, i.p.) was given 30 min prior to the administration of 6-HDA to block uptake of the toxin into noradrenergic terminals. This treatment produces a near-total loss of DA in the medial, periventricular zone of

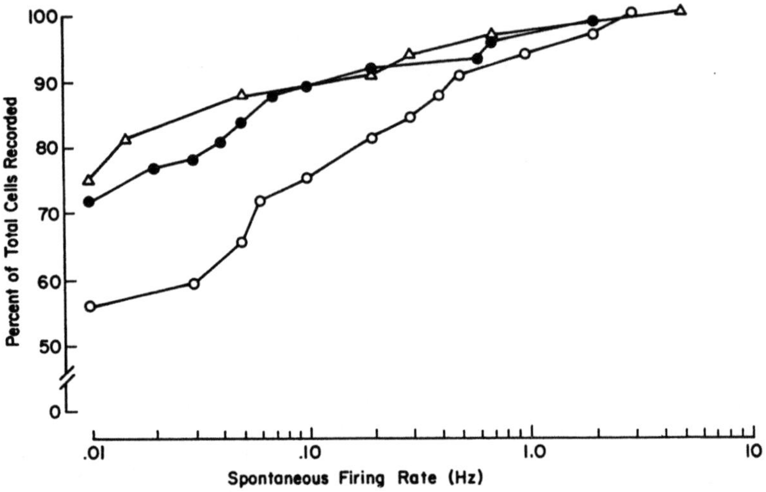

Fig. 1. Striatal activity in control and 6-HDA lesioned rats. Measurements were made 6-9 days after the administration of pargyline (50 mg/kg, i.p.) and either 6-HDA (180 µg, ivt.) or vehicle. Shown are recordings made in the medial (O—O, n=32) and lateral (Δ - Δ , n=32) striatum of 6-HDA-lesioned animals and in both striatal areas (●—●, n=73 for combined areas) of control animals. Subsequent biochemical analyses indicated that 6-HDA produced a 90-95% depletion of DA in the medial striatal area and a 60% depletion in the lateral area. (Unpublished observations of W.B. Orr and T.W. Berger)

the striatum, but residual terminals are present more laterally, with the number of terminals being inversely related to the amount of 6-HDA administered (Onn et al. 1982).

Approximately 1 week after administration of 6-HDA or vehicle solution (0.1% ascorbic acid in 0.9% NaCl), animals were anesthetized with chloral hydrate and the spontaneous activity of single striatal neurons was recorded extracellularly using glass pipettes. Striatal cells were identified by stereotaxic coordinates, action potential waveform (Skirboll and Bunney 1979), and their orthodynamic, excitatory response to neocortical stimulation. Neural activity was sampled in both the medial and lateral striatum of each rat. Thus, dopaminergic input to recorded striatal cells varied depending on the dose of 6-HDA used and the mediolateral location of the neuron within the striatum (Onn et al. 1982). Striatal cells recorded from normally behaving non-injected or vehicle-injected animals were silent, with 90% having firing rates below 0.2 Hz (Fig. 1). However, large doses of 6-HDA (250 µg) that denervated both medial and lateral regions of the striatum and produced behavioral akinesia, catalepsy, aphagia, and adipsia resulted in a tenfold increase in cellular activity (see also Schultz and Ungerstedt 1978). These results are consistent with previous reports suggesting that dopaminergic neurons normally exert an inhibitory influence on striatal activity (Connor 1970). However, this increase in striatal activity was observed only when DA depletions were near-total. Animals given smaller doses of 6-HDA (150 µg after 50 mg/kg of pargyline, i.p.) that depleted medial striatal areas by approximately 90-95%, but lateral striatal zones by only 60%, produced a significant increase in spontaneous firing rate in the medial striatum with no change at all in lateral striatal activity. These animals also did not exhibit the severe behavioral impairments seen in animals given the larger 6-HDA-induced lesion. Thus, our earlier observations that marked sensorimotor dysfunctions appear only when overall striatal DA depletions exceed 90-95% are paralleled by these cellular analyses. Collectively, these observations indicate that a relatively small number of DA terminals can perform at least some of the functions that normally are mediated by the full complement of terminals.

DA Efflux from Striatal Terminals After 6-HDA

One explanation for the resistance of striatal cells to disinhibition after damage to the nigrostriatal bundle is that the loss of dopaminergic afferents does not result in a proportional reduction in the amount of DA that is released into the synapse; in other words, the dopaminergic terminals spared by the lesion may provide more DA than they would in the intact animal. To examine this possibility, 300 μm coronal slices were prepared from the striatum of animals treated 2-4 weeks earlier with 6-HDA or its vehicle. Slices were superfused at 100 μl/min with Krebs-bicarbonate medium supplemented with ascorbic acid (1 mM) and tyrosine (10 μM). After an initial period of 15-30 min, slices were subjected to electrical field stimulation via platinum grid electrodes (2 ms 18 mA bipolar pulses at 2 Hz). The superfusion effluent was collected in 5-min aliquots, mixed with perchloric acid, and frozen prior to analysis of endogenous DA content by high performance liquid chromatography with electrochemical detection.

Using striatal slices from intact animals, we observed that DA efflux exhibited a biphasic response to field stimulation (Fig. 2). Initially, calcium-dependent DA release increased threeto fourfold. However, despite continued stimulation, this release declined within 10-20 min, usually achieving a steady-state rate of DA efflux that was approximately two times the basal levels. Pretreatment with alpha-methyltyrosine (50μM) had little effect on the initial rise in DA efflux while completely abolishing subsequent release (Stachowiak et al. 1984). Thus, we conclude that the decline in DA release with continuous electrical stimulation represents in the gradual exhaustion of DA stores and a resulting dependence on newly synthesized material.

Striatal slices prepared from 6-HDA-treated animals showed a different pattern of DA efflux. Instead of a rapid increase in DA release followed by a steady decline, these slices exhibited a progressive increase in DA efflux, often achieving near-control levels within 15 min (Fig. 2) (Stachowiak et al. 1984). Accordingly, when efflux was expressed as a fraction of the DA present in the slice at the onset of the experiment, the efflux from 6-HDA-lesioned slices was much greater than that from control tissue. Indeed, the fractional rate of DA release during depolarization increased exponentially with lesion size, reaching more than tenfold in animals with greater than 95% DA depletions. Both biochemical and histochemical studies have indicated that tissue catecholamine content in the brain-lesioned animal reflects the number of residual catecholaminergic terminals, and consequently the increase in fractional DA efflux suggests an elevation in efflux per residual DA terminal.

There are several mechanisms that could contribute to this increased DA efflux. For example, because DA is inactivated primarily by high

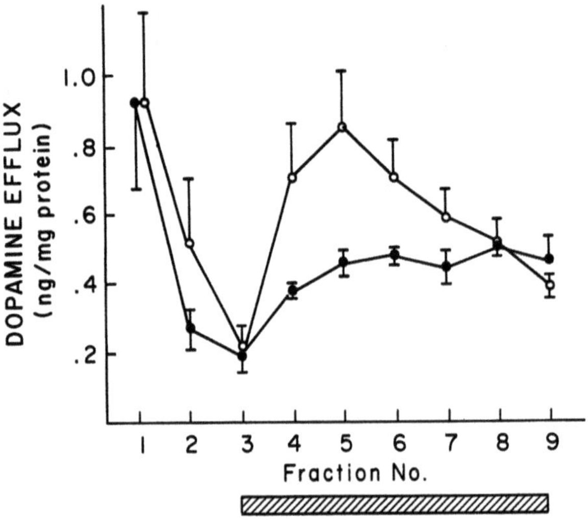

Fig. 2. Endogenous DA efflux from superfused striatal slices prepared from animals 1-2 months after intraventricular treatment with 250 µg 6-HDA (●—●, n=3) or vehicle (○—○, n=11). Shown are means ± SEM

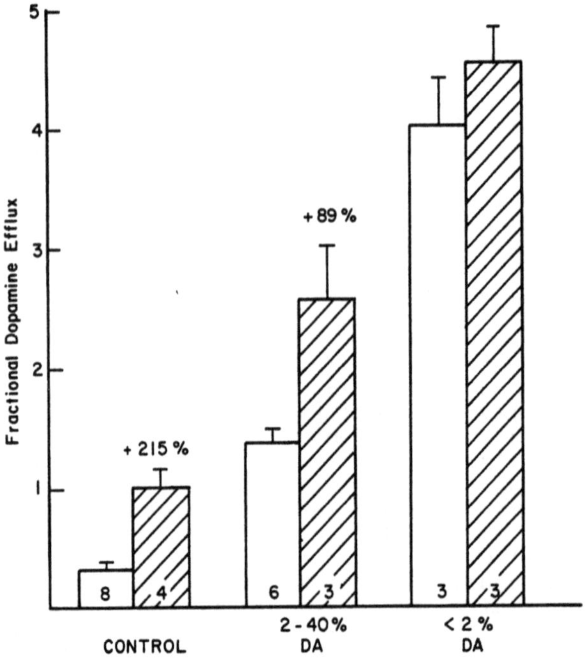

Fig. 3. Endogenous DA efflux from striatal slices prepared 1-2 months after 6-HDA or vehicle treatment. Slices were incubated under basal conditions (open bars) or in the presence of 10 µM nomifensin (hatched bars). Shown are the means ± SEM of the number of experiments indicated within each bar

affinity uptake into dopaminergic terminals, the degeneration of a large portion of these neurons might lead to an increased overflow. This appears to be the case. Nomifensin (10 μM), a drug that blocks DA uptake, increased DA efflux from control slices by two- to three-fold, whereas the drug had little or no effect on DA efflux from 6-HDA-lesioned tissue (Fig. 3). While these observations suggest that terminal loss plays a role in increasing the availability of DA after partial injury, further examination of the data indicates that this cannot by itself account for the increased fractional DA efflux, for several reasons. First, while depolarization of slices from either intact or 6-HDA lesioned animals is not associated with a depletion of tissue DA, depolarization of intact slices in the presence of nomifensin is accompanied by a depletion of DA to 45% of prestimulation levels. Second, even in the presence of nomifensin fractional DA efflux from intact slices was never as great as that seen in 6-HDA-lesioned slices without nomifensin (Fig. 3). We conclude that in addition to decreased reuptake, augmented synthesis and release of DA from residual terminals also contributes to the increase in overflow. This is consistent with our observation that there is a marked increase in the ratio of the metabolite dihydroxyphenylacetic acid to DA present in the striatum of rats after large 6-HDA-induced brain lesions (Zigmond et al. 1984).

Tyrosine Hydroxylase Activity After 6-HDA

An increased capacity for DA synthesis probably involves a change in the rate-limiting step in DA synthesis, the oxydation of tyrosine. This step is catalyzed by the enzyme tyrosine hydroxylase (TH). We examined enzyme activity in crude homogenates incubated with subsaturating concentrations of radiolabelled tyrosine and of the synthetic cofactor, 6-methyl-5,6,7,8-tetrahydropterin. While TH activity was found to be reduced in striatal homogenates prepared from 6-HDA treated animals, the decline was signifcantly less than was the loss of DA (Zigmond et al. 1984). The extent of this discrepancy could be expressed as the ratio of TH activity to DA content. This ratio increased exponentially with lesion size in a manner that paralleled the elevation in the fractional rate of DA release from our depolarized slices (Fig. 4). Thus, we believe that this increase in TH activity underlies the capacity of the residual terminals to release DA at a significantly faster rate without compromising tissue stores.

The phenomenon of increased transmitter biosynthesis after partial injury is one that we have seen now in many monoaminergic systems, including the serotonergic projections of the raphe nuclei (Stachowiak et al. 1983) and both peripheral and central noradrenergic projections (Acheson et al. 1980; Acheson and Zigmond 1981; Fluharty et al. 1984). It is with the latter system, the central noradrenergic projections of the locus coeruleus, that we have begun to examine the mechanism of the induced increase in transmitter biosynthesis.

In considering the molecular mechanism of increased transmitter biosynthesis, it seemed clear that an increased synthesis of TH protein, which occurs in the cell body, could not account for the rapid changes in enzyme activity that occurred in terminal fields (Fig. 3). Instead, an examination of TH shortly after the administration of 6-HDA suggested that several changes had occurred in existing enzyme molecules (Acheson and Zigmond 1981). For example, the affinity of the enzyme for its cofactor was markedly increased (Fig.

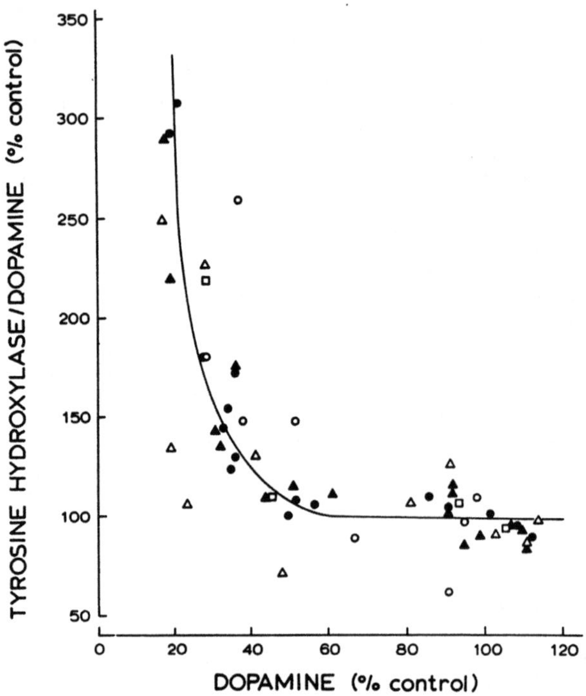

Fig. 4. Tyrosine hydroxylase (TH) activity as a function of striatal dopamine depletion. Animals were given intraventricular 6-HDA or vehicle and killed 3 (○), 5-6 (●), 12 (△), 21 (▲), or 35 (□) days later. TH activity was measured in the presence of a subsaturating concentration of cofactor and at a suboptimal pH. (Unpublished observations of A.L. Acheson and M.J. Zigmond)

5a). In addition, the pH profile of the enzyme had changed completely; instead of having a sharp peak at pH 6.2, it had a very broad pH optimum (Fig. 5b). Both of these effects could be mimicked in vitro by preincubating TH from intact tissue under cAMP-dependent phosphorylating conditions, whereas TH activity from 6-HDA lesioned tissue was not affected by these same conditions. For this and other reasons, we believe that the initial increase in NE biosynthesis probably results from the phosphorylation of existing TH molecules within residual terminals.

These alterations in the characteristics of TH were not long-lived. Gradually, the affinity of the enzyme for cofactor returned to normal (Fig. 5a) as did its pH profile (Fig. 5b). At the same time, the maximal velocity of the TH reaction increased. The time course of these changes depended on the distance of the terminal field from the cell bodies in locus coeruleus; while it took 21 days for the maximal velocity of hippocampal TH to reach a new steady state level, only 6 days were required for the same changes to occur in cerebellar TH (Acheson and Zigmond 1981). These observations are consistent with the gradual transport of new TH protein from cell bodies to residual terminals.

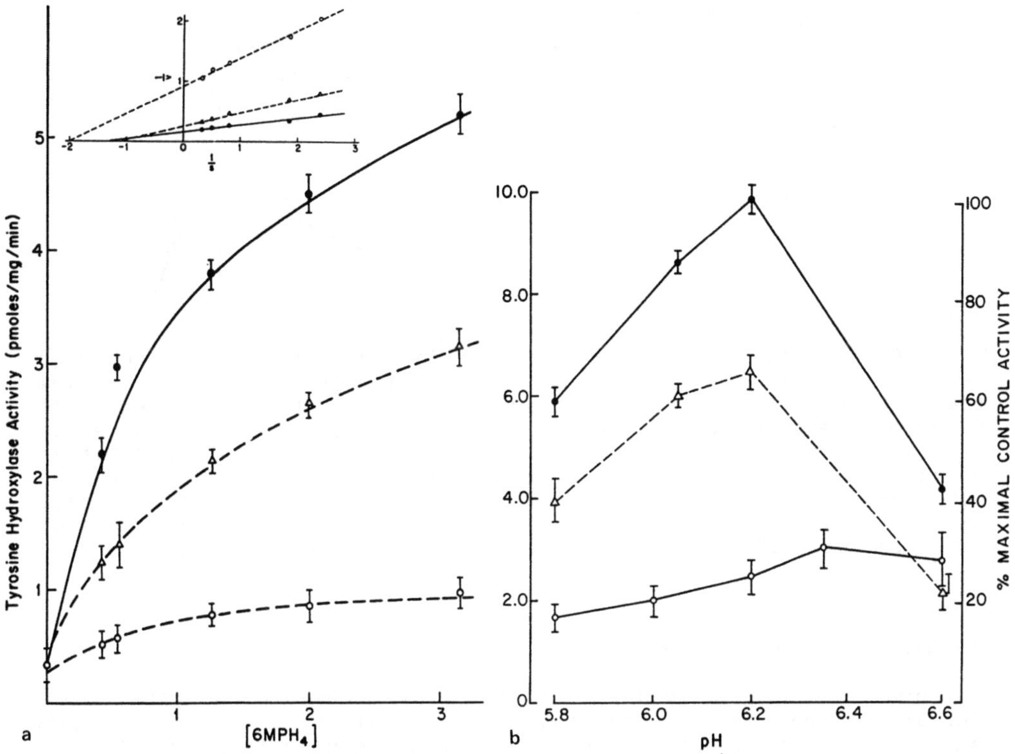

Fig. 5. Effect of 6-HDA-induced lesions on the characteristics of tyrosine hydroxylase (TH) in hippocampus. (a) TH activity as a function of the concentration of 6-methyl-5,6,7,8-tetrahydropterin ($6MPH_4$) in control animals (●) and in animals treated with 6-HDA 5 days (○) or 21 days (△) previously. Assays carried out at pH 6.5. (b) TH activity as a function of assay pH in control animals (●) and in animals treated with 6-HDA 1 1/2 days (○) or 21 days (△) previously. Assays carried out in the presence of 0.7 mM 6MPH . All values represent the means ± SEM of four or five animals. (Acheson and Zigmond 1981)

In short, there appear to be two, temporally distinct events responsible for the increased capacity of residual terminals to synthesize and release transmitter. First, there is a short-term activation of existing TH molecules resulting in increased activity under physiological conditions (i.e., pH above the optimum and low cofactor concentration). Second, there is a gradual increase in the number of TH molecules per terminal. This pattern appears to be a general one, as we have observed similar phenomena in the peripheral sympathoadrenal system (Fluharty et al. 1984) and in dopaminergic (Zigmond et al. 1984) and serotonergic (Stachowiak et al. 1983) systems in brain.

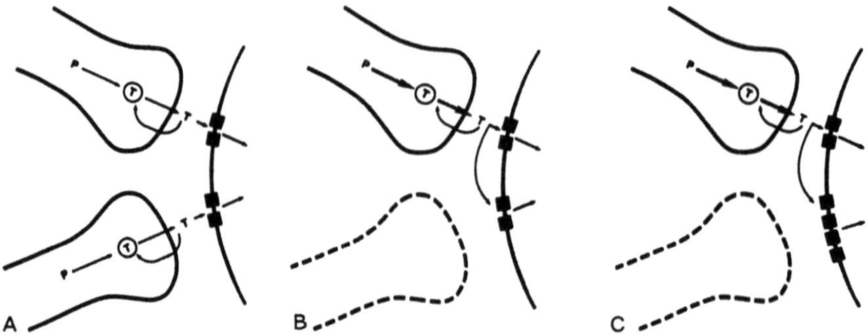

Fig. 6. Model for "biochemical reinnervation" after subtotal destruction of a monoaminergic system. (A) In the intact system transmitter (T) is formed from precursor (P), released onto receptors (■) and inactivated by high affinity reuptake. (B) After partial degeneration (dotted line) an increase in transmitter synthesis and release coupled with the loss of high affinity uptake sites permits T to act on denervated targets. (C) After still further degeneration, receptor proliferation serves to further increase the efficacy of available T. (Zigmond and Stricker, 1984b)

Summary and Conclusions

In summary, there is evidence at both the behavioral and the cellular levels that monoaminergic systems can withstand extensive injury and still maintain control over their postsynaptic targets. In this review we have described a mechanism by which this might occur involving the "biochemical reinnervation" of denervated neurons rather than any anatomical growth. We believe that this process may be in part a natural consequence of the synaptic homeostasis that appears to exist at monoaminergic synapses (Zigmond and Stricker 1984, 1985). According to this concept, decreased activation of receptors that monitor extracellular transmitter availability, as would occur as a result of injury, leads to an increase in impulse flow and an increase in transmitter release. This is accompanied by an increase in synthesis thereby maintaining presynaptic stores intact. The impact of these adaptive changes is further enhanced by the relative absence of the high affinity uptake sites normally responsible for transmitter inactivation (Fig. 6).

We have said nothing about the possible contribution to recovery of postsynaptic changes, such as increases in the number of transmitter receptors. That is because such alterations appear to occur only in response to very large depletions of transmitter (Zigmond and Stricker 1980; Creese 1981). Indeed the absence of increased DA receptor proliferation except after near-total degeneration of the dopaminergic neurons is consistent with our proposal that a small number of DA neurons are able to supply striatal targets with an adequate supply of transmitter for normal, basal function.

In closing, there is one further implication of our data that may be noted. Because relatively few hyperactive monoaminergic neurons may conceal a progressive degenerative disorder, it would seem important to develop tests that could be used to diagnose such conditions in their compensated, preclinical state (Zigmond and Stricker 1981). Such tests may be based on the demonstrated ability of monoaminergic antagonists (Heffner et al. 1977) and of stress (Cannon et al. 1929;

Stricker and Zigmond 1974; Snyder et al. 1982) to overwhelm the residual functional capacity of the injured system and thereby precipitate acute neurological dysfunctions.

Acknowledgments. This research was supported in part by grants from the American Parkinson's Disease Association and the U.S.P.H.S. (NS-19608 and MH-30915).

References

Acheson AL, Zigmond MJ (1981) Short- and long-term changes in tyrosine hydroxylase activity in rat brain after subtotal destruction of noradrenergic neurons. J Neurosci 1:493-504

Acheson AL, Zigmond MJ, Stricker EM (1980) Compensatory changes in cerebral tyrosine hydroxylase activity after intraventricular 6-hydroxydopamine. Science 207:537-540

Bernheimer H, Birkmayer W, Hornykiewicz O, Jellinger K, Seitelberger F (1973) Brain dopamine and the syndromes of Parkinson and Huntington: Clinical, morphological and neurochemical correlations. J Neurol Sci 20:415-455

Cannon WB, Newton HF, Bright EM, Menkin V, Moore RM (1929) Some aspects of the physiology of animals surviving complete exclusion of sympathetic nerve impulses. Am J Physiol 89:84-107

Connor JD (1970) Caudate nucleus neurones: correlation of the effects of substantia nigra stimulation with iontophoretic dopamine. J Physiol 208:691

Creese I (1981) Dopamine receptors. In: Yamamura HI, Enna SJ (eds) Neurotransmitter receptors, part 2, biogenic amines. Chapman and Hall, London, p 129

Fluharty SJ, Rabow LE, Stricker EM, Zigmond MJ (1984) Neurochemical adaptions after partial sympathectomy: adrenal and cardiac tyrosine hydroxylase activity. Fed Proc 43:1020

Heffner TG, Zigmond MJ, Stricker EM (1977) Effects of dopaminergic agonists and antagonists on feeding in intact and 6-hydroxymine-treated rats. J Pharmacol Exp Ther 201:386-399

Onn S, Zigmond MJ, Stricker EM, Berger TW (1982) Intraventricular 6-hydroxydopamine treatment in rats produce a gradient of dopamine depletions within the striatum. Soc Neurosci Abstr 8:743

Schultz W, Ungerstedt U (1978) Short-term increase and long-term reversion of striatal cell activity after degeneration of the nigrostriatal dopamine system. Exp Brain Res 33:159-171

Skirboll LR, Bunney BS (1979) The effect of acute and chronic haloperidol treatment on spontaneously firing neurons in the caudate nucleus of the rat. Life Sci 25:1419-1434

Snyder AM, Zigmond MJ, Stricker EM (1982) Stress-induced neurological deficits in dopamine-depleted rats and their reversal by dopamine agonists. Soc Neurosci Abstr 8:362

Stachowiak MK, Jacoby JH, Zigmond MJ, Stricker EM (1983) Increased tryptophan hydroxylase activity in central serotonergic neurons spared by 5,7-dihydroxytryptamine. Soc Neursosci Abstr 9:1222

Stachowiak M, Keller R, Stricker E, Zigmond M (1984) Increased dopamine release from terminals spared by 6-hydroxydopamine: An in vitro analysis. IUPHAR 9th International congress of pharmacology, London

Stricker EM, Zigmond EM (1974) Effects on homeostasis of intraventricular injection of 6-hydroxydopamine in rats. J Comp Physiol Psychol 86:973-994

Zigmond MJ, Stricker EM (1980) Increased response to dopaminergic agonists after intraventricular 6-hydroxydopamine: relation to lesion size. Experientia 36:436-438

Zigmond MJ, Stricker EM (1981) Behavioral tests for the detection of subclinical brain damage: an experimental model. In: Finger S (ed) Brain-behavior relationships. Lexington, Lexington p 93

Zigmond MJ, Stricker EM (1984) Parkinson's disease: studies with an animal model. Life Sci 35:5-18

Zigmond MJ, Stricker EM (1985) Adaptive properties of monoaminergic neurons and their functional implications. In: Lajtha A (ed) Handbook of neurochemistry, vol 9. Plenum, New York, pp 87-102

Zigmond MJ, Acheson AL, Stachowiak MK, Stricker EM (1984) Neurochemical compensation after nigrostriatal bundle injury in an animal model of preclinical Parkinsonism. Arch Neurol 41:856-861

Chemically Induced Homotypic Collateral Sprouting of Hippocampal Serotonergic Afferents

E. C. Azmitia and F. C. Zhou

Department of Biology, New York University, 100 Washington Square East, New York, NY 10003, USA

Introduction

The adult brain has a tremendous capacity to respond to injury. Damage in a localized region produces many responses, e.g., degeneration, regeneration, reactive synaptogenesis, proximal sprouting (damming effect), spared collateral sprouting (pruning effect), retrograde cellular reactivity, and compensatory collateral sprouting. It is expected that each response will have its own regulatory factors. Studies have shown that several specific soluble (Crutcher and Collins 1982) and membrane-bound (Braun et al. 1981) factors exist which have selective attractive properties. Thus despite intensive research extending into the last century, the analysis of the reparative growth response in the adult brain is still poorly defined. "Almost nothing is known of the steps which lead to this compensatory sprouting..." (Gall et al. 1979, p 539). One approach to this longstanding problem is to focus on a single cell type (serotonin) in a localized area (hippocampus) during a specific growth process (compensatory collateral sprouting) regulated by a known hormonal substance (corticosterone).

1. Single Cell Type - Serotonin-Producing Neurons. Serotonin has been recognized as a major brain chemical system for almost 3 decades. Although these few thousand neurons have their cell bodies confined to the raphe of the brainstem, their axons stretch from the olfactory bulb to the spinal cord (Azmitia and Gannon 1984) and they may be able to influence directly every cell in the cerebral cortex of the rat (Lidov et al. 1980). Serotonin is experimentally implicated in a large spectrum of physiological processes (e.g., blood flow, neurogenesis, temperature regulation, hormonal secretion, locomotor activity, pain, and sleep) and clinically implicated in disorders such as myoclonus, migraine, depression, epilepsy, and schizophrenia (Azmitia 1978, 1982; Calas 1981; Haber et al. 1981; Jacobs and Gelperin 1981).

2. Localized Area - Hippocampus. The hippocampus has been extensively studied both structurally and functionally (Ciba Foundation Symposium 1978; Isaacon and Pribram 1975). The main afferents and efferents of this structure were described by Ramon y Cajal (1911) and Lorenti de No (1934). Modern techniques have more precisely described the fine circuitry (Meibach and Siegal 1977; Swanson and Cowan 1977; Swanson et al. 1981). Serotonergic fibers were first described in the hippocampus using histochemical fluorescence (Fuxe

Experimental Brain Research, Suppl. 13

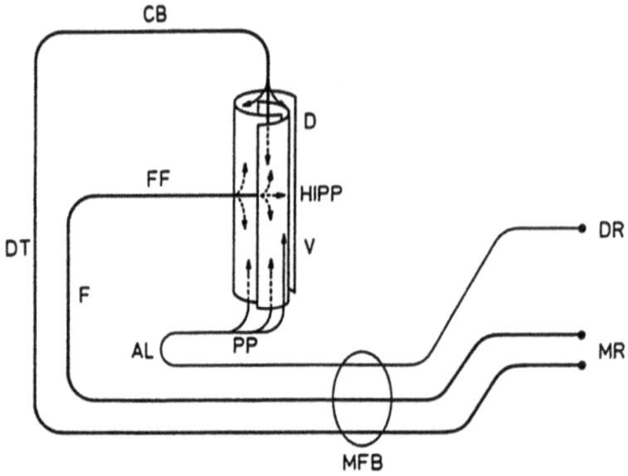

Fig. 1. This schematic diagram illustrates the main projections to the hippocampus (HIPP) from the median (MR) and dorsal (DR) raphe nuclei in the rat. The MR fibers enter the dorsal (D) hippocampus via the cingulum bundle (CB) and fornix-fimbria (FF). The DR innervates mainly the ventral (V) hippocampus via the ventral route which uses the ansa lenticularis (AL) and the perforant path (PP). DT, diagonal tract; MFB, medial forebrain bundle

1965), and using radioautographic tracing procedures three anatomically distinct pathways into this area were seen: the cingulum bundle (CB), the fornix-fimbria (FF), and a ventral entorhinal pathway (Fig. 1; Azmitia 1978; Azmitia and Segal 1978; Zhou and Azmitia 1983).

The hippocampus essentially acts as a comporator, comparing actual with expected stimuli (see Gray 1981). It has been firmly established that localized removal of serotonergic fibers from this limbic center by microinjections of specific neurotoxic drugs into the CB or FF disrupts normal hippocampal activity (Azmitia et al. 1978; McNaughton et al. 1980; Srebro et al. 1982; Williams and Azmitia 1981). The deficit produced in the animal's behavior - increased nocturnal locomotor activity - is negatively correlated with the amount of hippocampal reuptake of 5-HT (-0.89, $p<0.001$, Williams and Azmitia 1981). These studies substantiate the finding that 5-HT is an inhibitory transmitter on hippocampal neurons (Segal 1980).

3. Specific Growth Response-Homotypic Collateral Sprouting. The phenomenon of collateral sprouting has received a great deal of attention in the last decade. Two types have been studied in the adult brain: reactive synaptogenesis and homotypic collateral sprouting. The first type deals mainly with heterotypic innervation of a denervated target. This process has a fast onset, a short duration, favors those fibers most closely situated to the denervated sites (Raisman and Field 1973; Cotman and Lynch 1976; Murray and Goldberg 1974), and usually does not result in return of normal function. [An exception to this rule is the collateral sprouting induced by ipsilateral damage to the afferent fibers of a brain area innervated bilaterally (e.g., optic projections - Lund et al. (1973), Goodman et al (1973); entorhinal projections - Steward et al. (1974, 1976).] In this type of axonal sprouting, "little if any advantage is gained by

homology to the degenerating input....transmitter identity is not required to allow reactive synaptogenesis" (Cotman and Lynch 1976, p 73). The second type of collateral sprouting has a slow onset, long duration, and favors those fibers chemically identical to the denervated ones. This compensatory collateral sprouting has been well studied in the PNS (Edds 1953; Diamond 1979) and in the CNS (Azmitia et al. 1978; Gage et al. 1983). There are reports of specific heterotypic sprouting in the hippocampus of sympathetic noradrenergic fibers in response to central cholinergic denervation produced by electrolytic lesions of the septum (Loy et al. 1980). However, there is no evidence that this type of sprouting would compensate for loss of normal functions. To our knowledge, the chemical model we are developing is unique in that compensatory collateral sprouting is induced by localized removal of a single chemical pathway.

4. Hormonal Substance - Corticosterone. Corticosterone secretion into the plasma from the adrenal glands is under the control of ACTH, a peptide stored in the anterior pituitary and released during stress (Ganong and Hume 1954). Circulating corticosterone is the principal steroid in the rat and is selectively bound by neurons in the hippocampus (McEwen et al. 1970). Although this steroid has many stimulatory actions on hippocampal and serotonergic neurons and on glia cells, recent studies have shown that cortisol, a related glucocorticoid, inhibits reactive synaptogenesis in the adult brain (Scheff et al. 1980; Scheff and Cotman 1982). In contrast to this latter finding, there are reports that circulating adrenal steroids are required for both normal development (Sze 1980) and adult sprouting (Zhou and Azmitia 1982) of serotonergic axons. These differential effects of adrenal steroids raise the important possibility that reactive synaptogenesis and compensatory collateral sprouting may be regulated by separate growth factors, and the major benefit of our Hippocampal-5-HT-Steroid Model is that it permits a multidisciplinary analysis of a neuronal (chemically identified) circuit in a localized area of the brain known to preferentially concentrate a specific hormone. An experimental method has been developed to induce homotypic collateral sprouting of adult serotonergic neurons in normal rats.

Material and Methods

Axonal sprouting is induced by removing a single 5-HT-pathway using a specific neurotoxic injection procedure (Azmitia et al. 1978; Zhou and Azmitia 1984). Female rats (200-220 g) are injected intraperitoneally with desipramine (10 mg/kg) 45 min before receiving a microinjection of 3 μg/0.3 μl of free-base 5,7-dihydroxytryptamine into the cingulum bundle [co-ordinates (15°)-AP at bregma, 1.1 mm lateral, and 2.4 mm vertical]. The volume of injection is precisely controlled by use of a specially constructed glass micropipette device with a tip diameter between 50 and 80 μm. The 5,7-DHT solution is kept in the dark and on ice until the micropipette is filled with exactly 0.3 μl immediately before use. The solution is injected at a rate of 30 nl/min under visual inspection and the pipette is left in place for 5 min after all the solution has been injected. The animals are maintained on a 12 h light/dark cycle (8:00 and 20:00 h) with food and water provided ad libitum. The room is temperature and humidity controlled and all cages are routinely sterilized.

Immunocytochemical procedures with a specific 5-HT antibody are used to describe the fibers within the hippocampus. Rats are pretreated with L-tryptophan (50 mg/kg) and pargyline (50 mg/kg) (Azmitia and Gannon 1983; Frankfurt and Azmitia 1984). The rats are perfused through the ascending aorta with 4% paraformaldehyde in 0.1 M phosphate buffer (pH 7.4) containing 0.1% $MgSO_4$. The brains are coronally sectioned (50 µm) through the hippocampus with a vibratome (Oxford).

In the adrenalectomy studies 1 h after the injection of 5,7-DHT the animals are either bilaterally adrenalectomized (ADX) or sham operated. The use of chloropent (i.p. injection 0.25 ml/100 g body weight) allows sufficient time to perform both procedures. The replacement of physiological levels of corticosterone in the ADX rats are achieved by the implantation procedure (Meyer et al. 1979). Briefly, pure corticosterone (100 mg) is melted and poured into a 6 mm diameter hole drilled in a paraffin block. After the pellet has cooled, it is removed and implanted in the nape of the neck in a lightly anesthetized rat.

Results

A topographic description of the hippocampal 5-HT fibers using immunocytochemistry with an antibody against 5-HT is shown in Fig. 2.

The raphe neurons in the rat efferent to the hippocampus were located and quantified using HRP transport (DAB and TMB as substrate) (Zhou and Azmitia 1983), ^{3}H-5HT light and EM radioautography (intrahippocampal injections) (Azmitia and Gannon 1983), and a combination of HRP and ^{3}H-5HT as retrograde markers (Azmitia 1981b). Neurotoxin (5,7-DHT) lesions of the CB and FF were performed before HRP injections to identify the cells of origin of these pathways in the raphe nuclei (Zhou and Azmitia 1983). This specific lesioning procedure will reliably destroy the CB fibers within 24 h and within 1 week cause a decrease of 63% in hippocampal tryptophan hydroxylase activity, of 55% in ^{3}H-5HT high-affinity reuptake, and of 67% in the number of HRP labeled neurons in the midbrain raphe (Fig. 3).

Fig. 2 A - C. Dark-field photographs of the normal rat hippocampus ► stained with a specific 5-HT antibody and processed by the HRP-DAB method. The series of low-power views (8 x negative magnification) is from the same coronal section of the dorsal hippocampus. A The fibers are dense in the stratum lacunosum moleculare of CA_1. Three entry routes into the dentate gyrus are shown in this series. (1) Fibers enter from the fasciola cinerium to enter the apex of the dentate gyrus. (2) Fibers enter from LM to curve around the dorsal (fissure) arm of the dentate gyrus. These fibers past through the granular layer and are heaviest in the polymorphic region immediately beneath the granule cells. B A slightly more lateral view shows the third route. (3) Fibers enter from the fornix (shown in C). These fibers innervate mainly the ventral (ventricular) arm of the dentate gyrus. C A view of the immunoreactive fibers in fornix. LV, lateral ventricle; CC, corpus callosum

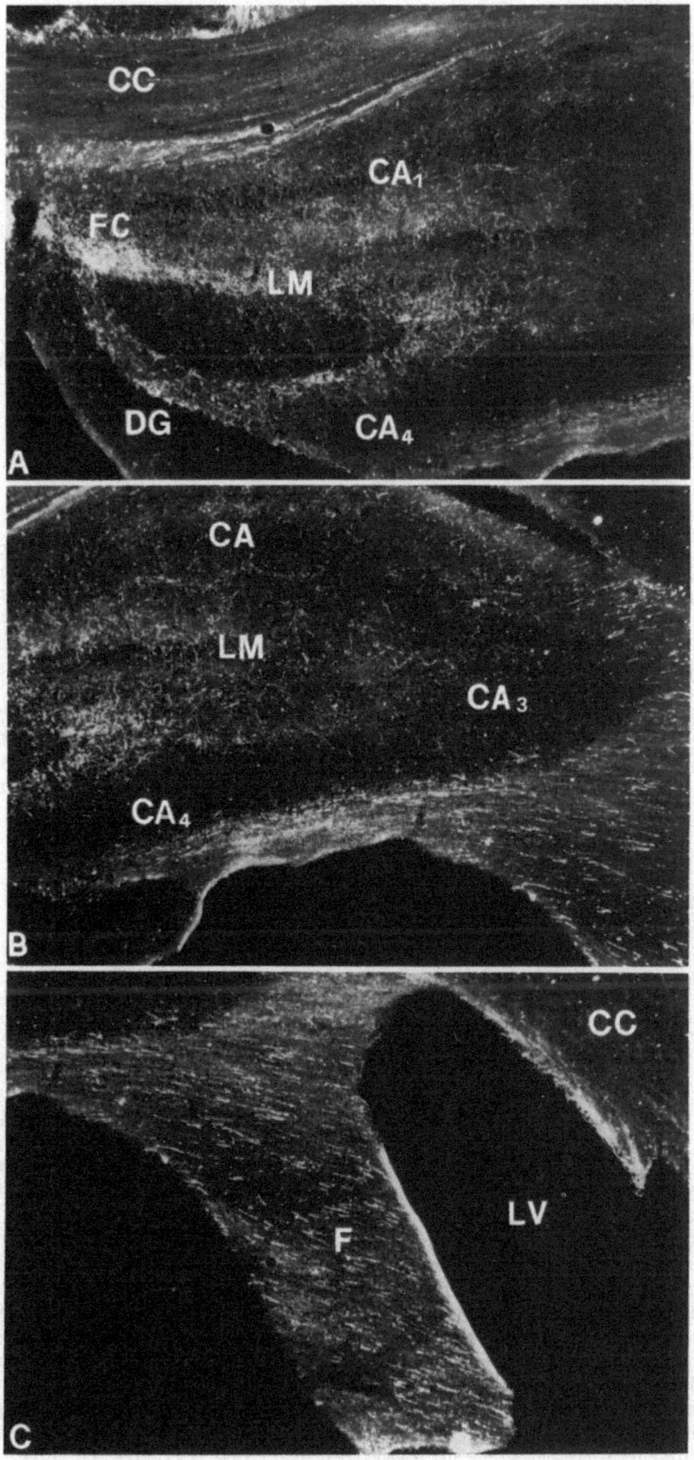
CC
CA1
FC
LM
DG
CA4
A
CA
LM
CA3
CA4
B
CC
F
LV
C

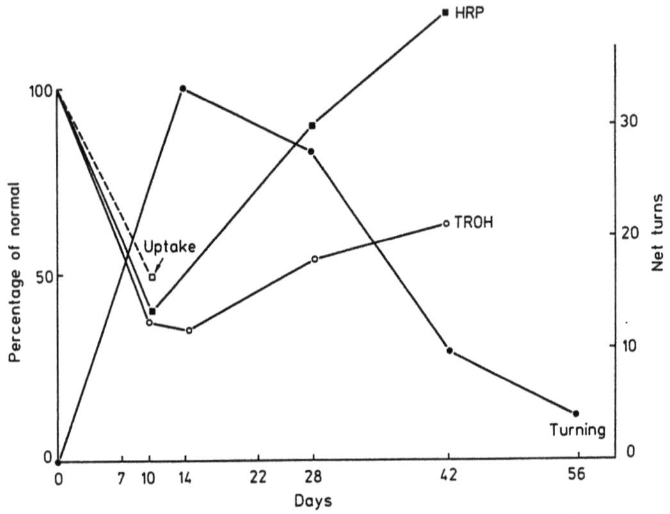

Fig. 3. Diagrammatic representation of four separate experiments showing changes in behavior [turning(●)], the activity of tryptophan hydroxylase [TROH(o)], the number of horseradish peroxidase containing cells in the MR after intra-hippocampal injections [HRP(■)], and the specific high-affinity uptake of ^3H-5-HT into hippocampal slices [uptake(□)]. The abscissa shows the number of days after the injection of 5,7-DHT (3-4 µg/0.3-0.4 µl Ringer's, solution) into the CB of desipramine (10 mg/kg)-pretreated female rats. The left ordinate shows the percentage change from normal for the HRP, TROH, and uptake experiments. The right ordinate shows the number of net turns from that seen in sham-injected rats (data from Azmitia et al. 1978; Clewans and Azmitia 1984; Zhou and Azmitia 1984)

Fig. 4. Dark-field photographs of dorsal hippocampus cut in cross► section 7 days after a unilateral injection into the cingulum bundle with 4 µg 5,7-DHT. A A low magnification (8 x negative magnification) picture of the lesioned (right side) dentate gyrus (DG) which contains fewer fibers than the normal DG (left-side). A marked reduction is also noted in the fasciola cinerea (fc) on the lesioned side. B A more anterior section than that shown in A. The fibers in the cingulum bundle (CB) are greatly reduced on the lesioned side (arrow with asterisk) compared to the control side (arrow). Again, a marked reduction of fibers is seen in the dentate gyrus (DG) on the lesioned side. Note that the dorsal fornix is equally stained on the two sides since the fibers in this tract enter from the fornix which was not damaged by the injection. C, D A higher magnification (15x negative magnification) of the CA_1 layer of the dorsal hippocampus on the control (C) and lesioned (D) side of the brain. Note the marked reduction of the 5-HT immunoreactive fibers in the pyramidal cell layer of CA_1 on the lesioned side

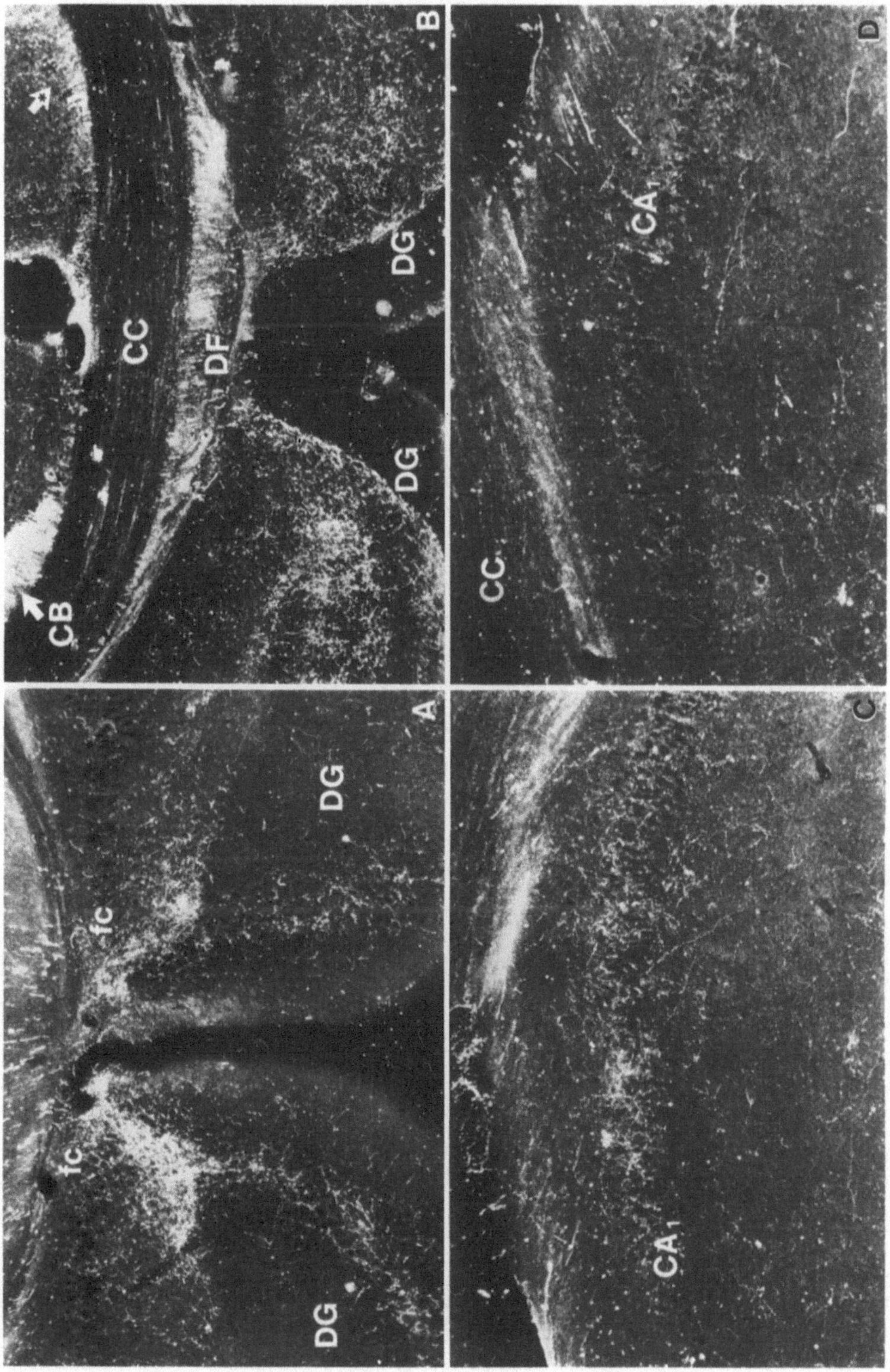
CB
CC
DF
DG
DG
B
fc
fc
DG
DG
A
CC
CA1
D
CA1
C

The effects of 5,7-DHT injections into the CB on the 5-HT immunoreactive fibers in the dorsal hippocampus are shown in Fig. 4. It can be seen that the fibers are mainly lost from the CB and in the ipsilateral hippocampus.

Evidence for collateral sprouting of the midbrain efferents was tested by a series of experiments where HRP was injcted into the dorsal hippocampus at various time points after CB 5,7-DHT lesions. The results show that within 28 days the number of labeled cells in the midbrain returns to normal levels. A subsequent injection into the FF resulted in the loss of these cells within 3 days. These results suggest that the neurons projecting from the midbrain to the dorsal hippocampus via the FF can reorganize following a specific 5,7-DHT injection into the CB (Zhou and Azmitia 1984).

Furthermore, an increase in the number of HRP-labeled cells in the raphe nuclei 21 days after unilateral 5,7-DHT microinjections into the CB was completely blocked by bilateral adrenalectomy (Zhou and Azmitia in press; Fig. 5). Both normal, bilateral adrenalectomy, and long-term, 5,7-DHT-lesioned normal rats showed approximately 300 labeled neurons in the midbrain raphe 24 h after HRP injection into the dorsal hippocampus. However, the short-term (5 days) 5,7-DHT-lesioned normals and long-term (21 days) 5,7-DHT-lesioned adrenalecto-

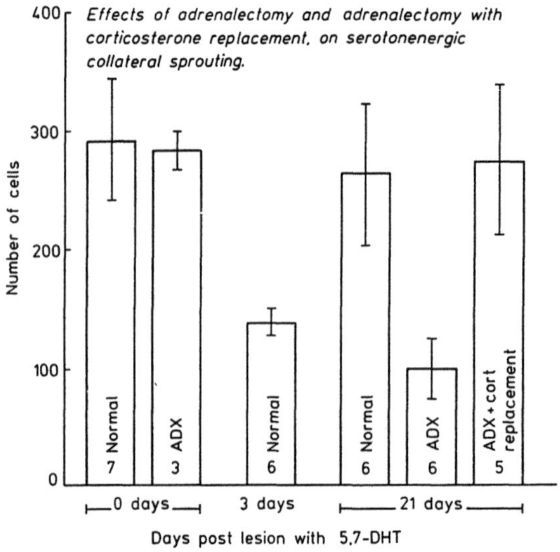

Fig. 5. Number of horseradish peroxidase (HRP) retrogradely labeled neurons (mean + SD) in the median raphe nucleus (ordinate) after 0,3, and 21 days post 5,7-DHT lesion of the CB serotonergic fiber pathway (abscissa). There was no difference between adrenalectomized (ADX) and normal (non-ADX) nonlesioned rat cell counts. However, the significant increase in the number of labeled cells seen in non-ADX + lesioned rats between 3 and 21 days ($P<0.01$) did not occur in ADX + lesioned rats (ADX vs non-ADX lesioned rats at 21 days, $P<0.001$). If a corticosterone pellet (CORT) was implanted into ADX + lesioned rats at the time of lesion, then a significant increase in labeled cells did occur (ADX vs ADX + CORT at 21 days, $P<0.01$)

mized rats showed a decrease from normal levels of between 56% and 63%. Studies are in progress to determine the effects of corticosterone replacement in these long-term 5,7-DHT adrenalectomized rats.

Conclusion

Two types of collateral sprouting have been described in the hippocampus after damage to the afferent fibers.
(A) Shows a hippocampal neuron receiving a 5-HT input from the CB and two other chemically distinct afferents (e.g., noradrenergic, GABA-ergic, cholinergic, glutaminergic, etc.). These neurons have also been shown to concentrate circulating adrenal steroids.
(B) The predicted sprouting occurring after lesion of the CB 5-HT fibers due to reactive synaptogenesis. Those fibers closest to the denervated sites sprout to occupy the vacated terminal sites. No chemical specificity is involved. The process begins within 2 days and is finished within 2 weeks. The chemical imbalance which results from this reorganization usually leads to dysfunction of the neuron.
(C) The predicted sprouting occurring after lesion of the CB 5-HT fibers due to compensatory collateral sprouting. The fibers homotypic (using the same neurotransmitter) to the lesioned fibers travel long distances in their growth response (possibly several millimeters) to reinnervate the denervated sites. The 5-HT fibers from the FF are primarily involved in this sprouting. The process

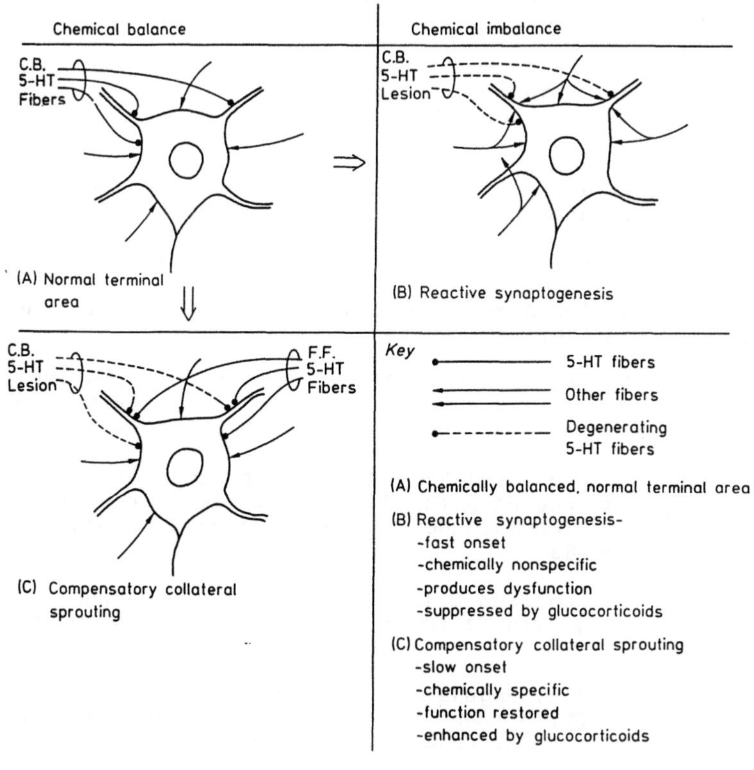

begins around 10-14 days and can continue for several months (>90 days). The chemical balance is restored, which in turn restores neuronal function.

References

Azmitia EC (1978) The serotonin-producing neurons of the midbrain median and dorsal raphe nuclei. In: Iversen LL, Iversen SD, Snyder SH (eds) Handbook of psychopharmacology, vol 9. Plenum, New York, chapt 5, pp 233-314

Azmitia EC (1981) Bilateral serotonergic projections to the dorsal hippocampus of the rat: simultaneous localization of retrogradely transported ^{3}H-5HT and HRP. J Comp Neurol 203:737-743

Azmitia EC (1982) Recent advances in serotonin methodology. J Histochem Cytochem 30:739-743

Azmitia EC, Gannon PJ (1983) The ultrastructural localization of serotonin immunoreactivity in myelinated and unmyelinated axons within the medial forebrain bundle of rat and monkey. J Neurosci 3: 2083-2090

Azmitia EC, Gannon PJ (1985) Serotonergic system in the primate and subprimate brain. Adv Neurol 33 (to be published)

Azmitia EC, Segal M (1978) An autoradiographic analysis of the differential ascending projections of the dorsal and median raphe nuclei in the rat. J Comp Neurol 179:641-659

Azmitia EC, Perlow MJ, Brennan MJ, Lauder (1981) Fetal raphe and hippocampal transplants into adult and aged C57BL/6N mice: a preliminary immunocytochemical study. Brain Res Bull 7:703-710

Braun SJ, Sweadner KJ, Patterson PH (1981) Neuronal cell surfaces distinctive glycoproteins of cultured adrenergic and cholinergic sympathetic neurons. J Neurosci 1:1397-1406

Calas A (1981) The serotonergic neuron. J Physiol (Paris) 77:147-524

Ciba Foundation Symposium (1978) Functions of the septo-hippocampal system. Elsevier, Amsterdam

Clewans C, Azmitia EC (1984) Compensatory changes in tryptophan hydroxylase after induced homotypic collateral sprouting of hippocampal serotonergic fibers. Brain Res 307:125-133

Cotman CW, Lynch GS (1976) Reactive synaptogenesis in the adult nervous system: the effects of partial deafferentation on new synapse formation. In: Barondes SH (ed) Neuronal recognition. Plenum, New York, pp 69-108

Crutcher KA, Collins F (1982) In vitro evidence for two distinct hippocampal growth factors: basis of neuronal plasticity? Science 217:76-78

Diamond J (1979) Regulation of nerve sprouting by extrinsic factors. In: Schmidt FO, Worden FG (eds) The neurosciences: fourth study program. MIT Press, Cambridge, p. 937

Edds MV Jr (1953) Collateral nerve regeneration. Quart Rev Biol 28: 260-276

Frankfurt M, Azmitia EC (1984) Regeneration of hypothalamic serotonergic fibers after unilateral intracerebral microinjections of 5,7-dihydroxytryptamine. Brain Res 298:273-282

Fuxe K (1965) Evidence for the existence of monoamine neurons in the central nervous system. IV. Distribution of monoamine nerve terminals in the central nervous system. Acta Physiol Scand 64, Suppl 247:39-85

Gage FH, Björklund A, Stenevi U (1983) Reinnervation of the partially deafferented hippocampus by compensatory collateral sprouting from spared cholinergic and noradrenergic afferents. Brain Res 268:27-37

Gall C, Rose G, Lynch G (1979) Proliferative and migratory activity of glial cells in the partially deafferented hippocampus. J Comp Neurol 183:539-550

Ganong WF, Hume DM (1954) Absence of stress-induced and compensatory adrenal hypertrophy in dog with hypothalamic lesions. Endocrinology 55:474-483

Goodman DC, Bogdasarin RS, Horel JA (1973) Axonal sprouting of ipsilateral optic tract following opposite eye removal. Brain Behav Evol 8:27-50

Gray JA (1981) The neuropsychology of anxiety: an enquiry into the function of the septo-hippocampal system. Oxford University Press, New York

Haber B, Gabay S, Issidorides MR, Alivisatos SGA (1981) (eds) Serotonin: current aspects of neurochemistry and function, Plenum, New York

Isaacon RL, Pribram CH (1975) (eds) The hippocampus, vols I and II. Plenum, New York

Jacobs B, Gelperin A (1981) (eds) Serotonin neurotransmission and behavior. MIT Press, Cambridge

Lidov HGW, Grzanna R, Molliver ME (1980) The serotonin innervation of cerebral cortex in the rat: an immunohistochemical analysis. Neuroscience 5:207-223

Lorenti de No R (1934) Studies on the structure of the cerebral cortex. II. Continuation of the study of the Ammonic system. J Psychol Neurol 46:113-177

Loy R, Milner TA, Moore RY (1980) Sprouting of sympathetic axons in the hippocampal formation: conditions necessary to elicit ingrowth. Exp Neurol 67:399-411

Lund RD, Cunningham TJ, Lund JS (1973) Modified optic projection after unilateral eye removal in young rats. Brain Behav Evol 8:51-72

McEwen BS, Weiss JM, Schwartz LS (1970) Retention of corticosterone by cell nuclei from brain regions of adrenalectomized rat. Brain Res 17:471-482

McNaughton N, Azmitia EC, Williams JH, Buchan A, Gray JA (1980) Septal elicitation of hippocampal theta rhythm after localized de-afferentation of serotonergic fibers. Brain Res 200:259-269

Meibach RC, Siegel A (1977) Efferent connections of the hippocampal formation in the rat. Brain Res 124:197-224

Meyer JS, Micco DJ, Stevenson BS, Krey LC, McEwan BS (1979) Subcutaneous implantation method for chronic glucocorticoid replacement therapy. Physiol Behav 22:867-870

Murray M, Goldberger ME (1974) Restitution of function and collateral sprouting in the cat spinal cord: the partially hemisected animal. J Comp Neurol 158:19-36

Raisman G, Field PM (1973) A quantitative investigation of the development of collateral reinnervation after partial deafferentation of the septal nuclei. Brain Res 50:241-264

Ramon y Cajal S (1911) Histologie du systeme nerveux de l'homme et des vertebres, vol 2. Maloine, Paris

Scheff SW, Cotman CW (1982) Chronic glucocorti therapy alters axon sprouting in the hippocampal dentate gyrus. Exp Neurol 76:644-654

Scheff SW, Bernardo LS, Cotman CW (1980) Hydrocortisone administration retards axon sprouting in rat dentate gyrus. Exp Neurol 68:195-201

Segal M (1980) The action of serotonin in the rat hippocampal slice. J Physiol (Lond) 303:423-439

Srebro B, Azmitia EC, Winson J (1982) Effect of 5-HT depletion of the hippocampus on the neuronal transmission from perforant path through dentate gyrus. Brain Res 235:142-147

Steward O, Cotman CW, Lynch G (1974) Growth of a new fiber projection in the brain of adult rats: reinnervation of the dentate gyrus by the contralateral entorhinal cortex following ipsilateral entorhinal lesion. Exp Brain Res 20:45-66

Steward O, Cotman CW, Lynch G (1976) A quantitative autoradiographic and electrophysiological study of the reinnervation of the dentate gyrus by the contralateral entorhinal cortex following ipsilateral entorhinal lesion. Brain Res 114:181-200

Swanson LW, Cowan WM (1977) An autoradiographic study of the organization of the efferent connections of the hippocampal formation in the rat. J Comp Neurol 172:49-84

Swanson LW, Sawchenko PE, Cowan WM (1981) Evidence for collateral projections by neurons in a horn, the dentate gyrus, and the ubiculum: a multiple retrograde labeling study in rat. J Neurosci 1:548-559

Sze PY (1980) Glucocorticoids as a regulatory factor for brain tryptophan hydroxylase during development. Dev Neurosci 3:217-223

Williams JH, Azmitia EC (1981) Hippocampal serotonin re-uptake and nocturnal locomotor activity after microinjections of 5,7-DHT in the fornix-fimbria. Brain Res 207:95-107

Zhou FC, Azmitia EC (1983) Effects of 5,7-dihydroxytryptamine on HRP retrograde transport from hippocampus to midbrain raphe nuclei in the rat. Brain Res Bull 10:445-451

Zhou FC, Azmitia EC (1984) Induced homotypic collateral sprouting of serotonergic fibers in hippocampus. Brain Res 308:53-62

Zhou FC, Azmitia EC (1985) Effect of adrenalectomy and corticosterone on the sprouting of serotonergic fibers in hippocampus. Neurosci Lett (to be published)

Spinal Plasticity After Nerve Injury: Mediolateral Localization of Rewired Cells

M. Devor[1] and P. D. Wall[2]

1 Life Sciences Institute, Hebrew University of Jerusalem, Jerusalem 91904, Israel
2 Cerebral Functions Group, Department of Anatomy and Embryology, University College London, London WC1E6BT, Great Britain

Introduction

There is now abundant evidence that lesions and environmental manipulations during the pre- and early postnatal period can affect neural connectivity in orderly and often predictable ways. What is more, in appropriate systems these changes can be shown to have specific and appropriate functional (behavioural) consequences. Some outstanding examples include the rearrangement of retinotectal connectivity and consequent changes in visual orientation behaviour (e.g. Schneider and Jhaveri 1974), experimental amblyopia after various unilateral manipulations of early visual experience (e.g. Hubel et al. 1977), and changes in odour-guided behaviour appropriate to altered connectivity of the lateral olfactory tract (LOT) following neonatal LOT injury (Devor 1975).

In general, both structural and functional lability in these preparations decline with age until in juveniles and adults they are no longer evident at all. However, in recent years several brain structures have been identified in which morphological and/or functional plasticity persists into adulthood. Among the more closely studied of these are the limbic forebrain, the red nucleus and the hippocampus (e.g. Raisman 1977; Tsukahara 1981; Cotman and Lynch 1976). Perhaps the best example in this category, however, are the various central way stations in the somatosensory pathway, including the spinal cord dorsal horn, the dorsal column nuclei, the trigeminal complex, the ventrobasal complex and the somatosensory cortex. In each of these structures, deafferentiation of the somatotopic representation of a part of the body leads to a functional invasion of the deafferented cells by sensory input from surrounding innervated skin.

Somatotopic Remapping

A typical case is respecification of the somatotopic representation of the forelimb in primary somatosensory cortex of the rhesus macaque. Immediately after amputation of the 3rd digit, cortical cells representing this digit (i.e. whose receptive field (RF) is located on the 3rd digit) can no longer be activated by stimulation of the skin. Over the following weeks and months, however, there is a gradual shift in the somatotopic map of skin represented by surround zones until most of the former 3rd digit cells acquire a new RF on the 3rd digit stump, or on the tips of the 2nd or 4th digits (Merzenich et al. 1984). A similar process has been documented in

Experimental Brain Research, Suppl. 13

the somatosensory cortex in rat, raccoon and cat, in the somatosensory thalamus in cat and rat, in the trigeminal nucleus in cat and rat, in the dorsal column nuclei in cat and in the dorsal horn in cat and rat (for references see Kaas et al. 1983; Devor 1983).

It is not known for certain, of course, whether a similar type of somatotopic rearrangement also occurs after partial deafferentation in humans, but a number of bizarre and previously unexplained sensory anomalies in amputees suggest that some such processes may be at work. For example, Cronholm (1951) described a large series of patients in whom cutaneous stimulation on the amputation stump or on adjacent torso, in addition to being felt locally, also evoked a sensation referred into the phantom limb. Such referred sensations can sometimes be quite painful (Howe 1983). A second example is the increase in sensory resolution on amputation stumps (Teuber et al. 1949; Haber 1955). If a larger mass of neural tissue is now devoted to proximal skin of the limb, it is expected that finer spatial discriminations will be possible.

Unlike many of the other examples of neural plasticity in adult mammalian CNS, the studies of cutaneous somatotopy are based on electrophysiological (RF) data rather than on neuroanatomical observations. Thus, although the functional effectiveness of newly formed connections is assured, the mechanism underlying the rewiring is not known. Two main classes of explanation have been proposed, "long distance" sprouting with the formation of new synaptic contacts, and strengthening of a previously existing but "relatively ineffective" synaptic channel (Wall 1977). A number of processes are capable of producing synaptic strengthening, including purely electrophysiological ones such as disinhibition (e.g. Wall and Devor 1981), and proliferative ones such as local sprouting and augmentation of pre-existing populations of terminals (e.g. Raisman 1977).

Spinal Somatotopy After Nerve Injury

Unfortunately, experimemts capable of deciding between these two possibilities are not easy to design. With this problem in mind, we (Devor and Wall 1978, 1981a,b) developed a new and particularly robust preparation which offers a number of technical advantages for the analysis of mechanisms underlying somatotopic reorganization. In this preparation we took advantage of the well defined somatotopic gradient represented along the mediolateral axis of the dorsal horn. Virtually all cells recorded in the medial half to two-thirds of the dorsal grey matter have RFs restricted to the foot. Most, in fact, are limited to the toes or even parts of individual toes. Moving laterally, RFs tend to increase in size and they shift proximally up the foot and calf and onto the thigh, perineum and lower back. Immediately upon transection of the peripheral nerves that innervate the toes and foot, the sciatic and saphenous nerves, virtually all cells in the medial part of the dorsal horn lose their toe-foot RF, although they can still be driven by electrical stimulation of the cut nerve end. However, when we examine the medial dorsal horn some time after the sciatic and saphenous nerves have been cut, many of the cells in this region turn out to have acquired a new RF in the field of distribution of the nearest neighbouring intact nerves, on the calf, thigh, perineum and lower back. This functional rewiring is first observed in cats about 2 weeks after nerve section, and it stabilizes after about 4 weeks. In rats, the time course is faster, with the process reaching completion 4 days postoperatively.

It is obvious that if the cut nerves managed to regenerate and reinnervate skin of the proximal hindlimb, then medially located dorsal horn cells would acquire a proximal RF. The principal argument against this trivial, peripheral explanation is that in each experiment the chronically cut sciatic and saphenous nerves were re-cut acutely on the day of the experiment. This would disconnect any putative regenerating sprouts in the periphery. Correspondingly, transection or block of nerves that normally innervate thigh skin does eliminate the novel RFs. A number of subsidiary control experiments that will not be reviewed here (see Devor 1983) exclude a range of more interesting peripheral possibilities. The phenomenon under consideration, therefore, involves a change in functional connectivity within the spinal cord proper. The nature of this change is currently under investigation. It goes without saying that an intensive investigation into mechanism calls for a reliable and hopefully robust plastic phenomenon. Somatotopic remapping in spinal dorsal horn following nerve section appears to meet these requirements on several grounds. First, it is one of a whole range of similar somatotopic remapping paradigms (see above) that have been observed in a number of somatosensory nuclei, in a number of species and by a number of independent investigators over a period of nearly 15 years. Second, we have observed somatotopically inappropriate RFs in medial dorsal horn in each and every one of the chronic neurectomized cats and rats that we have investigated. Third, the phenomenon has recently been replicated in cats (Lisney 1983), and with an interesting variation in rats (Markus et al. 1984) in other laboratories. We are concerned, however, by two recent studies in which somatotopic remapping was not found. In one, which has so far been published only in abstract form, Brown et al. (1983) failed to find somatotopically inappropriate RFs after peripheral neurectomy in cats. They appear, however, to have used a protocol quite different from our own, and particularly to have searched exclusively for spinocervical tract neurons. It is entirely possible that these cells, unlike many others, do not undergo functional rewiring. Publication of the full report may clarify the essential differences. The second report, by Pubols (1984), is of more concern because the experimental procedures followed ours more closely, and the map search procedure ought to have revealed somatotopically inappropriate RFs in cells of all classes. In her report Pubols suggested a number of possible explanations for this discrepancy, including animal strain differences, subtle differences in anaesthesia and other details of experimental procedure etc. We see no value in adding our speculations to this list. However, we feel obliged to deal with one seemingly more likely suggestion made by Pubols which, if correct, challenges the fundamental observation of somatotopic remapping.

Mediolateral Localization

The essential features of our experimental paradigm are that a sharp mediolateral somatotopy exists in the lower lumbar dorsal horn, and that the electrode penetrations in which we find novel proximal RFs indeed pass through the region of dorsal horn in which such RFs are somatotopically inappropriate, that is, through the medial half to two-thirds of the dorsal grey matter. In the present instance the underlying somatotopy of dorsal horn RFs is not in question. Pubols (1984) sees the same mediolateral map that we and all other investigators see. However, she raises the question as to whether we can be confident that our medial electrode penetrations were indeed medial and did not inadvertently pass through the lateral dorsal horn, a region in which proximal RFs are not somatotopically inappropriate.

Granted that the proportion of the dorsal width occupied by the RFs on toes and foot is large (half to two-thirds), the overall width of the dorsal horn is small (roughly 1.2 mm wide in adult cats and 0.7 mm wide in rats). Could we have made a serious and consistent error in mediolateral localization?

In our experiments (Devor and Wall 1978, 1981a,b) mediolateral localization was always determined in two ways. The first was by referring to superficial spinal landmarks, particularly the line of entry of the dorsal roots and the spinal midline, which is midway between the root entry zones on the two sides and is demarcated by a midline blood vessel or by the dorsal median spinal sulcus in many cases. The second, and more important, method of orientation was the use of landmarks linked to the actual underlying RF somatotopy. In a series of closely spaced vertical electrode penetrations, the location of the border between the dorsal column (white matter) and the dorsal horn (grey matter) was established bilaterally, based on signals coming off the electrode.

The glass electrode tips marking this border on the two sides were then cut off and left in place for later histological localization. In all of the cats and most of the rats, a second set of marker electrodes was then placed at the border between the toe-foot and the leg map representations. On the control side this was done by examining RFs in closely spaced vertical electrode penetrations. On the experimental (chronic neurectomized) side this was done by placing a marker electrode at the same distance lateral to the first marker as on the control side. Subsequent search penetrations were made between the two sets of marker electrodes, their positions with respect to the markers being read from the microdrive micrometer. Only occasionally were additional markers left. Pubol's (1984) suggestion is that since the intermediate search penetrations were not themselves marked, it remains possible that they were routinely deflected laterally into the region of spinal cord expected to have cells with proximal RFs.

The prospect of such radical deflections from the microdrive axis (at least 45°) and always in the same direction (laterally) seems, on the face of it, unlikely. This is all the more so as had similar deflections occurred on the control side, proximal RFs would have been found there too in electrode penetrations aimed at the medial dorsal horn. These and other arguments aside, Pubol's criticism can be answered directly by examining those few medial electrode penetrations in which the electrode tip was cut off and left in situ. These data, which were not given in the original publications, are shown in Figs. 1 and 2. The figures show projection drawings of cleared thick histological sections of the tissue in which the actual electrophysiological recordings were made. Each drawing includes the outline of the one or more glass electrode tips left in medial grey matter after the end of the recording penetration. For each such tip the position of cells encountered during the penetration that had proximal RFs is indicated. Depth positions were calculated from microdrive micrometer readings made at the time of recording (after correction for 15% tissue shrinkage in histology), and probably contain some measure of error. The 15% correction factor was determined by comparing micrometer readings with the location of dyespots injected through recording microelectrodes (Devor and Wall, 1981a). Mediolateral positions, on the other hand, are necessarily accurate. In Figs. 1 and 2 we show sections from the chronic cats (at least 28 days) and rats reported by Devor and Wall (1981a,b) in which histological data are still available and in which a medially placed electrode tip was present through which electrophysiological

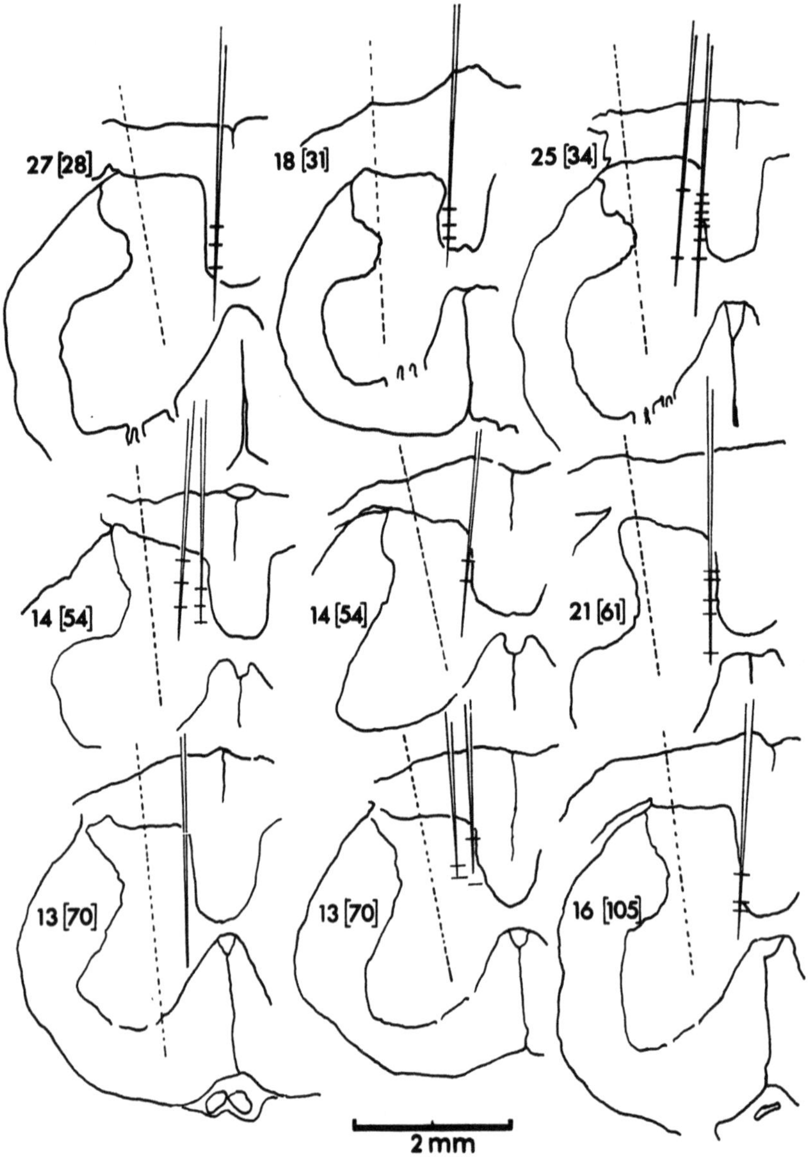

Fig. 1. Marked electrode penetrations in the L6-7 dorsal horn in chronic neurectomized cats. Dashed lines indicate boundary between toe-foot and leg representations transposed from the marker electrode on the contralateral side for each cat. Numbers refer to the animal code and survival time in days (in brackets)

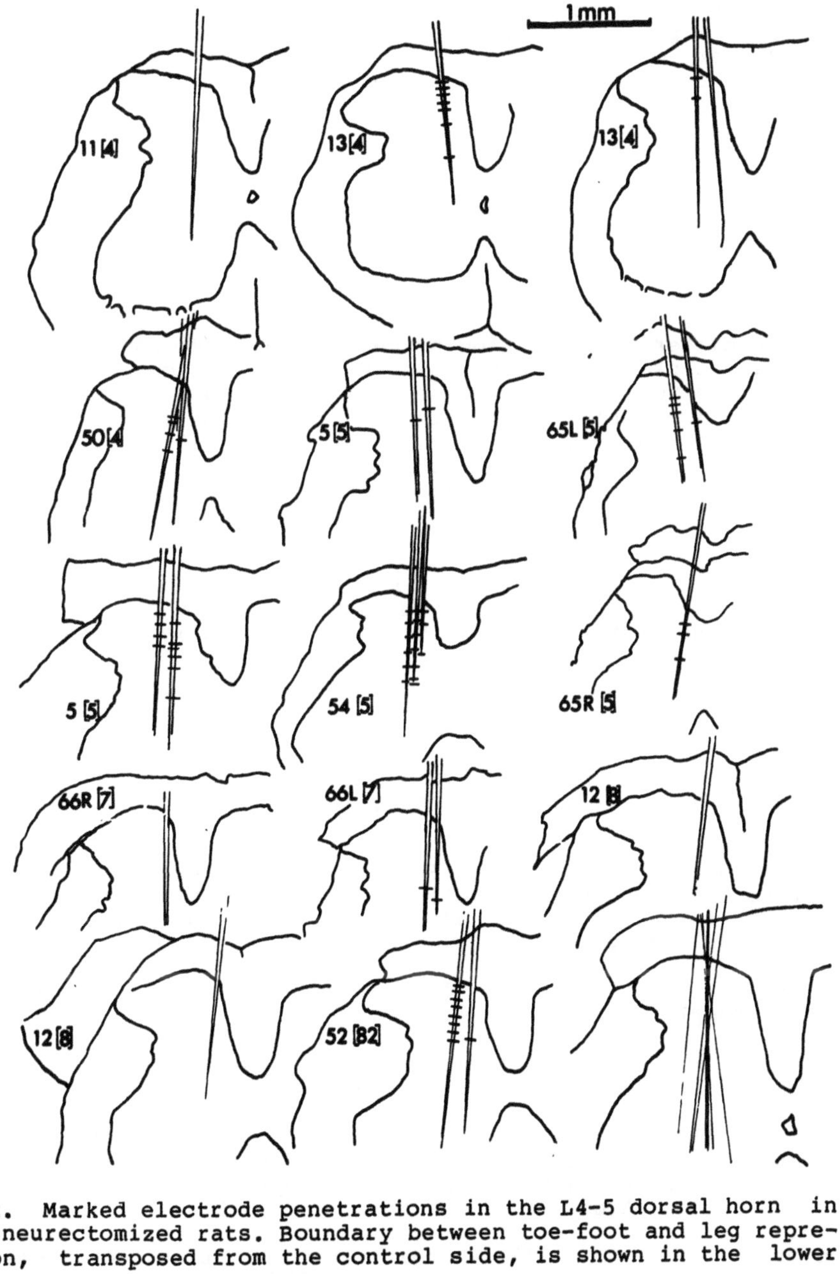

Fig. 2. Marked electrode penetrations in the L4-5 dorsal horn in chronic neurectomized rats. Boundary between toe-foot and leg representation, transposed from the control side, is shown in the lower right

observations were made. No selection was made. For clarity, marker electrode tips in the lateral part of the dorsal horn are not shown. The number of somatotopically inappropriate RFs encountered per electrode penetration in the marked electrode tracks in Figs. 1 and 2 was 2.9 ± 1.7 for cats and 2.7 ± 2.5 for rats. These values do not differ significantly ($p > 0.2$) from the overall figures based on all medial electrode penetrations, marked and unmarked (Devor and Wall 1981a,b). Similarly, RF properties were like those reported.

Conclusions

We conclude that the original reports of somatotopic remapping in the mediolateral axis following peripheral nerve injury could not be an artefact of erroneous mediolateral localization of electrode penetration as Pubols (1984) suggested. We do not know why she, unlike others (Lisney 1983; Markus et al. 1984), failed to replicate this finding. We suspect, however, that a detailed comparison of her experimental protocol with ours might prove instructive in understanding the specific conditions under which somatotopic remapping is initiated.

Acknowledgment. Support of the Israel Academy of Sciences and the International Spinal Research Trust is gratefully acknowledged.

References

Brown AG, Fyffe REW, Rowe MJ (1983) Effects of peripheral nerve section on dorsal horn somatotopy in the cat. Proc J Physiol (Lond) 339:36-37

Cotman CW, Lynch G (1976) Reactive synaptogenesis in the adult nervous system: the effects of partial deafferentiation on new synapse formation. In: Barondes SH (ed) Neuronal recognition, Plenum, New York, pp 69-108

Cronholm B (1951) Phantom limbs in amputees. Acta Psychiatr Neurol Scand [Suppl] 72:1-310

Devor M (1975) Neuroplasticity in the sparing or deterioration of function after early olfactory tract lesions. Science 190:998-1000

Devor M (1983) Plasticity of spinal cord somatotopy in adult mammals: involvement of relatively ineffective synapses. In: Haber B, Perez-Polo JR, Hashim GA, Giuffrida-Stella A-M (eds) Nervous system regeneration. Liss, New York, pp 287-314

Devor M, Wall PD (1978) Reorganization of spinal cord sensory map after peripheral nerve injury. Nature 275:75-76

Devor M, Wall PD (1981a) The effect of peripheral nerve injury on receptive fields of cells in the cat spinal cord. J Comp Neurol 199: 277-291

Devor M, Wall PD (1981b) Plasticity in the spinal cord sensory map following peripheral nerve injury in rats. J Neurosci 1:679-684

Haber WB (1955) Effects of loss of limb on sensory functions. J Psychol 40:115-123

Howe JF (1983) Phantom limb pain: a re-afferentation syndrome. Pain 15:101-107

Hubel DH, Wiesel TN, Le Vay S (1977) Plasticity of ocular dominance columns in monkey striate cortex. Philos Trans R Soc Lond [Biol] 278:377-410

Kaas JH, Merzenich MM, Killackey HP (1983) The reorganization of somatosensory cortex following peripheral nerve damage in adult and developing mammals. Annu Rev Neurosci 6:325-356

Lisney SJW (1983) Changes in the somatotopic organization of the cat lumbar spinal cord following peripheral nerve transection and regeneration. Brain Res 259:31-39

Markus H, Pomerantz BH, Krushelnycky D (1984) Spread of saphenous somatotopic projection map in spinal cord and behavioural effects after chronic sciatic denervation in adult rat. Brain Res 296:27-39

Merzenich MM, Nelson RJ, Stryker MP, Cynader MS, Schoppmann A, Zook JM (1984) Somatosensory cortical map changes following digit amputation in adult monkeys. J Comp Neurol 224:591-605

Pubols LM (1984) The boundary of proximal hindlimb representation in the dorsal horn following peripheral nerve lesions in cats. Somatosens Res (to be published)

Raisman G (1977) Formation of synapses in the adult rat after injury: similarities and differences between a peripheral and a central nervous site. Philos Trans R Soc Lond,[Biol] 279:349-360

Schneider GE, Jhaveri SR (1974) Neuroanatomical correlates of spared or altered function after brain lesions in the newborn hamster. In: Stein DG, Rosen JJ, Butters N (eds) Plasticity and recovery of function in the central nervous system. Academic, New York, pp 65-109

Teuber HL, Krieger HP, Bendell MB (1949) Reorganization of sensory function in amputation stumps: two point discrimination. Fed Proc 8:156

Tsukahara N (1981) Synaptic plasticity in the mammalian central nervous system. Annu Rev Neurosci 4:352-375

Wall PD (1977) The presence of ineffective synapses and the circumstances which unmask them. Philos Trans R Soc [Biol] 278:361-372

Wall PD, Devor M (1981) The effect of peripheral nerve injury on dorsal root potentials and on transmission of afferent signals into the spinal cord. Brain Res 209:95-111

Physiological Changes in Cortical Cells Following Partial and Complete Visual Cortex Deafferentation in Cats

U. Yinon, A, Hammer, and M. Podell

Maurice and Gabriela Goldschleger Eye Institute, Research Department, Tel-Aviv University, Sackler School of Medicine, Chaim Sheba Medical Centre, Tel-Hashomer 52621, Israel

Introduction

The primary visual cortex receives its main visual input via the direct geniculo-cortical and the indirect callosal pathways. The effectivity of visual information transfer through the callosal pathway has been already proven in behavioral and electrophysiological studies. It has been shown in behavioral studies that the posterior portion of the corpus callosum of cats is involved in interhemispheric transfer (Myers 1955; Berlucchi et al. 1978). It has also been proved electrophysiologically that individual axons of the posterior portion of the corpus callosum possess visual properties similar to those of cells in the primary visual cortex (Hubel and Wiesel 1967; Berlucchi et al. 1967; Shatz 1977).

More direct physiological evidence for callosal transfer of visual information was provided by several investigators who discovered neuronal activity in the primary visual cortex despite elimination of the crossing visual pathways (Berlucchi and Rizzolatti 1968; Cynader et al. 1981; Lepore and Guillemot 1982). In contrast, other investigators using the same experimental model have not found clear evidence for such a callosal transfer (Yinon and Hammer 1981, 1984; Yinon et al. 1982; Milleret and Buser 1983).

It was assumed that the high level of callosal transfer found in several of the above-mentioned studies is related to a compensational mechanism which may develop with time in the absence of the conventional input to the cortical cells. The corpus callosum may thus serve as a candidate pathway for substituting the absent channel. It was therefore studied whether the callosal pathway can be used as an alternative input channel for cortical cells deprived of their direct visual input. The induction of enhanced visual activity in the corpus callosum may in turn facilitate recovery of the visually deafferented cortical cells.

Experimental Procedures

Since our hypothesis was concerned with changes in the mature brain, adult animals (cats) were studied.

Two experimental approaches were adopted in order to examine physiologically the effects of callosal activity on recovery of cortical cells. In the first approach the activity of cells in the primary visual cortex of the completely deafferented hemisphere was

Experimental Brain Research, Suppl. 13

studied. This was done by electrophysiologically examining single cortical cells following unilateral optic tract sectioning (OTX). Visual activity is expected to originate in the intact hemisphere and to be callosally transferred to the deafferented one. For comparison we carefuly searched for visually responsive and unresponsive cells along the electrode penetration in the afferented hemisphere of the OTX cats and in the cortex of normal control cats. In order to assess the possible extent of cortical recovery with increasing postoperative survival time, cats were studied in both the acute and chronic phases (Table 1).

In the second approach the activity of cells in the primary visual cortex of the partially deafferented hemisphere was studied. The partial deafferentation was induced by sagitally sectioning the optic chiasm. Normal control adult cats were studied for comparison.

The OTX and the optic chiasm split (OCX) operations were modified from Myers (1955) [see also Yinon and Hammer (1981) and Podell et al. (1984)]. In order to observe changes with recovery time, various groups of cats with different postoperative survival periods ranging from acute to chronic (0-7 months) were studied.

Table 1. Properties of visual cortical cells of unilateral optic tract sectioned (OTX) adult cats

Experimental Condition (& No. of cats)	Hemisphere Recorded (& No. of cells)	% Selective Cells		% Unresponsive Cells (& % S Cells)[b]
		Orientation[a]	Direction[a]	
Normal (6)	Right (62)	80.7	74.2	14.5 (0)
	Left (126)	82.5	58.7	12.7 (0)
Acute Operated (2)	Intact (65)	41.6	27.7	44.6 (0)
	Deafferented (62)	0	0	93.6 (4.8)
Chronic Operated 3 months (2)	Intact (49)	49.0	28.5	34.7 (0)
	Deafferented (45)	0	0	91.1 (2.2)
Chronic Operated 6 months (1)	Intact (22)	50.0	27.3	27.3 (0)
	Deafferented (5)	0	0	80.0 (20.0)

[a] Cells selective to both orientation and direction (% of total visually active cells). Orientation bias cells are not included. Other cells were nonselective or visually unidentified

[b] Cortical cells with unclear response to visual stimulation

The OTX and OCX cats which were not studied physiologically immediately following the operation were all observed behaviorally. These behavioral tests included examination of the size of the visual field, following movements of the eye and the head, general visuomotor coordination, and pupillary reactions.

A whole week was usually needed for recovery, as observed from the typical visual behavior found after this period. Furthermore, both pupils of the OCX cats were widely dilated for most of the postoperative periods, as previously observed (Caulfield et al. 1968). In the OTX cats only the pupils contralateral to the sectioned optic tract were found to be dilated, as previously described (Antonini et al. 1979).

Action potentials were recorded using tungsten microelectrodes from the boundary of area 17/18, where the callosal axons terminate (Innocenti 1980) and where the vertical meridian is retinotopically represented (Tusa et al. 1979). The location of the vertical meridian was directly determinated in accordance with the position of the optically projected area centralis and by calculating its distance from the area centralis (Vakkur et al. 1963).

Other details with regard to anesthesia of the cats, their surgical preparation for the electrophysiological sessions, unit recording, optical techniques, visual stimulation, receptive field mapping, and data analysis were as previously described (Yinon 1980; Yinon and Hammer 1981).

At the end of the experiments, histological examinations were carried out to verify the site of penetration of the microelectrode and the sectioning of the optic tract and chiasm.

Results

Complete Cortical Deafferentation - OTX Cats

We will first consider, for comparison, the responses in the afferented intact (contralateral to the OTX) hemisphere of the OTX cats. The histograms of the ocular dominance distribution of cortical cells of the intact hemisphere revealed a high proportion (80.8%) of binocularly driven cells. The proportion of these cells was similar to that of our own (Yinon et al. 1982) and of Hubel and Wiesel's (1962) normal cats. Furthermore, it was similar to the proportion of these cells previosly found in the intact hemisphere of OTX cats (Yinon et al. 1982; Lepore et al. 1983). In particular, the percentage of binocularly driven cells in the intact hemisphere was 88.9% for the acute group, 82.7% for the chronic 3 month group, and 53.8% for the chronic 6 month group (Fig. 1).

Another characteristic of the intact hemisphere in the OTX cats was the high percentage of visually unresponsive cells as compared to that in the normal cats (these cells were recognized according to their spontaneous activity and the absence of response to visual stimulation). The highest proportion of the visually unresponsive cells was found in the acute group and the lowest proportion in the chronic 6 month group (Fig. 1; Table 1). Thus, there is a trend toward recovery as expressed by the diminution in the visually unresponsive cells in the intact hemisphere with increasing survival time.

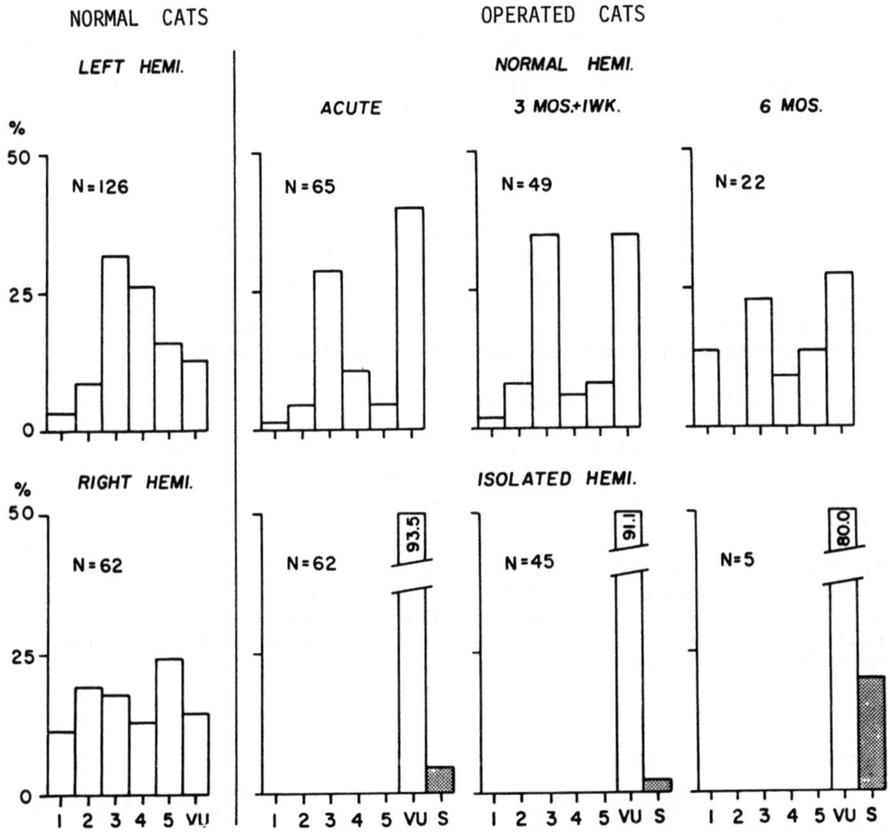

Fig. 1. Ocular dominance histograms and responsiveness of cortical cells in cats with unilaterally sectioned optic tract (OTX). 1,5 = cells driven monocularly by the right and the left eye, respectively; 2,4 = cells binocularly driven with dominance of the right and left eye, respectively; 3 = cells equally binocularly driven from both eyes. VU = visually unresponsive cells (hatched blocks); S = cells with unclear visual activity (stippled blocks); N = number of cells. Normal Hemi. = hemisphere contralateral to the sectioned tract; Isolated Hemi. = hemisphere ipsilateral to the sectioned tract

It is important to note here that the relative absence of responsiveness in the intact hemisphere was consistently found in all cortical layers of the OTX cats, including layers III and IV, where Lepore and Guillemot (1982) discovered callossally driven cortical cells with clear receptive field properties.

Changes were also found in the receptive field properties of cortical cells in the intact hemisphere of the OTX cats. A cell was classified as being selective if it was responsive - either monocularly or binocularly - within its receptive fields to a certain orientation and/or direction of the moving light bar. Nonselective cells included those which responded within their receptive field to all orientations and directions.

A marked decrease in the proportions of both orientation and direction selective cells was found (Table 1). It was 26.3% of the responsive cells for the acute OTX cats, 26.2% for the chronic 3 month cats, and 27.8% for the chronic 6 month cats. In addition, many cells of the intact hemisphere of the OTX cats were widely tuned in their range of orientation response in comparison to cells of the normal control cats.

In comparison, in our normal cats only a small proportion of cells were visually unresponsive (Table 1). Furthermore, a high proportion of cells from both hemispheres (56.0%) were selective for both orientation and direction. In addition, many cells in the normal cats were narrowly tuned to the preferred orientation. The results with regard to orientation and direction selectivity were similar for both the right and the left hemispheres (Table 1).

In the deafferented hemisphere of the OTX cats, no one cell had a clear response to visual stimulation; only a very small proportion of cells (S cells) were suspected of exhibiting visual activity (Fig. 1; Table 1). This visual activity had indicated some direct as well as indirect callosal activation. These cells displayed asynchronous firing to visual stimulation and their action potential counts were inconsistent from sample to sample. Furthermore, the receptive fields of these cells, if existing, could not be mapped. However, the absolute majority of the cells in the deafferented hemisphere were visually unresponsive. In addition, no recovery was found in the deafferented hemisphere as a function of the survival time.

Partial Cortical Deafferentation - OCX Cats

In this experimental paradigm only a partial deafferentation of cortical cells was induced. The only input missing for the binocularly or monocularly driven cortical cells in the OCX cats was that from the contralateral eye. This is the case if the corpus callosum is not involved.

Considering the summed results for the OCX cats with the various survival times, the main finding was that the majority (87.1%) of the visually responsive cells reacted monocularly following stimulation of the ipsilateral eye. Only a few cells in the OCX cats showed a callosal transfer by either reacting binocularly (2.5%) or monocularly (1.2%) to stimulation of the contralateral eye. In comparison, in the normal control cats, only 10.9% of the cells reacted to stimulation of the ipsilateral eye, while 74.5% were binocularly and 11.1% were contralaterally driven (Fig. 2).

A very high proportion of cells were visually unresponsive in the OCX cats (summed results for all groups: 53.9%) in comparison to the normal cats (12.3%) (Fig. 2). The considerable number of unresponsive cells found in these cats is not attributed to a surgical effect on the uncrossed chiasmal fibers since the latter were visually active and their receptive fields were found, as expected, in the center of the field. Furthermore, the receptive fields of these cells had even crossed the vertical meridian to the opposite hemifield. Thus, no evidence for a central scotoma was found. In addition, visually responsive cells were equally found in the left (43.7%) and in the right (43.8%) hemispheres of the OCX cats, indicating no unilateral damage (similar results were also obtained for the single OTX cats).

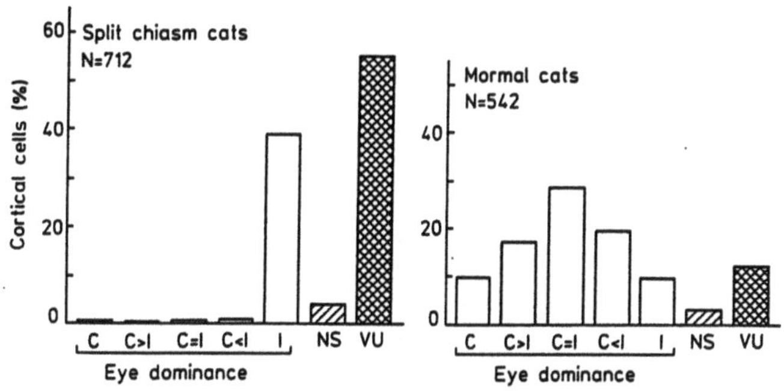

Fig. 2. The ocular dominance distribution of cortical cells in the operated split chiasm (OCX) adult cats and in the normal control cats. Cells were classified into eye dominance groups according to whether they were monocularly contralaterally (C) or ipsilaterally (I) driven, or driven binocularly with dominance of either the contralateral (C>I) or the ipsilateral (C<I) eye. C = I represents cells equally driven through either eye. Percentage of all groups of cells was calculated from the total number of identified cells (N) recorded in all the OCX cats. NS = nonspecific cells; VU = visually unresponsive cells

The high proportion of unresponsive cells in operated OCX cats is attributed to the fact that in these cats nearly two-thirds of the optic nerve fibers are transected (Polyak 1968; Duke-Elder 1976). This practically includes all the crossing (contralateral) optic nerve fibers.

Unidentified cells, i.e., cells with unclear visual properties and cells nonspecific in their response to visual stimulation, were more common in the OCX cats (Table 2). We have also found in the OTX cats cells which exhibited typical habituation behavior although being selective to visual stimulation. Not one cell of this type was encountered in our normal control cats. The increase in the proportion of the above-mentioned groups of "weak" cells may be explained by the absence of a converging facilitatory binocular input on these cells. Despite this, nearly 40% of the cells were stimulus selective in the OCX cats, in contradiction to Berlucchi and Rizzolatti's (1968) results for OCX cats.

There was no consistent change in cortical cell responsiveness with post-operative survival time. The proportion of visually responsive cells was 46.1% of the total number of cells in the acute group (0-1 weeks survival time), 54.7% in the chronic 3.5-6 weeks group and 33.0% in the most chronic group (6.25-7 months). The relationship between the proportion of visually responsive cells and the length of the survival time gave a correlation coefficient of -0.45.

There was neither a consistent change with survival time in the proportion of nonspecific and unidentified cells, nor in the proportion of monocularly driven cells. However, a slight increase was found in the percentage of binocularly driven cells, from 0% in the acute group to 2.1% in the most chronic group of cats, a finding which may indicate some callosally mediated recovery of cortical cells. The situation is different if all the callosally activated cells are considered binocularly (those driven as well as

Table 2. Experimental conditions of the operated split chiasm adult cats[a] (OCX) and percent of visually responsive and unresponsive cortical cells (of total no. of cells)

Experimental Condition	Survival time	No. of cats (& total no. of cells)	Responsive Cells			Visually Unres-ponsive cells	Uniden-tified cells
			Monocularly driven	Binocularly driven	Nonspecific		
Chiasm Split (OCX)	0-1 wks	4 (191)	41.9	0	4.2	51.3	2.6
	3.5-6 wks	4 (247)	47.0	0.8	6.9	44.5	0.8
	6.25-7 mos.[b]	4 (291)	29.5	2.1	1.4	63.6	3.4
Total		12 (729)	38.7	1.1	4.0	53.9	2.3
Normal Control		13 (544)	19.3	65.1	2.9	12.3	0.4

those monocularly driven). They comprise 5.0% of the cells with clear ocular dominance in the cats with 0-1 weeks survival time, 1.7% in the 3.5-6 week cats, and 6.5% in the 6.25-7 month cats. This result does not seem to indicate a clear trend toward recovery of cortical cells in the OCX cats.

Discussion

A complete elimination of binocular retinal input to one hemisphere takes place following unilateral sectioning of the optic tract. It leaves the deafferented hemisphere without its direct visual input and makes it dependent solely on the callosal visual input. Our results showed that the isolated hemisphere was practically deprived from its direct visual input. Furthermore, almost no callosal input from the intact, afferented hemisphere was found, even after long survival periods such as 6 months (Yinon et al. 1982, 1984; Podell et al. 1984).

Our findings are in contrast with those of previous investigations on OTX cats (survival time unknown) which have shown effective callosal activation in the deafferented hemisphere (Choudhury et al. 1965; Vesbaesya et al. 1967). The situation is different in the afferented hemisphere of the OTX cats. This hemisphere was found to be visually active, unlike the deafferented hemisphere. However, diminution was found in orientation and direction selectivity of the cells and in their responsiveness to visual stimulation. A possible explanation for the changes in selectivity is that the cortical cells are normally dependent upon facilitatory callosal input that possesses those specific properties. The removal of the callosally mediated visual input thus results in the loss of this facilitation.

As mentioned above, a considerable reduction in visual responsiveness was also found in the afferented hemisphere. Deafferentation thus seems to induce inactivation of the corpus callosum, which in turn affects the overall degree of visual responsiveness in both hemispheres. This may be a result of the absence of a facilitatory interhemispheric action, as mentioned above. The reduction found in responsiveness was not due to transitional dysfunctioning because of the operation, since it was also found in the chronic cats with survival periods of 3 and 6 months.

The afferented hemisphere remained quite normal with regard to the binocular retinal input, regardless of its deprivation from the deafferented hemisphere. This is in contrast to what would be expected in view of the fact that callosal inactivation must result in binocularity diminution, as proved following callosotomy (Payne et al. 1980). However, concerning the general reduction in responsiveness found by us, we suggest a relatively equal diminution in all classes of binocularly and monocularly driven cells. This diminution does not greatly affect the profile of the ocular dominance distribution: There was even a slight bias in favor of the binocularly driven cells. This is in keeping with recent findings showing that callosotomy does not affect the ocular dominance distribution of cortical cells (Yinon and Weiser 1984, unpublished results).

The possible involvement of a callosally mediated interhemispheric compensatory action was also examined by us on animals with partial cortical deafferentation. Partial deafferentation was induced by surgically eliminating the strong contralateral pathways (OCX); the remaining weak ipsilateral pathways were left intact. Following this procedure only a very small activation of the contralateral pathways was seen, indicating some callosal contribution (Yinon and Hammer 1981; Yinon et al. 1982, 1984). This is also supported by the findings of Milleret and Buser (1983) in OCX cats (survival time in all of the above-mentioned studies: 1-6 weeks).

On the other hand a relatively high proportion of callosally driven cells were found in the visual cortex of OCX cats by other investigators (Berlucchi and Rizzolatti 1968; Cynader et al. 1981; Lepore and Guillemot 1982) (a survival time of 2-4 weeks is known only for Berlucchi and Rizzolatti's study). We do not yet know how to explain the discrepancies between our results and those of the above-mentioned investigators. One possible difference could be the extent of the survival time given. It was assumed that the corpus callosum, which mediates a certain level of binocularity, will become more active with increasing survival time. This question has never been examined previously.

Despite long survival periods in our study - up to 7 months - we found no physiological recovery in the OCX cats. The elimination of the binocular contralateral visual input was not compensated for by an input from the alternative intact callosal pathway. A small proportion of cells reacted via the callosal pathway, but with no consistent trend toward recovery as a function of increase in survival time.

It is important to mention here that the level of responsiveness in the OCX cats was markedly reduced, as shown by the difference between the expected and the actual properties of responsive cells in these cats. This is based on calculating the proportion of binocularly (65.1%) and monocularly (9.5% ipsilateral; 9.7% contralateral) driven cells in our normal cats (total sum: 84.4%). Following chiasm split, each hemisphere might be expected to retain the original 9.5% ipsilaterally monocularly driven cells and 65.1% binocularly driven cells with ipsilateral input only (total sum: 74.6%). This is a much higher proportion than the actual proportion of these cells (43.8%) obtained in the OCX cats.

One of the explanations for the unexpectedly large reduction of neuronal responsiveness in the OCX cats is the absence of binocular convergence of visual inputs. This is true especially if the contralateral and the ipsilateral inputs are tuned for similtaneous action. This may be the explanation not only for the low responsiveness level but also for the increase in the proportion of nonspecific and unidentified cells in these cats. The latter may be correct especially for the cells which had a weak input from the ipsilateral eye prior to their disengagement from the contralateral eye.

Another explanation for the unexpectedly high proportion of visually unresponsive cells is that part of the ipsilateral fibers cross in the optic chiasm. Thus, some of the cells receive their input from these ipsilateral fibers via the corpus callosum (Ramachandran et al. 1981; Diao et al. 1983). If this condition is also applicable to cats it would explain the very considerable reduction in visual activity following optic chiasm transection.

It was concluded that deafferentation of visual cortex cells in adult cats does not lead to recovery of the deprived cells by using visual input from the alternative callosal pathway. Only a small proportion of cells were callosally driven in these cats, but we could not deduct from this a process of intensification of axonal reactivity in the corpus callosum.

Although cellular changes in the visual cortex resulted following the induced trauma to the optic nerve and tract fibers, we could not suggest a positive effect following partial or complete cortical deafferentation. In the present study it was shown that even long survival periods did not seem to induce additional neuronal changes either in terms of recovery or degeneration of the cells concerned.

The prevention of direct visual input from cortical cells during adulthood somehow weakened them more than would be expected on the basis of anatomical considerations, because of the absent interhemispheric interaction. It is probably different in the developing animals, where neural plasticity is much more common. This subject is under study in our laboratory.

Summary

Isolation of visual cortex cells from their major input was performed in adult cats in order to study possible recovery of these cells by using input from the alternative, callosal visual pathway.

Partial deafferentation of the cells from their main retinal input was performed by cancellation of the contralateral pathways following optic chiasm split (OCX). Complete cortical deafferentation was induced by a unilateral transection of the optic tract (OTX).

Unit recording was acutely and chronically (survival time: 0-7 months) performed in the boundary of visual areas 17/18 in order to characterize electrophysiologically changes in the ocular dominance, responsiveness level, and receptive field properties of the cells.

It was found that the distribution of the cortical cells according to their ocular dominance under either partial (OCX) or complete (OTX) visual cortex deafferentation was as expected from anatomical considerations. However, the level of responsiveness of cortical cells and their selectivity to orientation and direction were reduced. Only a few monocularly and binocularly driven cortical cells showed a callosal transfer of visual activity under these experimental paradigms.

It was concluded that deafferentation of visual cortex cells in adult cats does not intensify axonal reaction in the alternative pathway (corpus callosum) to compensate for the missing direct visual input. Oppositely, the deafferentation seems to weaken the normal callosal transfer and thus to prevent some mutual interhemispheric activity which was present prior to the deafferentation.

Acknowledgment. Research supported by the Israel Institute for Psychobiology, Charles E. Smith Family Foundation, Jerusalem, Grant No. 7/83 and by the Pinchas Sapir Foundation of Mifal Hapayis, Tel-Aviv, to U. Yinon. Many thanks to the Jack Haberrer Trust Fund of the First International Bank of Israel, Tel-Aviv, for the support given to A. Hammer

References

Antonini A, Berlucchi G, Marzi CA, Sprague JM (1979) Behavioral and electrophysiological effects of unilateral optic tract section in ordinary and Siamese cats. J Comp Neurol 185:183-202

Berlucchi G, Rizzolatti G (1968) Binocularly driven neurons in the visual cortex of split-chiasm cats. Science 159:308-310

Berlucchi G, Gazzaniga MS, Rizzolatti G (1967) Microelectrode analysis of transfer of visual information by the corpus callosum. Arch Ital Biol 105:583-596

Berlucchi G, Buchtel HA, Lepore F (1978) Successful transfer of visual pattern discriminations in split-chiasm cats with section at the intertectal and posterior commissures. Physiol Behav 20: 331-338

Caulfield TH, Robinson JS, Voneida TJ (1968) Pupil response in cats after chiasm section. Psychol Sci 12:289-290

Choudhury BP, Whitteridge D, Wilson ME (1965) The function of the callosal connections to the visual cortex. Q J Exp Physiol 50: 214-219

Cynader M, Lepore F, Guillemot JP (1981) Interhemispheric competition during post-natal development. Nature (Lond) 290:139-140

Diao Y-C, Wang Y-K, Pu M-L (1983) Binocular responses of cortical cells and the callosal projection in the albino rat. Exp Brain Res 49:410-418

Duke-Elder S (1976) System of ophthalmology, vol I. The eye in evolution. Kimpton, London, p 487

Hubel DH, Wiesel TN (1962) Receptive fields, binocular interaction and functional architecture in the cats' visual cortex. J Physiol (Lond) 160:106-154

Hubel DH, Wiesel TN (1967) Cortical and callosal connections concerned with the vertical meridian of visual fields in the cat. J Neurophysiol 30:1561-1573

Innocenti GM (1980) The primary visual pathway through the corpus callosum: Morphological and functional aspects in the cat. Arch Ital Biol 118:124-188

Lepore F, Guillemot JP (1982) Visual receptive field properties of cells innervated through the corpus callosum in the cat. Exp Brain Res 46: 413-424

Lepore F, Samson A, Molotchnikoff S (1983) Effects on binocular activation of cells in visual cortex of the cat following the transection of the optic tract. Exp Brain Res 50:392-396

Milleret C, Buser P (1983) Patterns of unit responses in the visual cortex (area 18) of the adult cat after chiasmotomy. C R Seances Acad Sci 297:75-80

Myers RE (1955) Interocular transfer of pattern discrimination in cats following section of crossed optic fibers. J Comp Physiol Psychol 48:470-473

Payne BR, Elberger AJ, Berman N, Murphy EA (1980) Binocularity in the cat visual cortex is reduced by sectioning the corpus callosum. Science 207:1097-1099

Podell M, Yinon U, Hammer A (1984) Properties of visual cortical cells of the intact and the deafferented hemisphere of unilateral optic tract sectioned acute and chronic adult cats. Exp Brain Res 55:91-96

Polyak S (1968) The vertebrate visual system. University of Chicago Press, Chicago, p 788

Ramachandran VS, Bravo H, Pettigrew JD (1981) Binocular neural mechanisms in goats and sheep. Invest Ophthalmol Vis Sci [Suppl] 20: 73 (ARVO Abstracts)

Shatz C (1977) Abnormal interhemispheric connections in the visual systems of Boston Siamese cats: a physiological study. J Comp Neurol 171:229-246

Tusa RJ, Rosenquist AC, Palmer LA (1979) Retinotopic organization of areas 18 and 19 in the cat. J Comp Neurol 185:657-678

Vakkur GJ, Bishop PO, Kozak W (1963) Visual optics in the cat, including posterior nodal distance and retinal landmarks. Vision Res 3:289-314

Vesbaesya C, Whitteridge D, Wilson ME (1967) Callosal connections of the cortex representing the area centralis. J Physiol (Lond) 191:79-80

Yinon U (1980) Monocular deafferentation effects on responsiveness of cortical cells in adult cats. Brain Res 199:299-306

Yinon U, Hammer A (1981) Physiological mechanisms underlying responsiveness of visual cortex neurons following optic chiasm split in cats. In: Flohr H, Precht W (eds) Lesion induced neuronal plasticity in sensorimotor systems. Springer, Berlin Heidelberg New York, pp 360-368

Yinon U, Hammer A (1985) Optic chiasm split and binocularity diminution in cortical cells of acute and chronic operated adult cats. Exp Brain Res (to be published)

Yinon U, Weiser Z (1985) Ocular dominance in the cat visual cortex following sectioning of the corpus callosum. (to be published)

Yinon U, Hammer A, Podell M (1982) The hemispheric dominance of cortical cells in the absence of direct visual pathways. Brain Res 232:187-190

Yinon U, Podell M, Goshen S (1984) Deafferentation of the visual cortex: The effect on cortical cells in normal and in early monocularly deprived cats. Exp Neurol 83:486-494

Transneuronal and Peripheral Mechanisms for the Induction of Sprouting and Synapse Formation in Amphibian and Mammalian Motor Nervous System

S. Rotshenker, M. Tal, G. Ring, H. Sugarman, and F. Reichert

Department of Anatomy and Embryology, Hebrew University-Hadassah, Medical School, P.O. Box 1172, Jerusalem 91010, Israel

Introduction

The peripheral motor nervous system responds to injury in two ways: (a) axotomized motoneurons regenerate their axons and re-establish synaptic connections with denervated muscle fibers, and (b) intact motor neurons sprout their axons and form additional synaptic connections. Sprouting of intact motor axons can be subdivided into two categories: sprouting into denervated areas and sprouting in normally innervated muscles. An example of the former mode of sprouting is that following partial denervation. After cutting some of the axons innervating a muscle, the remaining intact axons grow new processes that innervate denervated muscle fibers (e.g., reviews by Edds 1953; Brown et al. 1981). Sprouting and synapse formation in normally innervated muscles occur after denervating homologous muscles on the contralateral side of the body (Rotshenker 1979, 1982; Rotshenker and Reichert 1980; Rotshenker and Tal 1984; Ring et al. 1983; Elizalde et al. 1983).

As previously suggested (Rotshenker 1982), mechanisms for the induction of sprouting and synapse formation can be divided into two major groups: peripheral and transneuronal. Peripheral mechanisms involve the direct influence of a growth-promoting factor (e.g., trophic or inactivity) on the peripheral extensions of the motoneuron (the intramuscular axons, preterminal branches and nerve endings) that responds with sprouting. The transneuronal mechanism involves the induction of axonal sprouting of intact motoneurons by other nerve cells within the spinal cord (motor, sensory, or others).
The purpose of the present communication is to summarize the evidence obtained by us for the existence of both transneuronal and peripheral mechanisms, and to provide evidence suggesting that muscle fibers can regulate axonal growth at the level of the cell body of the motoneuron. Since it is beyond the scope of this paper to provide a complete description of all experimental procedures and data, the reader is advised to consult the original reports.

Morphological and Electrophysiological Identification of Sprouting and Synapse Formation Following Contralateral Axotomy

We examined the pattern of motor innervation to individual muscle fibers of the cutaneous-pectoris and sartorius muscles of the frog, and the peroneal and extensor digitorium longus muscles of the mouse. Some degree of sprouting and synapse formation was observed

Experimental Brain Research, Suppl. 13

in all intact muscles of normal animals. Sprouting was enhanced in intact muscles by axotomizing the neurons to the contralateral, homologous muscles (contralateral axotomy).

Frog cutaneous-pectoris muscles were stained with a combined silver cholinesterase stain that visualizes the entire neuromuscular junction: axons, preterminal branches and nerve endings (silver), and end-plate regions (cholinesterase) (Rotshenker 1982) (Fig. 1). The majority of muscle fibers are innervated by a single myelinated axon. In some instances, two myelinated axons innervate a single muscle fiber at a single site. Occasionally, single muscle fibers receive innervation from two different sources: a myelinated axon and a sprout that arises either from another axon or from nerve endings of a neighboring neuromuscular junction (Fig. 1). The incidence of muscle fibers demonstrating the latter pattern of innervation was 1.5±0.6% in intact muscles of normal frogs. This incidence of sprouting and synapse formation increased to an average value of 16.9±3.0% after denervating contralateral cutaneous-pectoris muscles. Thus muscle fibers that were previously innervated by a single motoneuron (via one axon) became polyneuronally innervated (via the sprout).

Polyneuronally innervated muscle fibers were also identified electrophysiologically by recording multiple end-plate potentials in response to graded intensities of stimulation applied to the motor nerve (Rotshenker 1979, 1982). The incidence of polyneuronal innervation detected in intact muscles of normal frogs was about 16%, the anatomical basis being the innervation of single muscle fibers by two myelinated axons. The incidence of polyneuronal innervation increased 2.5-fold following contralateral axotomy. The additional number of muscle fibers that became polyneuronally innervated is similar to the additional number of muscle fibers that became innervated by sprouts (Rotshenker 1982). Thus, sprouting and synapse formation that follow contralateral axotomy result in an increased pattern of polyneuronal innervation. (For additional light and electron microscopical evidence for sprouting and synapse formation in intact cutaneous-pectoris muscles following contralateral axotomy, see Rotshenker and Reichert 1980).

Sartorius muscles of the frog were studied with the light microscope after they were stained with a combined silver-cholinesterase stain (Ring et al. 1983). The incidence of sprouting and synapse formation observed in intact muscles of normal frogs was 6.0±0.8%. It increased more than three-fold to 20.2±2.6% after contralateral axotomy.

Peroneal and extensor digitorum longus (edl) muscles of the mouse were also examined with the light microscope after neuromuscular junctions were visualized by a combined silver-cholinesterase stain (Rotshenker and Tal 1984).

In intact muscles of normal mice, all fibers were innervated by a single axon. In a small proportion of cases (7.0±0.9%) axons gave rise to sprouts, and both parent axon and sprout innervated the same muscle fiber (Fig. 2). This incidence of sprouting and synapse formation increased almost threefold to an average value of 20.7±2.1% after contralateral axotomy.

Thus contralteral axotomy resulted in sprouting and synapse formation in both amphibian and mammalian muscles. However, in the mouse, sprouting did not result in a polyneuronal pattern of innervation as it did in the frog.

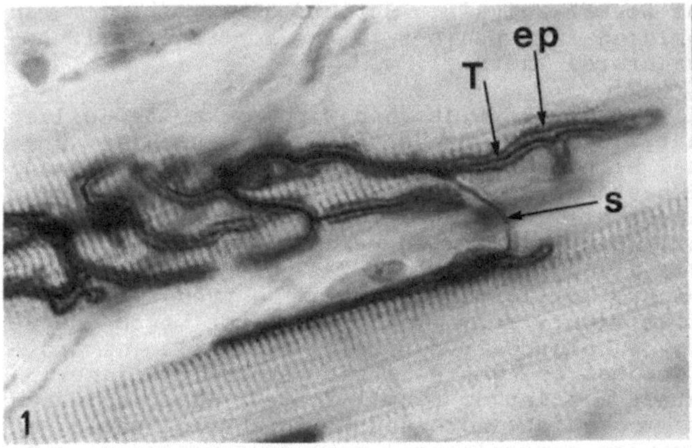

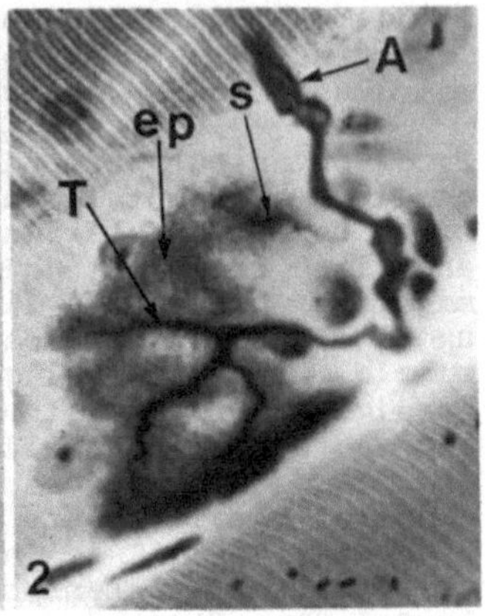

Fig. 1. A combined silver-cholinesterase stain visualizes myelinated axons (A), nerve endings (T), sprouts (S), and end-plate regions (ep) in a frog preparation. The sprout arising from a neuromuscular junction on one muscle fiber innervates a neighboring muscle fiber

Fig. 2. A combined silver-cholinesterase stain of a peroneal muscle of the mouse. The sprout (S) that arises from the myelinated axon (A) contributes to the innervation of the muscle fiber innervated by the parent axon

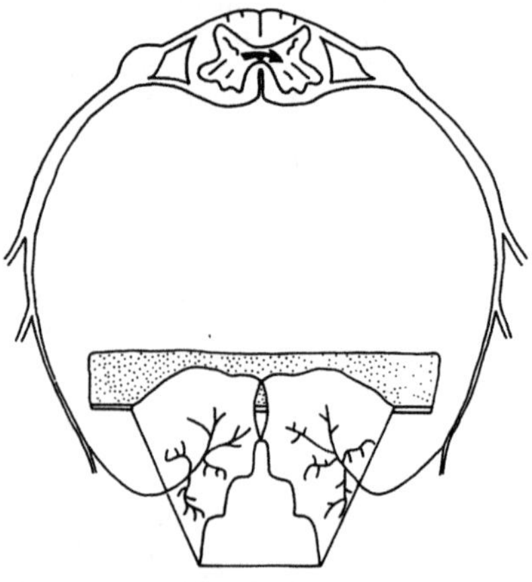

Fig. 3. A schematic representation of the paired cutaneous-pectoris muscles of the frog and their motor innervation that arises from the 2nd spinal segment. Axotomy of the nerve to one muscle was followed by sprouting and synapse formation in the opposite intact muscle. Similar observations were made in sartorius muscles of the frog, and peroneal and edl muscles of the mouse

The Transneuronal Mechanism for the Induction of Sprouting and Synapse Formation

A transneuronal mechanism has been formulated to explain the induction of sprouting and synapse formation in intact muscles after contralateral axotomy . (Rotshenker 1979, 1982). We suggest that axotomy initiates a signal for growth in cell bodies of axotomized motoneurons. The signal is then transferred transneuronally across the spinal cord to the intact motoneurons, which respond with sprouting and synapse formation (Fig. 3). The experimental results that led to the formulation of the transneuronal mechanism did so (a) by ruling out alternative mechanisms, and (b) by being in accord with predictions of the suggested mechanism.

A) Possible alternatives to the transneuronal mechanism are that sprouting was induced by: (a) products of degeneration or denervation, or (b) inactivity of the denervated muscles.

a) Products of degeneration and denervation that arise from the denervated muscle could reach the opposite muscle by local or systemic distribution, and induce neuronal sprouting. If this were the case, then the surgical removal of one cutaneous-pectoris muscle should not be followed by contralateral sprouting since the sources for products of degeneration and denervation would be removed. Such was not the case; sprouting and synapse formation developed in intact cutaneous-pectoris muscles after removing the homologous muscle on the opposite side (Rotshenker 1979). Conversely, denervation of one hind limb, which would allow the systemic distribution of products of degeneration and denervation, produced sprouting in contralateral hind limb muscles (Elizalde et al. 1983; Ring et al. 1983; Tal and Rotshenker 1984) but not in the chest cutaneous-pectoris muscles (Rotshenker 1979).

The possible role of products of degeneration and denervation was further ruled out by the observation that axotomy of the 2nd spinal nerve at its exit from the spinal column did not produce sprouting in intact opposite muscles in about one-half of the animals. Sprouting was always observed after contralateral axotomies that were placed at distances of 3 mm or more from the spinal column (Ring et al. to be published). We do not understand why contralateral axotomy that was placed in too close proximity to the spinal cord failed to produce sprouting. However, one conclusion is clear: products of degeneration and denervation could not produce sprouting. At the same time, this experimental result does not contradict the transneuronal mechanism. It can be argued, based on extensive data in the literature (e.g., Lieberman 1971, 1974), that the response of the motoneuron cell body to axotomy is different for injuries which are placed either very close to the cell body or at some distance from it (e.g., cell death following axotomy is more prominent after proximal axotomy). The sprouting response, which is suggested to be dependent on changes in cell bodies of contralateral, axotomized motoneurons, could also differ in conjunction with differences in the soma reaction to axotomy.

b) It is possible that inactivity of the denervated muscles could produce sprouting in opposite muscles. The experimental results that contralateral axotomy of the 2nd spinal nerve did not produce sprouting, whereas more distant axotomies always did, suggest that inactivity could not be a major factor in the induction of sprouting. In other experiments, we altered the sensory input to the spinal cord by producing a tenotomy of one cutaneous-pectoris muscle (Rotshenker 1979) or by cutting the 2nd dorsal root through which sensory information from the muscle reaches the spinal cord (Ring et al. to be published). Neither of these procedures resulted in transneuronal sprouting and synapse formation.

B) One prediction of the transneuronal mechanism theory is that the delay with which new synapses develop in intact muscles should decrease when the contralateral axotomy is placed closer to the spinal cord. This prediction is based on the assumption that sprouting is dependent on changes in cell bodies of contralateral axotomized motoneurons. It is known that retrograde postaxtomy changes (e.g., chromatolysis, increase in RNA production) develop more rapidly when the site of axotomy is placed closer to the soma (Lieberman 1971, 1974). The predicted results were observed in both frogs and mice. In the frog cutaenous-pectoris mucle (Rotshenker 1979, 1982), sprouting and synapse formation occurred about 1, 2.5, and 4 weeks after contralateral axotomies that were placed, respectively, 5 mm from the spinal column, about midway between the spinal cord and the muscle, and at the site of the nerve entry into the muscle. In the mouse (Rotshenker and Tal 1984) a difference of 35 mm in the distance of the contralateral axotomies from the spinal cord resulted in a difference of 5 to 7 days in the development of sprouting and synapse formation. The variation in the time to onset of sprouting was not due to the difference in the number of axons that were involved in the injuries at the different sites.

The Possible Role for a Muscle-Derived Trophic Substance as a Regulator of Growth at the Level of the Nerve Cell Body

The transneuronal mechanism formulated above is composed of two steps: (a) the initiation of a signal for growth by the axotomy in the injured nerve cell bodies, and (b) the transneuronal transfer of the signal to intact motoneurons. In what way did axotomy initiate a signal for growth?

Axotomy blocks normal trophic interaction between nerve and muscle. We suggest the possibility that axotomy prevented the normal supply of a muscle-derived trophic substance to the motoneurons' cell body. A reduced concentration of such a substance in the cell body could initiate in it a signal for growth.

We tested our working hypothesis by designing two experimental conditions that are likely to inhibit the retrograde supply of a trophic substance from the periphery of the nerve cell to its soma: (a) the inhibition of the retrograde axonal transport, on whose integrity the retrograde transport of trophic substances is dependent, and (b) the removal of muscle fibers, which are the source of the hypothetical trophic substance, without injuring the innervating axons.

a) The inhibition of the retrograde axonal transport in one cutaneous-pectoris muscle was achieved by the local application of colchicine to the nerve in doses that did not produce nerve degeneration or muscle denervation (Rotshenker 1982). Anatomical and electrophysiological examination of the intact cutaneous-pectoris muscle on the opposite side revealed the development of sprouting and synapse formation 2.5 weeks after drug application to the nerve at the axilla. The development of new synapses in the intact muscle following the contralateral inhibition of axonal transport mimicked that resulting from contralateral axotomy at the same site in magnitude and timing.

b) The removal of muscle fibers of one cutaneous-pectoris muscle was produced by the local application of the myotoxic, local anesthetic carbocaine (Tal and Rotshenker 1984). Drug application was followed by partial degeneration of the muscle. Macrophages phagocytized degenerating muscle fibers, leaving behind vacated spaces enclosed by a basal lamina sheath. Motor axons and nerve endings remained structurally intact and retained their ability to recycle synaptic vesicles as normal (Tal and Rotshenker 1983). Thus intact motoneurons became partially deprived of their target muscle fibers and of trophic substances derived from them. Cutaneous-pectoris muscles were examined electrophysiologically. Sprouting and synapse formation developed in intact muscles about 4 weeks after the application of the drug to opposite muscles. The response was similar in magnitude and timing to that obtained after cutting the cutaneous-pectoris nerve close to the muscle. Sprouting and synapse formation developed also in the drug-treated muscles but about 2.5 weeks after drug application. These results agree well with predictions from our working hypothesis. A signal for growth is expected to develop in the cell bodies of motoneurons that have been deprived of a muscle-derived trophic substance. Therefore, these motoneurons should be the first to sprout. After the signal for growth is transferred transneuronally across the spinal cord to intact motoneurons, the recipient cells should respond with sprouting, too, but with some delay.

Peripheral Mechanisms for the Induction of Sprouting

We define peripheral mechanisms for the induction of sprouting as those that involve the direct effect of a growth-promoting factor on the peripheral extensions of the axon. Such factors could be trophic factors (e.g., NGF), products of denervation, products of degeneration, inactivity, toxins, and others (e.g., review by Edds 1953; Brown et al. 1981). The evidence, in in vivo systems, that peripheral mechanisms exist is rather indirect and circumstantial. Recently, we were able to provide direct evidence for the possibility that such mechanisms exist.

In experiments previously described, colchicine was applied to one cutaneous-pectoris nerve at the axilla. Sprouting and synapse formation developed in the intact muscle on the opposite side after 2.5 weeks. Surprisingly, new synapses were also formed by the nerve that was exposed to colchicine, as little as 2 days after drug application (Rotshenker 1981, 1982). The rapid response to drug application (2 days) raised the possibility that colchicine induced sprouting by affecting nerve endings and axons directly without the immediate involvement of the cell body. To test this possibility, we examined the ability of axons that were separated from their cell bodies to sprout. In the frog, degeneration that follows axotomy is rather slow. Failure of synaptic transmission in cutaneous-pectoris muscles occurred first on the 4th day after cutting the 2nd spinal nerve close to the spine. The pattern of innervation to single muscle fibers did not change from normal during the first 3 days after surgery. Thus, it was possible to cut the 2nd spinal nerve and apply colchicine to the axillary nerve at the same time, and then test the muscle 2 and 3 days later. Anatomical and electrophysiological examination provided evidence for sprouting and synapse formation. Thus, axons that were separated from their cell bodies were still able to respond with sprouting and synapse formation.

Conclusions

Our studies on the amphibian and mammalian motor system suggest that sprouting of intact motoneurons and synapse formation can be regulated by two mechanisms: peripheral and transneuronal. Peripheral mechanisms provide the means of a direct mode of interaction between the periphery of the nerve cell and the target to determine the extent of target innervation. A transneuronal mechanism provides a vehicle by which the pattern of innervation of a muscle can be altered by nerve cells that do not themselves innervate the muscle but are an integral part of the entire system. In addition, we also provide evidence that the target muscle can regulate growth by a feedback mechanism that involves the effect of a muscle-derived trophic substance at the level of the cell body.

References

Brown MC, Holland RL, Hopkins WG (1981) Motor nerve sprouting. Annu J Rev Neurosci 4:17-42

Edds MV (1953) Collateral nerve regeneration. Q Rev Biol 28:260-276

Elizalde AM, Huerta M, Stefani E (1983) Selective reinnervation of twitch and tonic muscle fibers of the frog. J Physiol 340:513-524

Lieberman AR (1971) The axon reaction: a review of the principal features of perikaryon responses to axon injury. Int Rev Neurobiol 14:49-124

Lieberman AR (1974) Some factors affecting retrograde neuronal responses to axonal lesions. In: Bellair R, Bray EG (eds) Essays on the nervous system. Clarendon, Oxford

Ring G, Reichert F, Rotshenker S (1983) Sprouting in intact sartorius muscles of the frog following contralateral axotomy. Brain Res 60:313-316

Ring C, Sugarman H, Rotshenker S (1985) The role of motor and sensory nerve cells in the transneuronal induction of sprouting. To be published

Rotshenker S (1979) Synapse formation in intact innervated cutaneous-pectoris muscles of the frog following denervation of opposite muscles. J Physiol 292:535-547

Rotshenker S (1981) Sprouting and synapse formation by motor axons separated from their cell bodies. Brain Res 223:141-145

Rotshenker S (1982) Transneuronal and peripheral mechanisms for the induction of motor neuron sprouting. J Neurosci 2:1359-1368

Rotshenker S, Reichert F (1980) Motor axon sprouting and site of synapse of formation in intact innervated skeletal muscle of the frog. J Comp Neurol 193:413-422

Rotshenker S, Tal M (1984) The transneuronal induction of sprouting and synapse formation in intact mouse muscles. J. Physiol (to be published)

Tal M, Rotshenker S (1983) Recycling of synaptic vesicles in motor nerve endings separated from their target muscle fibers. Brain Res 270:131-133

Tal M, Rotshenker S (1984) Sprouting and synapse formation produced by carbocaine. J Neurosci 4:458-463

Changes in Target Tissue After Trauma

Protein Synthesis Under Dendritic Spine Synapses During Lesion-Induced Synaptogenesis: Evidence for Regulation of Reinnervation by the Target Cell

O. Steward

Departments of Neurological Surgery and Physiology, University of Virginia, Medical Center, Box 212, Charlottesville, VA 22908, USA

Introduction

The formation of interconnections between neurons during either normal development or compensatory growth following trauma is thought to involve two processes: (1) the growth of the presynaptic axon, and (2) the formation of the specialized synaptic contact between pre- and postsynaptic partners. Crucial to the understanding of the cellular mechanisms of the formation of connections is the determination of the cellular participants and their roles. The growth of the axon is largely a presynaptic process, although it can certainly be regulated by extrinsic factors. The evidence for this conclusion is solid; neurons grown in culture are capable of extending axons for considerable distances without directly interacting with potential postsynaptic targets or other cells. This of course does not mean that neurons grow without cellular contact in vivo, or that these interactions are not important. The data simply indicate that an axon can be generated through processes which are intrinsic to the presynaptic neuron. Because axon growth seems to be largely a presynaptic process, the major focus of reserach on the cellular mechanisms of the formation of interconnections has been on the presynaptic neuron and its axon. This has been particularly true in studies of the response of neurons to trauma. Since axonal damage in the vertebrate CNS rarely leads to effective axonal regeneration, a guiding assumption has been that the principal impediment to the reconstruction of circuitry was at the stage of axon growth.

Perhaps because of the concentration on the presynaptic neuron and the growth of its axon, the postsynaptic cell has often been viewed as a passive participant in growth processes, even when growth leads to the formation of synaptic connections. For example, when neurons which lose synaptic input as a consequence of lesions are reinnervated, it is often thought that the postsynaptic cell is simply stripped of the debris of degenerating synapses by glia, thereby leaving sites available for reinnervation (see for example, Raisman 1969; Matthews et al. 1976). In this conception, the postsynaptic cell need only preserve synaptic sites. By implication, the most active neuronal participant in the growth response is the presynaptic neuron which reinnervates these sites. Clearly, however, the phase of synaptogenesis is a cooperative interaction between pre- and psostsynaptic partners. At a minimum, the postsynaptic cell must maintain sites for synaptic contact, and therefore plays a permissive role. Even in such a passive role, the postsynaptic cell is likely to select between potential presynaptic partners by being unreceptive to some synapses. The postsynaptic cell may also select between synapses after initial contacts are formed through processes of synapse stabilization and/or repression (Changeux 1983).

Experimental Brain Research, Suppl. 13

Fig. 1 A-D. Polyribosomes under dendritic spines in mid-proximo-distal dendrites of granule cells of the dentate gyrus. A illustrates a collection of polyribosomes (arrow) which appear to be closely associated with a membranous cistern which extends into the neck of the spine. B illustrates a segment of a dendrite with three clusters of polyribosomes under mounds in the dendrite. Serial sections reveal that such mounds usually represent the base of spines which extend out of the plane of the section. C and D illustrate examples of polyribosomes under spines in the denervated dentate gyrus (6 days post-lesion). s = spine head, t = presynaptic terminal, den = dendrite, d = degenerating terminal. A, C, and D are reproduced from Steward, 1983a, and B from Steward and Levy, 1982 ►

There is an emerging body of evidence, however, which suggests that the postsynaptic cell may play an even more active role. For example, studies of invertebrates suggest that pre- and postsynaptic cells must be mutually growing for some types of synapse to form (Hadley et al. 1983). Since axonal and dendritic growth go hand-in-hand during normal development and at least some examples of lesion-induced growth in vertebrates as well (Caceres and Steward 1983), mutual pre- and postsynaptic growth during synaptogenesis may be the rule. In an even more active mode, there is growing evidence that the postsynaptic cell may regulate the growth of the presynaptic axon. In the peripheral autonomic nervous system, the targets of sympathetic axons produce nerve growth factor. In addition to its role in maintaining the presynaptic neuron, nerve growth factor may also induce or guide the presynaptic axon during development. (For reviews, see Levi-Montalcini 1976; Thoenen et al. 1979.) Similar target-derived growth and survival factors may exist for CNS neurons (see for example, Crutcher and Davis 1981, 1982; Crutcher and Collins 1982).

Given this evidence for a regulatory role for the postsynaptic cell in synaptogenesis, the question becomes how might the neuron accomplish this. Regulation of synaptic contacts would seem to be a particular problem for CNS neurons, which may receive tens of thousands of synaptic contacts on their dendrites. Furthermore, different portions of dendrites may receive different types of synaptic contact, suggesting a local specialization of the receptive surface of postsynaptic cells. From this perspective, we have been particularly interested by our discovery of a selective localization of polyribosomes under dendritic spines of CNS neurons, particularly during times of synaptogenesis (see Fig. 1 and Steward and Levy 1982; Steward and Fass 1983; Steward 1983a,b). The present chapter will develop the hypothesis that postsynaptic neurons regulate the formation of synapses along their receptive surface by producing specific proteins in cytoplasmic microdomains under sites of synaptic contact. The hypothesis derives from studies of lesion-induced synapse growth and synapse formation during normal development which reveal: (1) that polyribosomes are particularly prominent under developing spine synapses and increase substantially during lesion-induced synapse growth, and (2) that in parallel with the increase in polyribosomes, there are increases in the incorporation of protein precursors in neuropil regions, which appear to reflect protein synthesis by the spine-associated polyribosomes.

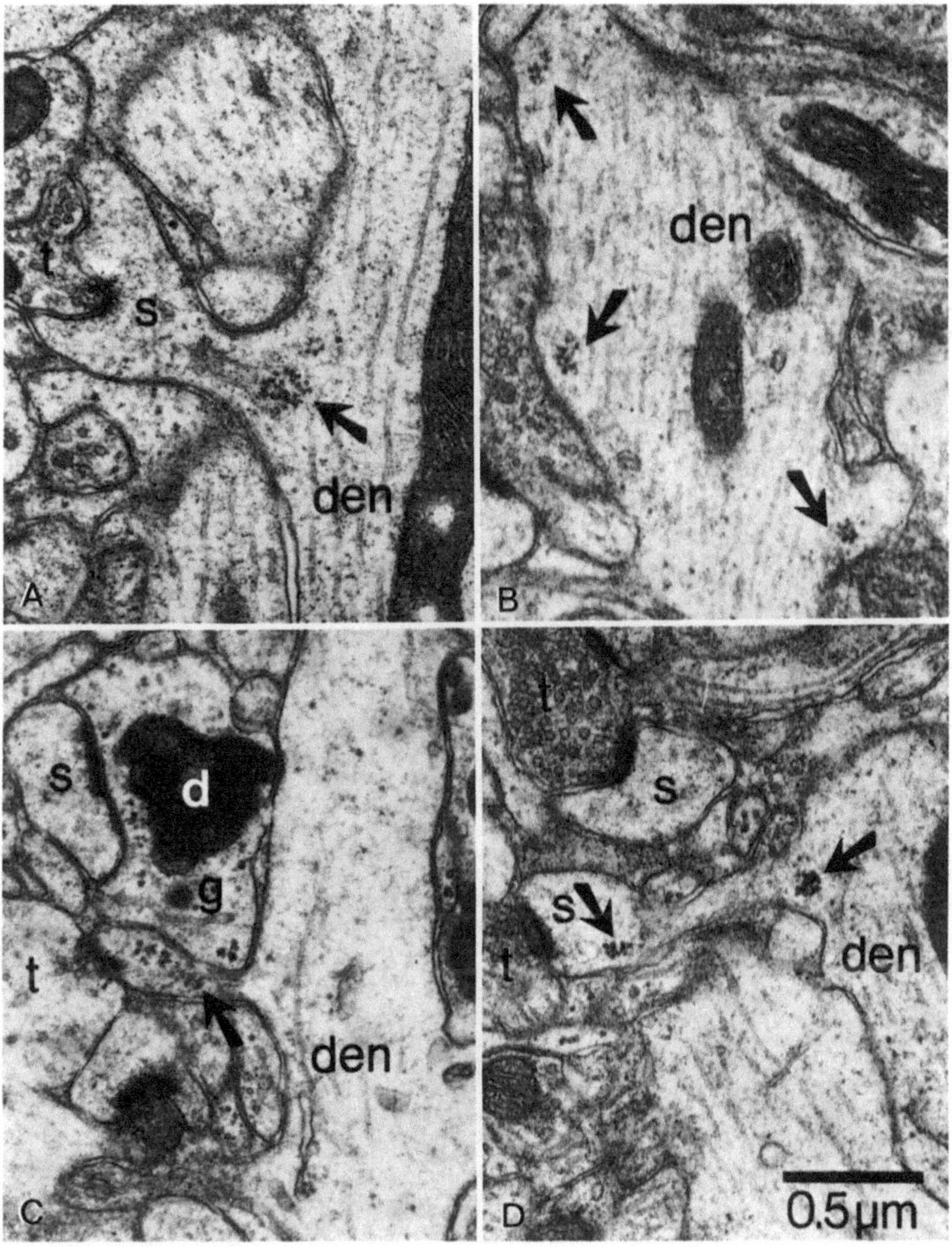
t
s
den
A
den
B
s
d
g
t
den
C
t
s
s
t
den
0.5 µm
D

Lesion-Induced Synaptogenesis in the Dentate Gyrus of the Adult Rat

The initial evidence that polyribosomes under dendritic spines might play a role in regulating synapse growth came from studies of lesion-induced synaptogenesis in the dentate gyrus. Lesions of the entorhinal cortex result in the degeneration of more than 90% of the synapses on distal dendrites of dentate granule cells ipsilateral to the lesion (Steward and Vinsant 1983; and see Matthews et al. 1976 and Lee et al. 1977 for slightly different numerical estimates). These lost synapses are rapidly replaced; there is a proliferation of presynaptic process within the denervated zone (terminal proliferation) and formation of new synaptic connections (reactive synaptogenesis). The time course of the terminal proliferation and reactive synaptogenesis has been defined by counting intact presynaptic processes and synapses within the denervated zone at various times following the lesion (Steward and Vinsant 1983). As illustrated in Fig. 2A, most of the presynaptic growth and synapse formation seems to occur between about 6 and 14 days post lesion. Caution must be exercised in interpreting the details of the time course of synapse formation because of the difficulty of identifying synapses early in their formation. Nevertheless, these results define the time in which to look for the cellular processes which might be involved in the growth.

Polyribosomes Under Dendritic Spines, and Their Changes During Lesion-Induced Synaptogenesis

It has long been known that dendrites, unlike axons, do contain a few polyribosomes. In general, the highest concentrations of ribosomes are found in proximal dendritic segments, and the number decreases with distance from the soma. What caught our attention about the polyribosomes was their preferential localization under the base of dendritic spines. Initially, our observations were limited to granule cells in the dentate gyrus, but we have found similar selectivity in all cell types we have evaluated (see Steward 1983b). In single sections, polyribosomes are most often found either under profiles which can be unequivocally identified as spine bases (Fig. 1A) or under mounds in the dendrite which serial sections reveal to be the base of spines which extend out of the plane of section (Fig. 1B). Quantiative analyses revealed that at least 81% of the polyribosomes in mid proximodistal dendrites of dentate granule cells lay under positively identified or presumed spine bases (the mounds) (Steward and Levy 1982). Very rarely, polyribosomes were also observed in spine necks or heads, although in normal material these clusters usually contained only two to three ribosomes. An evaluation of the incidence of polyribosomes in positively identified spines (where the spine head and the neck connecting it with the dendrite are visible an electron micrograph) revealed a relative incidence of 12% in mid proximodistal dendrites of granule cells (Steward and Levy 1982; and see Fig. 2B). Serial section reconstruction of a limited collection of spines revealed that the actual incidence is somewhat higher (closer to 30%).

During the lesion-induced growth which follows destruction of the input from the entorhinal cortex to the dentate gyrus, there are striking changes in the distribution of polyribosomes within granule cell dendrites. Beginning at about 6 days postlesion, polyribosomes were much more frequent under spine bases, and also often were observed in spine necks and heads (see Fig. 1C, D). Quantitative studies (Steward 1983a) revealed substantial increases in the inci-

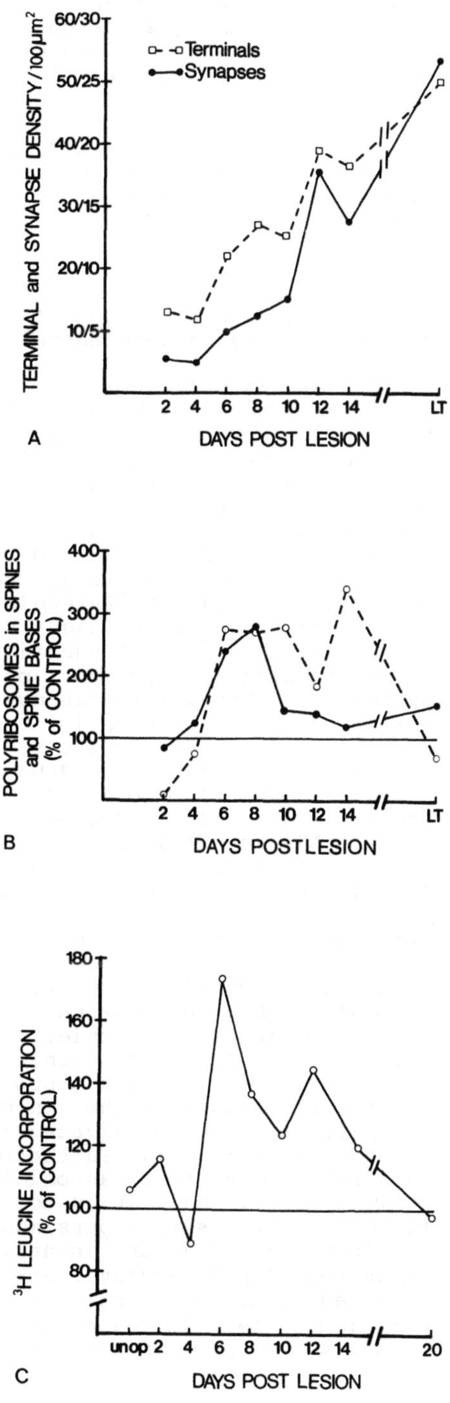

Fig. 2 A-C. Relationship between reactive growth in the denervated dentate gyrus, changes in polyribosomes, and changes in incorporation of protein precursors in the denervated neuropil. A) Time course of the increase in terminal and synapse density in the denervated neuropil of the dentate gyrus after unilateral entorhinal cortical lesions. The numbers on the ordinate refer to terminal density (numbers before slash) and synapse density (numbers after slash) per 100 µm² of neuropil. LT, long-term postlesion (7 months). (Redrawn from Steward and Vinsant 1983). B) Time course of increases in polyribosomes under the base of dendritic spines (solid line) and increases in polyribosomes spike necks and heads (broken line) in the denervated neuropil, from the same animals as analyzed in A (Steward 1983a). C) Time course of increases in incorporation of protein precursors (³H-leucine) in the denervated neuropil of the dentate gyrus after unilateral entorhinal cortical lesions (Fass and Steward 1983). Note that the increase in polyribosomes and the increase in protein precursor incorporation occur at about the same time (peaking at 6-8 days postlesion), which corresponds to the initial period of terminal proliferation and reactive synaptogenesis

dence of polyribosomes under spine bases and in the number of polyribosome-containing spine heads per area of neuropil. These changes were most prominent between 6 and 10 days postlesion (Fig. 2B). During this interval, the spines with polyribosomes were apposed to both degenerating and intact presynaptic elements. Occasional profiles were observed with polyribosomes both at the base and within the spine head. The "intact" synapses often appeared quite immature, and occasionally contained relatively large vesicles reminiscent of those found in presumed axonal growth cones in developing animals (Vaughn and Sims 1978).

Increases in Incorporation of Protein Precursors in the Denervated Neuropil During Lesion-Induced Growth

The changes in polyribosomes associated with dendritic spines suggest that there may be a local regulation of protein synthesis in the denervated dendrites during reinnervation. In fact, our evaluation of polyribosomes which led to the discovery of the selective localization of these elements under spines was initially carried out because of a separate study on the synthesis of proteins in the denervated neuropil (Fass and Steward 1983). We used an autoradiographic approach similar to Droz and Leblond (1963). At various times following a unilateral entorhinal lesion, rats were injected intravenously with ^{3}H-leucine. The animals were perfused 30 min later, and prepared for light microscopic autoradiography. Presumably, the 30-min postinjection survival interval is too short to permit substantial transport of labeled proteins from their sites of synthesis. Thus, the presence of label is thought to indicate the extent of protein synthesis, and the sites at which this synthesis occurs. We found that while there were no increases in labeling over the lamina containing granule cell somata, there were sizable increases in incorporation in the denervated neuropil (Fass and Steward 1983). The increases first appeared at 6 days postlesion, and persisted until day 14. Thus, the increases in incorporation paralleled almost exactly the changes in polyribosome distribution in the denervated dendrites (compare Fig. 2C and B).

The parallels between the increases in polyribosomes under spines and the increases in incorporation suggest that the two events are related. In addition, the available evidence suggests that the other elements within the denervated zone do not contribute significantly to the increases. The chief constituents of the denervated neuropil are the dendrites of the dentate granule cells, presynaptic processes, and glia. Presynaptic processes are not likely to make a detectable contribution to the incorporation, particularly since so few are present in the denervated zone during the time of maximal incorporation (see Fig. 1). Furthermore, in the autoradiograms, the label did not appear to be associated with glial cell bodies (Fass and Steward 1983). While it is not possible to exclude some contribution by glial processes, a parsimonious hypothesis is that the increase in incorporation in the denervated zone is a direct result of the increases in polyribosomes associated with spines. Even if spine-associated ribosomes do not account for all of the increase in incorporation, their increased incidence and redistribution indicate a regulation of protein synthetic machinery during synapse growth.

Polyribosomes Under Growing Synapses in Developing Animals

The replacement of synapses following lesions occurs simultaneously with degenerative processes involving the removal of degenerating presynaptic elements and perhaps the deterioration of the postsynaptic specialization, the spine, and the dendrite. One interpretation of the increase in spine-associated polyribosomes is that they are involved with the degradation of old sites of contact rather than the building of new ones. Strong evidence that the polyribosomes are involved in synapse growth rather than degradation comes from studies of developmental synaptogenesis. The dentate gyrus is an optimal site to study developmental synaptogenesis. Most granule cells undergo their final mitosis after birth, and thus most synaptic connections develop postnatally. In animals prepared for electron microscopy at 7 days of age, when most synapses in the dentate molecular layer must be either recently formed or in the process of forming, polyribosomes were quite prominent under spines (Fig. 3). Not only do most of the visible spines have polyribosomes; most have more than one cluster. Polyribosomes are also more frequent within the dendrite proper, but they still seem to be preferentially localized under spine bases. We evaluated the incidence of polyribosomes during synapse development using quantitative electron microscopic techniques. Rats were prepared for routine electron microscopy at 7, 10, 15, 20, and 28 days of age (two animals at each time point, except at 15 days, when three animals were analyzed). From a collection of 10 000 X electron micrographs covering about 2100-2500 μm^2 of the middle molecular layer of the dorsal blade of the dentate gyrus, we determined (1) the density of synapses/100 μm^2 of neuropil, and (2) the proportion of spines with underlying polyribosomes. The values obtained from the developing animals were compared with the average values from a collection of adult animals which had been analyzed similarly. Figure 4 illustrates the increases in synapse density and the changes in the incidence of polyribosomes during development. We found the highest incidence of polyribosomes early in development; at 7 days of age an average of over 60% of the synapses had underlying polyribosomes. As is also the case in mature animals, this is probably an underestimate, since not all of the base of a given spine is contained in a single thin section. As synapse density increased, the proportion of spines with polyribosomes decreased toward levels characteristic of mature animals. Presumably, as synaptogenesis proceeds, the proportion of synapses which are recently formed or are in the process of forming decreases. The high incidence of polyribosomes in the youngest animals, and the decreases in the proportion of spines with polyribosomes as synaptogenesis proceeds again, suggest a role for the polyribosomes in synapse growth.

Association Between Polyribosomes and Other Organelles Under the Spine

Given the prominence of polyribosomes under growing spine synapses, it is of considerable interest to determine what types of proteins these elements might be producing. Some hints about this can be derived from the relationship between the polyribosomes and intracellular membranes. In general, polyribosomes which are bound to intracellular membranes (such as those forming the rough endoplasmic reticulum) synthesize membrane proteins or proteins destined for packaging and release from the cell (Palade 1975). Proteins produced by membrane bound polyribosomes are cotranslationally inserted into the membrane proper, or into the lumen of the endoplasmic reticulum. In

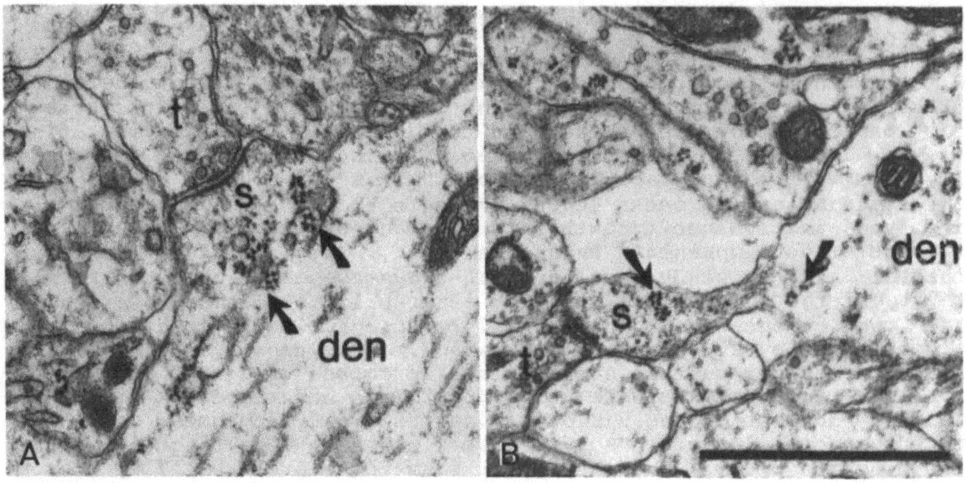

Fig. 3 A, B. Examples of polyribosomes under spines of 7-day-old rats, where most synapses are either growing or recently formed. Note the polyribosomes both under the base of the spine and within the head. The unlabeled calibration bar = 1 μm

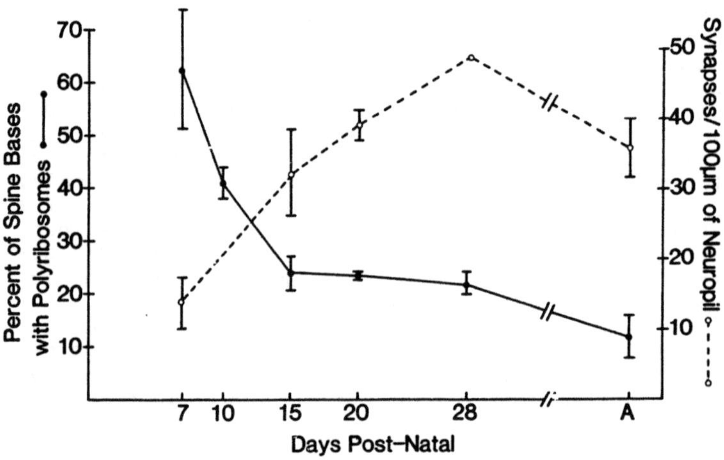

Fig. 4. The time course of synapse development in the dorsal blade of the dentate gyrus is revealed by plotting synapse density/100 μm^2 at each age (broken lines). The solid line indicates the incidence of polyribosomes under spine bases (percent of bases with polyribosomes) in the same series of animals. Vertical bars indicate the standard deviation (each point is the average of two animals, except at 15 days postlesion, when three animals were analyzed time point,

contrast, polyribosomes which are not membrane bound (free polyribosomes) tend to synthesize proteins for utilization within the cell (cytoplasmic or cytoskeletal proteins). There may be exceptions to this generalization, of course; some proteins seem to become associated with membranes after their synthesis (Bloebel 1981). Nevertheless, the rule is true for most proteins, and as a generalization can provide an indication of the probable type of protein being produced by different classes of polyribosomes.

Viewing of individual thin sections does not suggest a prominent relationship between spine-associated polyribosomes and intracellular membranes, although occasional profiles can be found which do suggest an association. For example, in the spine illustrated in Fig. 1A, the cluster of polyribosomes lies at the base of a membranous cistern which lies within the neck of the spine. Nevertheless, analysis of serial sections and thick sections viewed in stereo suggests that polyribosomes are often membrane associated. It becomes particularly apparent when viewing relatively thick sections (150-200 nm) in stereo that polyribosomes are often distributed around the end of thin cisterns which lie under the base or within the neck of the spine (see Fig. 5). These cisterns in turn can sometimes be seen to be in continuity with a spine apparatus when one is present. A quantitative estimate of the association between polyribosomes and cisterns was obtained by evaluating a collection of stereo photographs of 289 randomly selected polyribosomes from mid proximodistal dendrites of dentate granule cells (Steward 1983b). Of this collection, about half (49%) lay within one microtubule diameter of a sub-synaptic cistern. This is almost certainly an underestimate of the actual proportion of polyribosomes associated with cisterns. Because the cisterns tend to be quite small and because

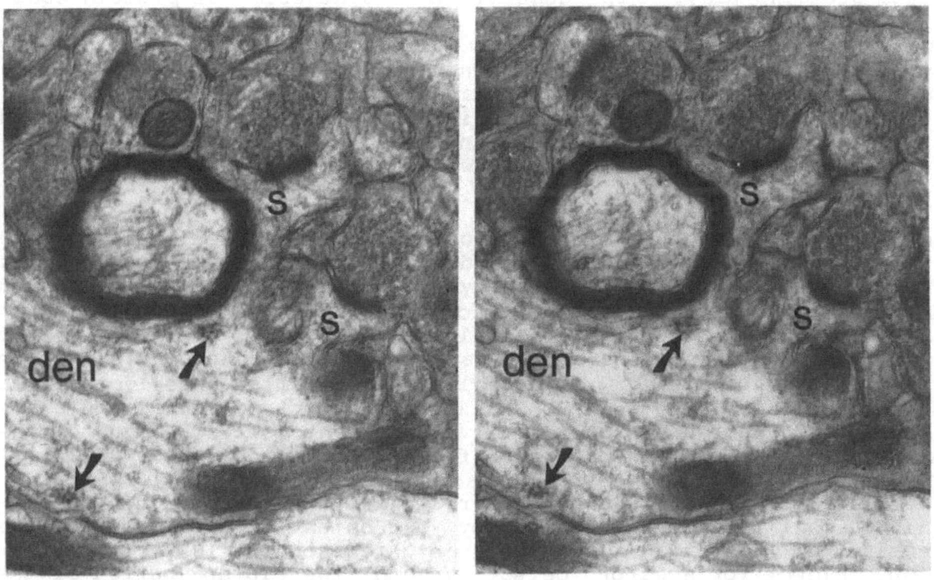

Fig. 5. Stereo pairs of spines from 200 nm thick sections prepared from tissue that had been perfused with tannic acid containing fixatives. Note the association between polyribosomes (arrows) and membranous cisterns under the spine base. Magnification, x 30 000

Fig. 6 A,B. Effects of intradentate injections of cycloheximide on ▶ the sprouting of AChE-containing septo-hippocampal fibers. A illustrates the increase in AChE staining in the denervated zone of the dentate gyrus in control animals which had received intradentate injections of saline. B illustrates the lack of intensification of AChE staining in animals treated with cycloheximide. AD = area dentata, RS = regio superior, arrows indicate the denervated zone. The long arrows indicate the edge of the defect produced by the cannula

the polyribosomes often appear to be distributed around the terminus of the cistern (see Fig. 5), section angle is critical for observing an association. Clusters on cisterns which are oriented perpendicular to the plane of section would not be seen to be associated with the membranes unless the section was fortuitous. Thus many clusters may well lie adjacent to membranous cisterns which are just out of the plane of a single section. For this reason, the proportion of polyribosomes associated with membranous cisterns may be much higher than the 50% estimate obtained from an analysis of thick sections. More accurate estimates will require serial reconstruction of a large number of spines.

Inhibition of Protein Synthesis in the Denervated Target Area: Effects on Lesion-Induced Growth

The increases in spine-associated polyribosomes during lesion-induced growth, and their prominence under developing synapses in young animals suggest that the proteins which they produce may be important in synapse formation. The crucial question is what role these proteins play. In general the synthesis of protein could either be ancillary to the process of synapse formation or necessary for some phase of it. We have attempted to explore this question by evaluating the effects of protein synthesis inhibitors on lesion-induced synaptogenesis in the dentate gyrus. If protein synthesis within denervated target cells is important for lesion-induced growth, then inhibiting protein synthesis within the denervated target area should block the growth within the area of effective inhibition.

Adult rats received unilateral entorhinal cortical lesions, and were simultaneously implanted with indwelling cannulae in the rostral dentate gyrus. Between 6 and 9 days postlesion the animals received daily intradentate injections of cycloheximide (10 μg cycloheximide in 1 μl saline). Control animals with similar lesions received injections of the saline vehicle alone, and a second control group without lesions received comparable cycloheximide injections. Ten days postlesion, the animals were prepared for acetylcholinesterase (AChE) histochemistry to reveal the extent of sprouting of the AChE-containing septo-hippocampal fibers (see Lynch et al. 1972).

As illustrated in Fig. 6A, the sprouting of AChE-containing septo-hippocampal fibers is revealed by a substantial increase in the density of AChE staining in the denervated zone at 10 days postlesion. Except for a minor defect around the tip and barrel of the cannula, the pattern of staining for AChE in control animals which had received saline injections resembled that seen in animals without cannulae. In contrast, there was a much less dramatic increase in staining in the animals which had received the intradentate cycloheximide injections between 6 and 9 days postlesion. The

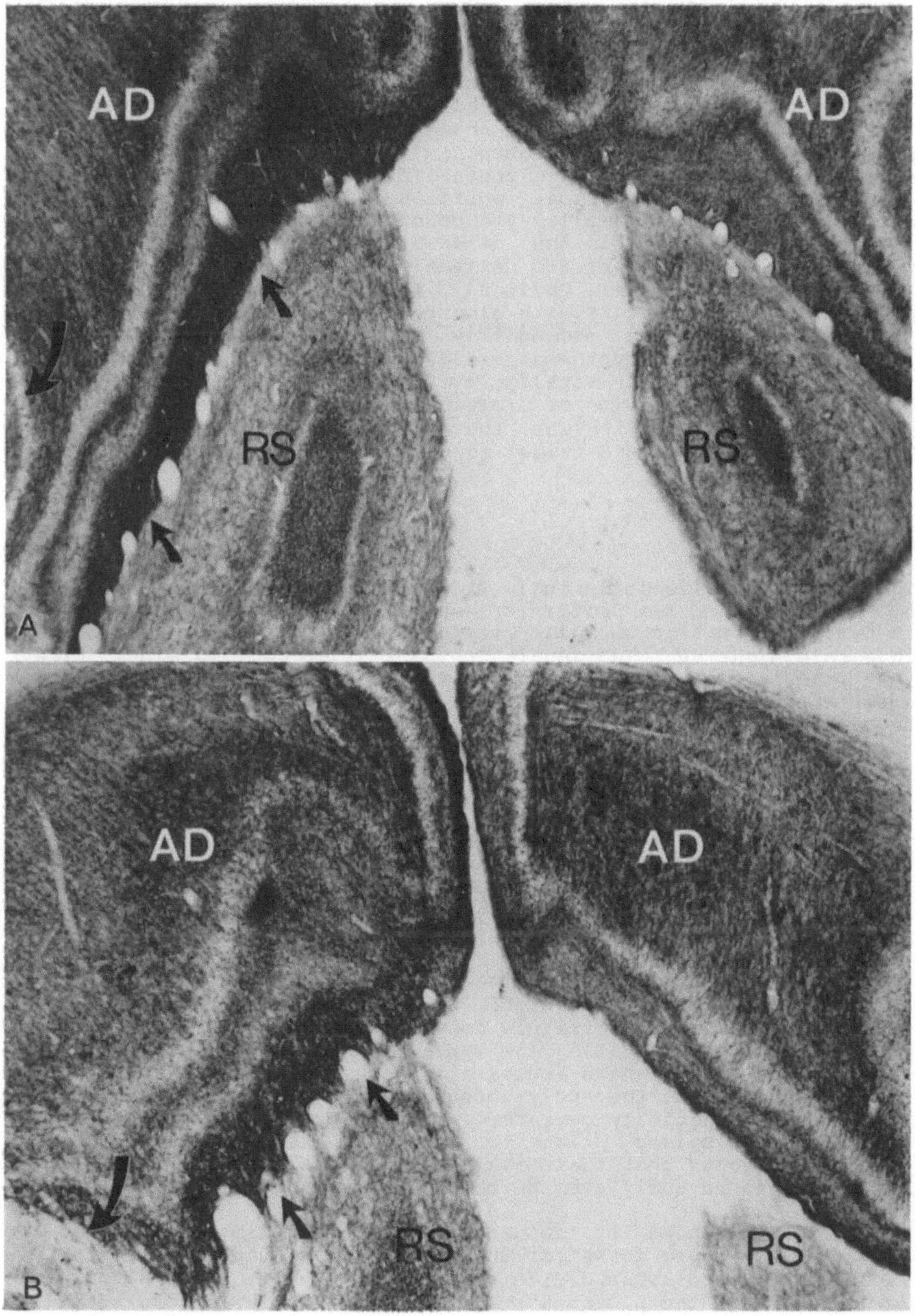
AD
AD
RS
RS
A
AD
AD
RS
RS
B

cycloheximide appeared to affect the lesion-induced growth, but not the normal pattern of innervation. The nondenervated laminae exhibited patterns of AChE staining which were comparable to those on the control side. Furthermore, cycloheximide injections in animals without lesions did not appreciably alter the pattern of AChE staining in the dentate gyrus (data not shown). The apparent inhibition of sprouting was a local phenomenon; the increase in AChE staining appeared to be unaffected in zones distant from the injection site (for example, in the caudal portions of the denervated dentate gyrus). All of these results are consistent with the interpretation that protein synthesis in the denervated target plays a role in the sprouting response. They are certainly not conclusive evidence for this hypothesis, however. Obviously, the daily inhibition of protein synthesis could interfere with almost any aspect of granule cell metabolism which could secondarily affect the sprouting response. Furthermore, protein synthesis would also be inhibited in other cell types, including glia. Finally, cycloheximide certainly has effects other than the inhibition of protein synthesis. At best, these data are suggestive. Nevertheless, the results are exactly what would be predicted if there was a target regulation of lesion-induced growth.

Discussion

The combined evidence strongly suggests that the spine-associated polyribosomes and the protein(s) which they produce play some role in synapse formation on spine-bearing neurons. The question is what this role might be. Two general possibilities come to mind. First, the presynaptic axon could induce the accumulation of polyribosomes under the site of synaptic contact. This would be a case of afferent regulation of the target. Second, polyribosome distribution could be regulated by the postsynaptic cell, and the proteins produced by these elements could induce, permit, or otherwise regulate the formation of the synaptic contact. This would be an example of target regulation of the afferents.

As we have argued elsewhere (Steward 1983a), the temporal relationships between the increases in polyribosomes and synapse formation seem more consistent with the interpretation that the accumulations of polyribosomes are regulated by the postsynaptic cell. In the case of lesion-induced growth, the increases in polyribosomes under spines are maximal just at the onset of the presynaptic growth response (6-8 days postlesion), before reinnervation has occurred (Steward 1983a). This certainly does not indicate that the afferents play no role in regulating the synthetic activity of the polyribosomes. Indeed, we have argued elsewhere (Steward and Levy 1982; Steward 1983b) that the polyribosomes seem optimally positioned to be affected by activity over the synapse. Nevertheless, their accumulation under spines prior to the time that reinnervation has occurred suggests that their distribution is altered by the postsynaptic cell as an early step in the reinnervation process.

Protein(s) produced by spine-associated polyribosomes could be involved in synapse formation in a number of ways. As noted previously (Steward 1983a,b), in the case of synapse replacement following lesions, the proteins produced by the polyribosomes could be involved in the degradation of old areas of membrane specialization, or other processes related to the deterioration of the synapse, the spine, or the dendrite. These hypotheses for a role of the polyribosomes in degradative processes are made much less likely by the observation that polyribosomes are even more prominent under growing spine synapses in developing animals. Thus, it seems more likely that the polyribosomes are involved in synapse construction. In this regard, again a number of possibilities exist. The polyribosomes could be involved in the construction of new postsynaptic specializations, or in the production of some protein associated with the site of the synaptic contact. They could also be involved in the rebuilding of the cytoskeletal scaffolding associated with the synapse. There is growing evidence that the cytoskeleton of the spine is quite different from that of the parent dendrite (Caceres et al. 1983, 1984); how the differential composition is produced is at present not clear. Obviously, one possibility is the local synthesis of some component of the spine cytoskeleton.

Another very attractive possibility is that the proteins produced by the polyribosomes might be involved in inducing the presynaptic growth or the synapse formation by the presynaptic processes. One way that this could take place is through the production of a protein for the postsynaptic specialization. Such a protein could form part of a recognition or adhesion complex which would induce or permit the formation of the synapse. Alternatively, the protein(s) produced by the polyribosomes might be packaged and released from the spine to affect the growth of nearby presynaptic axons. The fact that protein synthesis inhibitors block postlesion growth would be consistent with either of these interpretations. For the latter possibility, it is interesting to note that there is a large increase in the amount of neuronotrophic factors (NTFs) in the hippocampal formation after entorhinal cortical lesions in addition to an accumulation of factors in the wound cavity (Nieto-Sampedro et al. 1983). The time course of the increases in the concentration of at least some of these NTFs is quite similar to the time course of the increases in polyribosomes under spines.

Whatever the specific relationship between the formation of synapses, the changes in polyribosomes, and the changes in protein synthesis within the postsynaptic cell, it is clear that the postsynaptic cell is not simply a passive participant in synaptogenesis. During developmental synaptogenesis, it is certainly growing in concert with the presynaptic axons. During postlesion growth, the postsynaptic cell appears to actively restructure its dendritic specializations (particularly spines) and alters the cytoplasmic microenvironment of the spine during the period of reinnervation. All of these results suggest that the target cell regulates synapse formation along its receptive surface. We believe that one way that it may accomplish this is through the positioning of the polyribosomes under synaptic sites under construction. In this way, it becomes more reasonable to think of the postsynaptic cell regulating the construction of individual synapses at various sites, even if tens of thousands of synapses are involved, and even though many of these lie far away from the principal synthetic machinery in the cell body.

While the present evidence clearly suggests an involvement of the spine-associated polyribosomes in synapse growth, these elements are also present under spines in mature animals. Indeed, they are a ubiquitous component of spines on all CNS neurons which have been evaluated (Steward 1983b). The presence of polyribosomes under spines in mature animals does not contradict the suggestion that they are involved in synapse growth. Indeed, there is a growing consensus that synapse growth is an ongoing process in the undamaged CNS, and that growth following lesions is simply an amplification of this process (see Cotman et al. 1981). In fact, synaptic connections, like their constituent proteins, may be continually turning over. As we have noted elsewhere (Steward and Levy 1982; Steward 1983b) the localization of polyribosomes under spine bases seems ideal for the production of proteins for the individual synapse with which they are associated. Furthermore, such a localization also seems optimal for a regulation of synthesis by functional activity over the synapse. This would seem to represent an ideal substrate for activity-dependent modification of the synapse, a process that certainly is important for the fine-tuning of synaptic connections during development, and for the modifcation of connections in response to experience in the mature nervous system. Thus, the elucidation of the nature of the proteins being produced at the synapse, and the definition of their role, may provide insights not only into the cellular and molecular mechanisms of synapse growth but also into the mechanisms of synapse maintenance and modification throughout the life of the organism.

Acknowledgments. I thank S. Vinsant and P.M. Falk for their excellent technical help in the electron microscopic studies of synapse replacement following lesions (S. Vinsant) and synapse growth during development (P.M. Falk). Thanks also to Hakon Heimer for his excellent help in the studies of the effects of cycloheximide on sprouting, to P. Palmer for technical help in the development of the tannic acid procedures which were used for preparing tissue for stereo viewing, to Lee Snavely for photographic help, and to M.P. Janssen for excellent secretarial help. This work was supported by National Institute of Health Grant NS 12333. The author was the recipient of a Research Career Development Award NS 00325 during the time that much of this work was undertaken.

References

Blöbel G (1981) Regulation of intracellular protein traffic. Cold Spring Harbor Symp Quant Biol 46:7-16

Caceres A, Steward O (1983) Dendritic reorganization in the denervated dentate gyrus of the rat following entorhinal cortical lesions: a Golgi and electron microscopic analysis. J Comp Neurol 214:386-403

Caceres A, Payne MR, Binder LI, Steward O (1983) Immunocytochemical localization of actin and microtubule associated protein (MAP2) in dendritic spines. Proc Natl Acad Sci USA 80:1738-1742

Caceres A, Binder LI, Payne MR, Bender P, Rebhun L, Steward O (1984) Differential subcellular localization of tubulin and the microtubule associated protein MAP2 in brain tissue as revealed by immunocytochemistry. J Neurosci 4:394-410

Changeux J-P (1983) Concluding remarks: On the "singularity" of nerve cells and its ontogenesis. In: Changeux J-P, Glowinski J, Imbert M, Bloom FE (eds) Molecular and cellular interactions underlying higher brain functions. Prog Brain Res 58:465-478

Cotman CW, Nieto-Sampedro M, Harris EW (1981) Synapse replacement in the nervous system of adult vertebrates. Physiol Rev 61:684-784

Crutcher KA, Collins F (1982) In vitro evidence for two distinct hippocampal growth factors: Basis of neuronal plasticity. Science 217:67-68

Crutcher KA, Davis JN (1981) Sympathohippocampal sprouting is directed by a target factor. Brain Res 204:410-414

Crutcher KA, Davis JN (1982) Target regulation of sympathetic sprouting in the rat hippocampal formation. Exp Neurol 75:347-359

Droz B, Leblond CP (1963) Axonal migration of proteins in the central nervous system and peripheral nerves as shown by radioautography. J Comp Neurol 121:325-346

Fass B, Steward O (1983) Increases in protein precursor incorporation in the denervated neuropil of the dentate gyrus during reinnervation. Neuroscience 9:633-664

Hadley RD, Kater SB, Cohan CS (1983) Electrical synapse formation depends on interaction of mutally growing neurites. Science 221:466-468

Lee KS, Stanford EJ, Cotman CW, Lynch GS (1977) Ultrastructural evidence for bouton proliferation in the partially deafferented dentate gyrus of the adult rat. Exp Brain Res 29:475-485

Levi-Montalcini R (1976) The nerve growth factor: its role in growth, differentiation and function of the sympathetic adrenergic neuron. Prog Brain Res 45:235-258

Lynch G, Matthews DA, Mosko S, Parks T, Cotman C (1972) Induced AChE-rich layer in rat dentate gyrus following entorhinal lesions. Brain Res 42:311-318

Matthews DA, Cotman C, Lynch G (1976) An electron microscopic study of lesion-induced synaptogenesis in the dentate gyrus of the adult rat. I: Magnitude and time course of degeneration. Brain Res 115:1-21

Nieto-Sampedro M, Manthorpe M, Barbin G, Varon S, Cotman CW (1983) Injury-induced neuronotrophic activity in adult rat brain: correlation with survival of delayed implants in the wound cavity. J Neurosci 3:2219-2229

Palade G (1975) Intracellular aspects of the process of protein synthesis. Science 189:347-358

Raisman G (1969) Neuronal plasticity in the septal nuclei of the adult rat. Brain Res 14:25-48

Steward O (1983a) Alterations in polyribosomes associated with dendritic spines during the reinnervation of the dentate gyrus of the adult rat. J Neurosci 3:177-188

Steward O (1983b) Polyribosomes at the base of dendritic spines of CNS neurons: their possible role in synapse construction and modification. Cold Spring Harbor Symp Quant Biol 48:745-759

Steward O, Fass B (1983) Polyribosomes associated with dendritic spines in the denervated dentate gyrus: evidence for local regulation of protein synthesis during reinnervation. In: Changeux J-P, Glowinski J, Imbert M, Bloom FE (eds) Molecular and cellular interactions underlying higher brain functions. Prog Brain Res 58:131-136

Steward O, Levy WB (1982) Preferential localization of polyribosomes under the base of dendritic spines in granule cells of the dentate gyrus. J Neurosci 2:284-291

Steward O, Vinsant SL (1983) The process of reinnervation in the dentate gyrus of the adult rat: a quantitative electron microscopic analysis of terminal proliferation and reactive synaptogenesis. J Comp Neurol 214:370-386

Thoenen H, Otten U, Schwab M (1979) Orthograde and retrograde signals for the regulation of neuronal gene expression: the peripheral sympathetic nervous system as a model. In: Schmitt FO, Worden FG (eds) The neurosciences fourth study progam. MIT Press, Cambridge, pp 911-928

Vaughn JE, Sims TJ (1978) Axonal growth cones and developing axonal collaterals from synaptic junctions in embryonic mouse spinal cord. J Neurocytol 7:337-363

Acetylcholine Receptor Turnover Rates at Innervated and Denervated Vertebrate Neuromuscular Junctions

M. M. Salpeter

Section of Neurobiology and Behavior, W113 Seeley Mudd Hall, Cornell University, Ithaca, NY 14853, USA

Junctional receptors at innervated adult muscle are present at high density (~18 000 sites/μm^2) in a specialized region on the top 2000 Å of the postjunctional folds. They are essentially immobile (Fertuck et al. 1975; Axelrod et al. 1976) and have a slow turnover time ($t_{1/2}$~ 8 - 10 days; see review by Fambrough 1979). These and other properties of the junctional acetylcholine receptor (AChR) are significantly different from those of extrajunctional AChR on embryonic or denervated muscles, which are diffusely located (~400 sites/μm^2), are mostly freely mobile in the plane of the membrane, and have a half-life of 1 day as well as having other distinguishing physiological and pharmacological properties (see also review by Frank and Fischbach 1979 and recent review by McArdle 1983).

Work in several laboratories (including our own) has shown that the properties which distinguish the mature "junctional" from "extrajunctional" or embryonic receptors arise during development in an asynchronous manner (e.g., Burden 1977) and are not all equally affected by denervation. For example, the high-density receptor clustering at the specialized region of the postsynaptic membrane, which is initially induced by the nerve, does not revert after denervation (Loring and Salpeter 1980) and may respond to cues permanently imparted to the basal lamina (Burden et al. 1979). Metabolic stability, on the other hand, requires the presence of the nerve to be maintained, takes longer to develop, and is reversed after denervation but in a complex manner. In this talk I will describe some studies from our laboratory dealing with aspects of receptor turnover rates at innervated and denervated neuromuscular junctions (nmj's).

Using quantitative EM autoradiography and gamma counting, we first showed that after denervation there is a progressive increase in the degradation half-life of junctional AChRs and that by 16 days after denervation the half-life was about 30 h and thus close to the value for embryonic extrajunctional receptors (Levitt et al. 1980). We then asked whether the progressive increase in half-life reflects a continuous increase in turnover rate of all the junctional AChRs, or whether it reflects a changing ratio of two populations of receptors: (1) the "original" receptors preexisting at the time of denervation, which degrade at the slow junctional rate, and (2) the "new" receptors, which are inserted at the denervated junction to replace the degrading original receptors and which have the same turnover rate as embryonic receptors. By labeling the receptors before cutting the nerve, we first established that the turnover rate of the original receptors changed very little, remaining essentially stable with a half-life of about 8 days, for about 10 days after denerva-

Experimental Brain Research, Suppl. 13

tion and then accelerated to a half-life of about 3 days (Levitt and Salpeter 1981). Similar results were also reported from other laboratories (Stanley and Drachman 1981; Brett et al. 1982) and subsequently confirmed more elegantly by Bevan and Steinbach (1983), who found that even during the first 10 days after denervation there was a slight acceleration in the degradation rate of original receptors.

After modeling the information regarding the degradation of the original receptors and that of the total receptors reported by us earlier (Levitt et al. 1980), we concluded that the progressive increase in turnover rate after denervation does indeed result in a replacement of the original receptor with a population of new receptors that has a half-life of 1 day. This suggests that receptors inserted at the denervated adult nmj never become metabolically stable in the absence of the nerve (Levitt and Salpeter 1981). Subsequently we confirmed this finding directly: just before denervating the muscle, we inactivated all junctional receptors with nonradioactive α-bungarotoxin (BTX) [the irreversible inhibitor of the acetylcholine binding site of the AChR (Chang and Lee 1963).] We then used 125-BTX to label only the new receptors that had been inserted in the membrane during a 6-day period. These new receptors did indeed have a half-life of 20 h.

The studies described above showed that clustering and turnover were independently controlled in the adult nmj as they are during development, and that although metabolic stabilization required the presence of the nerve, once receptors had become stabilized they remained stable for a considerable time after denervation. Furthermore, since more stable (original) and less stable (new) receptors can coexist in a denervated nmj, metabolic stability is probably a property of the individual receptor or its microenvironment.

Recently Tami Levitt-Gilmour and I have examined the effect of reinnervation on receptor stabilization. We found that at the time of reinnervation receptors already in the membrane can be metabolically stabilized by a regenerating nerve.

The denervated nmj thus provides a model system for the study of receptor turnover and stabilization. It is morphologically easy to identify and visualize both in vivo and in vitro; it has a very high concentration of receptors at all times after denervation independent of its turnover rate, and it has a precise timetable which governs the turnover rates of two different populations of receptors. In addition it is easy to label the total population of receptors or each subpopulation separately, allowing each to be studied independently. Future studies using this system should clarify our general understanding of the membrane receptors and how they can be controlled.

References

Axelrod D, Ravdin P, Koppel DE, Schlessinger J, Webb WW, Elson EL, Podleski TR (1976) Lateral motion of fluorescently labeled acetylcholine receptors in membranes of developing muscle fibers. Proc Natl Acad Sci USA 73:4595-4598

Bevan S, Steinbach JH (1983) Denervation increases the degradation rate of acetylcholine receptors at endplates in vivo and in vitro. J Physiol (Lond) 336:159-177

Brett RS, Younkin SG, Konieczkowski M, Slugg RM (1982) Accelerated degradation of junctional acetylcholine receptor-α-bungarotoxin complexes in denervated rat diaphragm. Brain Res 233:133-142

Burden S (1977) Acetylcholine receptors at the neuromuscular junction: developmental change in receptor turnover. Dev Biol 61:79-85

Burden SJ, Sargent PB, McMahan UJ (1979) Acetylcholine receptors in regenerating muscle accumulate at original synaptic sites in the absence of the nerve. J Cell Biol 82:412-425

Chang CC, Lee CY (1963) Isolation of neurotoxins from the venom of Bungarus multicinctus and their modes of neuromuscular blocking action. Arch Int Pharmacodyn Ther 144:241-257

Fambrough DM (1979) Control of acetylcholine receptors in skeletal muscle. Physiol Rev 59:165-227

Fertuck HC, Woodward WW, Salpeter MM (1975) In vivo recovery of muscle contraction after α-bungarotoxin binding. J Cell Biol 66:209-213

Frank E, Fischbach GD (1979) Early events in neuromuscular junction formation in vitro. J Cell Biol 83:143-158

Levitt TA, Salpeter MM (1981) Denervated endplates have a dual population of junctional acetylcholine receptors. Nature 291:239-241

Levitt TA, Loring RH, Salpeter MM (1980) Neuronal control of acetylcholine receptor turnover rate at a vertebrate neuromuscular junction. Science 210:550-551

Loring R, Salpeter MM (1980) Denervation increases turnover rate of junctional acetylcholine receptors. Proc Natl Acad Sci USA 77:2293-2298

McArdle JJ (1983) Molecular aspects of the trophic influence of nerve on muscle. Prog Neurobiol 21:135-198

Stanley EF, Drachman DB (1981) Denervation accelerates the degradation of junctional acetylcholine receptors. Exp Neurol 73:390-396

Models for Axonal Growth

Neuronal Interactions Between Peripheral Sensory and Central Monoamine Neurons Grafted to the Anterior Eye Chamber

Å. Seiger and C. Ayer-LeLievre

Department of Histology, Karolinska Institutet, Box 60400, 10401 Stockholm, Sweden

Introduction

The scope of this presentation is to describe ways by which central monoamine neurons have been made to interact with various sources of peripheral nerve entities. Spatial and temporal nerve fiber interaction and possible reciprocal innervation of sequentially transplanted nervous tissues have been studied. Strong attractive forces for nerve fiber outgrowth and pattern formation have been revealed by these studies.

The Eye Chamber Model

Transplantation of nervous tissues to the anterior eye chamber has been performed for more than 50 years (Sartori 1926). It has become a useful model for studies of a variety of developmental and regenerative processes in both peripheral and central neurons. Upon grafting, the nervous tissue attaches to the anterior surface of the host iris and becomes revascularized within a couple of days. Neuronal tissue survives for as long as the recipient lives: up to years.

The iris has a neuronal input constituted of the adrenergic sympathetic and cholinergic parasympathetic parts of the autonomic ground plexus as well as sensory trigeminal innervation. Grafted neuronal tissue can be made to interact with all these endogenous iris nerves or any specific set of nerves thereof after selective denervations of the iris. Histological analysis of fiber growth from grafted neurons can be performed in sections of the graft neuropil and in large details in the host irides prepared as whole mounts. In such conditions the nerve fibers in the thin iris tissue can be visualized as a two-dimensional plexus. Against the background of intrinsic fibers, the outgrowth of fibers from the graft can be recognized as a halo of radiating fibers and/or network of fibers and axon bundles of varying patterns. Innervation of the grafts by intrinsic iris nerves can be studied in sections.

As an alternative, when sequential grafting of two or more neuronal tissues, peripheral or central, is performed, their reciprocal innervation or interactions on the iris can be studied. For details on the intraocular transplantation technique, see Olson et al. (1983).

Experimental Brain Research, Suppl. 13

Sensory Ganglia

Various sensory ganglia from fetal, neonatal, and adult donors have been grafted. Their neurons can survive for several months in the anterior eye chamber. General fiber outgrowth and cell morphology have been visualized by Linder's silver staining (Linder 1978) both in sections of grafts and in whole-mount preparations of stretched irides (Ayer-LeLievre et al. 1984). Ganglia containing neurons of neural crest and/or placodal presumed origins have been grafted (dorsal root, superior jugular and trigeminal or nodose and petrous ganglia). No obvious difference in the fiber outgrowth or neuron survival has been noted.

After intraocular maturation most neuronal perikarya present a normal appearance. Large and smaller cell bodies are present, some being richer in silver-stained neurofilaments as in normal sensory ganglia. It can be observed in isolated neurons or whole mounts of irides that some neurons retain their pseudo-unipolar morphology (Figs. 1, 3). The number of neurons surviving after transplantation is generally higher when the ganglia are taken from the adult or neonatal rats than at very early fetal stages, though in such cases a limited number of neurons may produce a relatively large outgrowth of fibers.

All types of ganglia can produce the same kind of nerve fiber outgrowth which is superimposed on the intrinsic sensory innervation, producing sensory hyperinnervation locally.

Close to the ganglia a dense halo of thin and thick fibers is generally observed. Thick myelinated fibers form fascicles and large bundles (Fig. 2) running for large distances in parallel and diverging at the periphery. These fascicles preferentially grow towards the sphincter, where they spread along the sphincter margin (Fig. 5). The graft origin of such a plexus of fibers can be confirmed experimentally. They are the only silver-stained fibers left on the host iris after trigeminal lesion of the host, which results in degeneration of intrinsic myelinated fibers. Alternatively, short-term (2-day) intraocular transplantation of the grafted irides results in their complete intrinsic denervation.

In sensory ganglia a varying small percentage (<20%) of neurons are immunoreactive to substance P antisera (SPLI), as evidenced by indirect immunofluorescence technique (Coons 1958). In grafted sensory ganglia a minor proportion of surviving neurons are also SPLI (Fig. 3). They produce a sparse plexus of delicate smooth and varicose fibers (Fig.4). These thin fibers form a halo around the graft and may form a network on the dilator plate but do not present the same pattern as the plexus of myelinated fibers from the graft.The graft-derived SPLI fibers are best visualized in hosts treated systemically with capsaicin before the operation (Fig. 6). This results in the disappearance of endogenous SPLI nerves (Jessell et al. 1978) of the host iris and allows quantitative studies of SPLI nerve fiber outgrowth from the graft in various experimental conditions.

Interaction Between Central Monoamine Neurons and the Intrinsic Iris Innervation

Fetal noradrenergic neurons from the nucleus locus coeruleus (LC), dopaminergic neurons from substantia nigra, and serotonergic neurons from the nucleus raphe dorsalis all survive intraocular grafting and

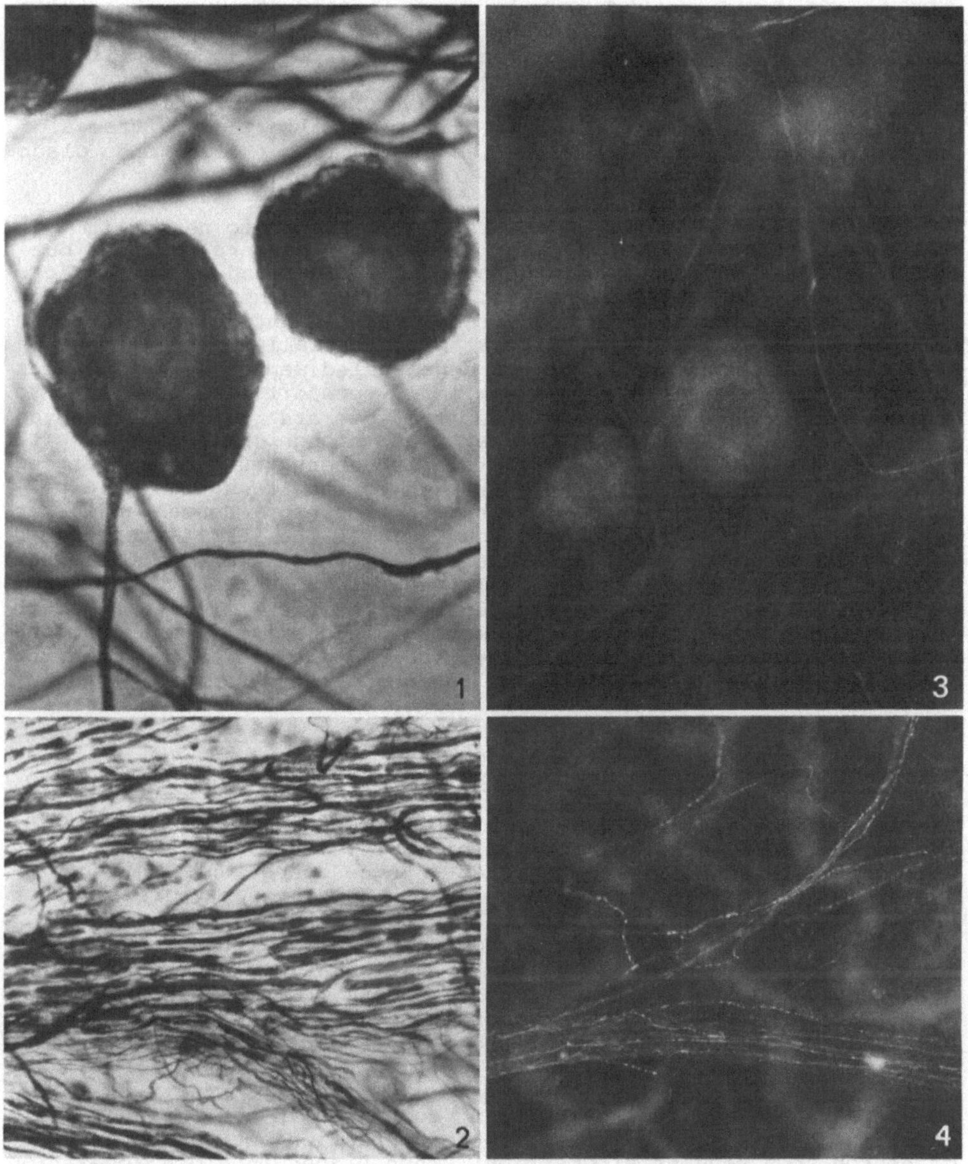

Fig. 1. Large sensory cell bodies from a superior-jugular ganglion maturated in oculo. Iris whole mount, silver-stained, X 1300

Fig. 2. Mixed fascicles of thick myelinated and thin unmyelinated fibers from cografted fetal trigeminal ganglion and locus coeruleus. Iris whole mount, silver-stained, X 540

Fig. 3. Two SPLI neurons from a grafted nodose ganglion. The larger cell appears pseudo-unipolar. Immunohistochemistry, fluorescein, X 700

Fig. 4. Sparse plexus of thin SPLI fibers (smooth and varicose) from a grafted nodose ganglion. The host had been capsaicin treated; intrinsic SPLI fibers have disappeared. Immunohistochemistry, rhodamine, X 330 produce a halo of nerve fibers in the host iris close to the brain graft attachment, as evidenced by Falck-Hillarp histofluorescence (Falck et al. 1962; Corrodi and Jonsson 1967). The fiber growth will continue linearly for 5-6 weeks, after which it ceases. A varicose plexus partly resembling the pattern of autonomic nerves is formed, intermingled with a few fascicles of axons following the tracks of the endogenous sensory nerve bundles of the iris (Seiger and Olson 1977a).

The presence of autonomic adrenergic or cholinergic nerves in the host iris does not influence the amount or pattern of fiber outgrowth from the central monoamine neurons (Seiger and Olson 1977c). Upon a sensory denervation, however, maturated LC neurons are reinitiated to grow in the iris and spread over the entire iris surface. The growth is rapid and the fibers preferentially form large axon fascicles along the tracks of the degenerated sensory axon bundles (Fig. 8) (Olson et al 1978). A similarly spread fiber ingrowth into denervated irides can be achieved if irides are grafted into eyes already carrying a maturated LC graft and brought into contact with the brain tissue. The iris graft, totally denervated by the grafting procedure, becomes rapidly reinnervated by the LC neurons in a pattern largely mimicking the distribution of sensory nerve bundles in the iris (Fig. 9) (Seiger and Olson 1977b). The reinitiated growth is directed only into the denervated iris graft, leaving the fiber halo in the host iris unchanged. In view of these results it seems highly probable that the sensory denervation itself elicits a growth stimulation of the LC neurons, directing the growing fibers into the denervated tissue and guiding the individual fibers to form axon fascicles, following the tracks of the degenerated sensory axons.

Combination of Monoamine Neurons and Sensory Ganglia (Fig. 11)

The directed reinitiated growth of LC fibers into an iris transplant placed on an LC graft can be inhibited if this iris transplant is not completely sensorily denervated by the operation. This can be achieved if the iris transplant itself has previously received a trigeminal ganglion graft (1 month before). This will maintain a secondary sensory innervation in some areas of the iris transplant. Several weeks after this last iris transplantation, secondarily sensory innervated areas did not receive fibers from the cografted LC. Reinitiation of the growth or noradrenergic fibers took place but was directed only into areas of the iris transplant lacking trigeminal fibers (Seiger 1980).

Locus coeruleus grafts have been combined in oculo with sensory ganglion grafts to study the direct interaction between both types of neuron upon sequential grafting. When placed in close apposition to each other, the halo of LC fibers growing out in the area of the iris where maturated trigeminal neurons have already formed fascicles will change distribution entirely. The LC fibers follow the routes of the trigeminal fibers, as evidenced by Falck-Hillarp histofluorescence (Fig. 7) followed by Linder's silver staining of the same specimens (cf. Fig. 5). In areas of the iris where trigeminal fascicles are not present, the distribution of LC fiber outgrowth is not altered.

The trigeminal ganglion grafts became densely innervated by the LC-derived fibers. Individual perikarya were approached by several fluorescent varicose nerves and the myelinated axon bundles traversing the ganglion graft were all accompanied by LC fibers. This het-

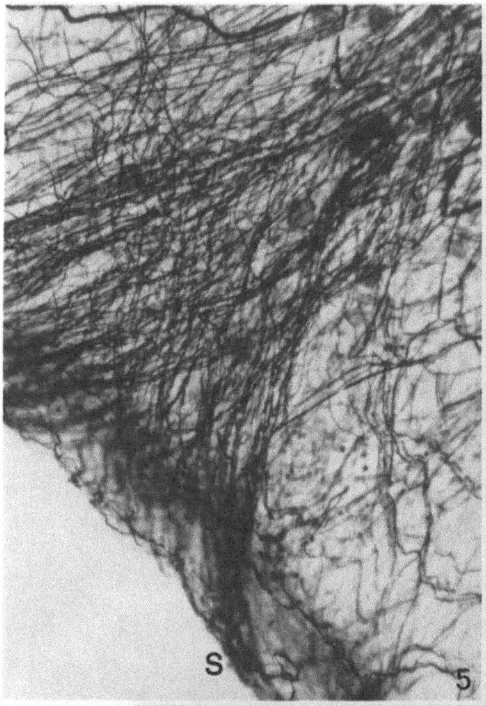

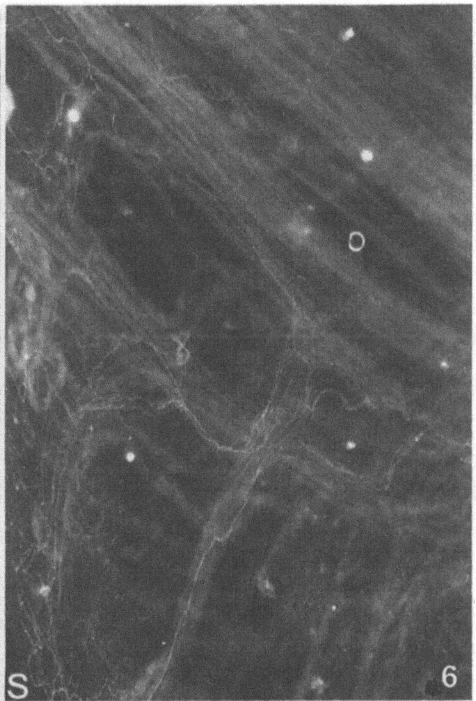

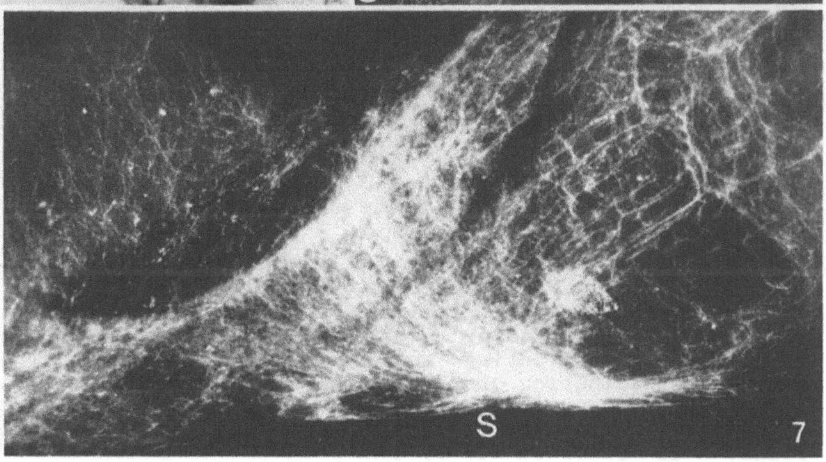

Fig. 5. Dense plexus of fibers from a grafted trigeminal ganglion. Large fascicles have grown towards the sphincter and spread along its margin(s). Silver staining, X 160

Fig. 6. Sparse plexus of SPLI fibers from a grafted nodose ganglion on the iris of a capsaicin-treated rat. The pattern of thin SPLI fibers in the sphincter area is different from that of thick myelinated fibers seen in background. Immunofluorescence, rhodamine, X 160

Fig. 7. Outgrowth of monoaminergic fibers from an LC cografted with a nodose ganglion. Parallel fluorescent thin fibers from the ganglion are oriented into the sphincter area, thus mimicking the outgrowth pattern of the sensory ganglion on the iris (cf. Fig. 5). Falck-Hillarp fluorescence, X 200

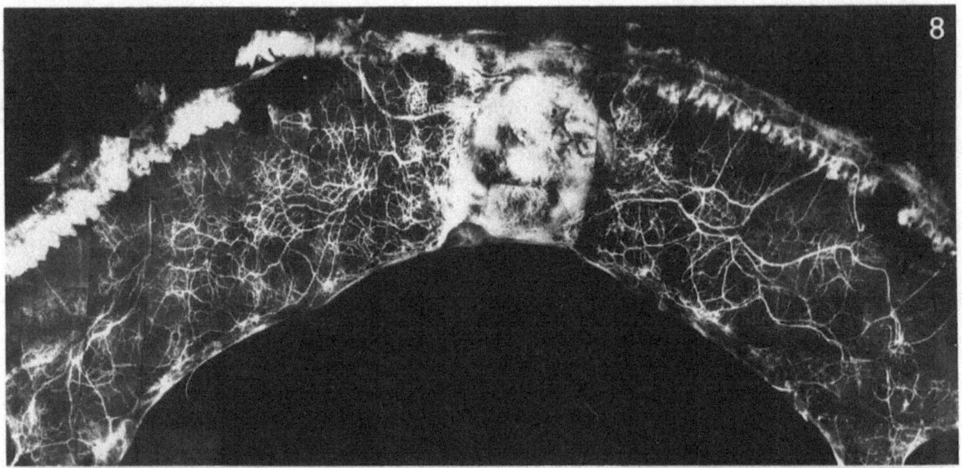

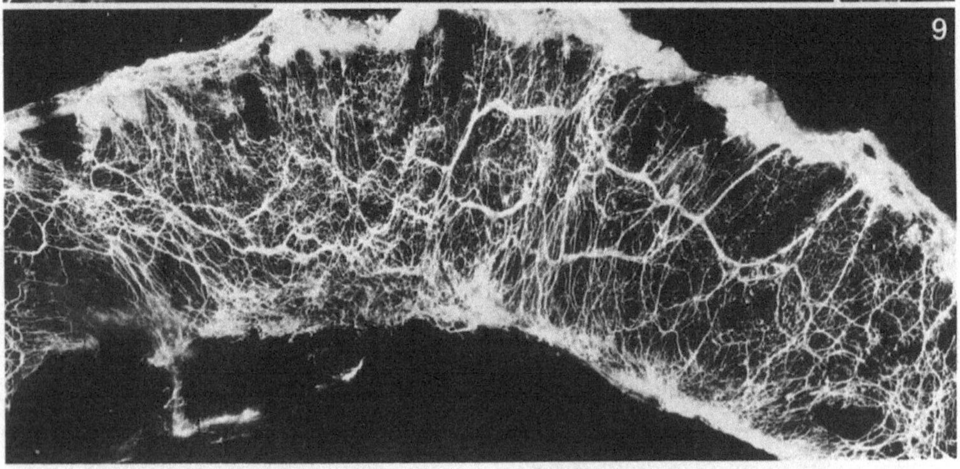

Fig. 8. Survey of an iris whole mount. Brightly fluorescent area (center) represents the LC transplant with heavily packed fluorescent noradrenaline-containing nerve fibers. Three weeks after trigeminotomy the fluorescent monoamine nerve fibers are here seen to spread out over the entire surface of the host iris. This differs from the distribution of the control specimens, where a restricted halo of fluorescent fibers is found centered around the CNS tissue attachment. Montage of approximately 200 fluorescence microphotographs, X 15 (Olson et al. 1978)

Fig. 9. Overview of a whole iris transplant that has been attached to a mature LC transplant for 1 month in oculo. The entire iris transplant surface is covered with a dense terminal plexus of varicose fibers intermingled with many thick axon bundles which partly mimics the normal sensory pattern of innervation. Montage of 120 fluorescence microphotographs of an iris whole mount, X 30 (Seiger and Olson 1977b)

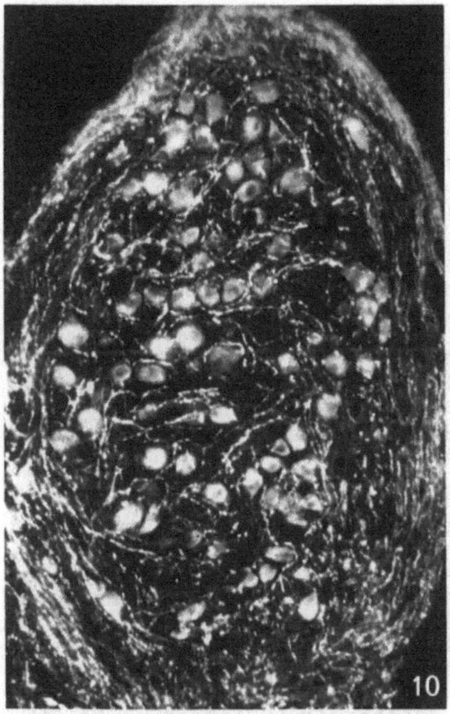

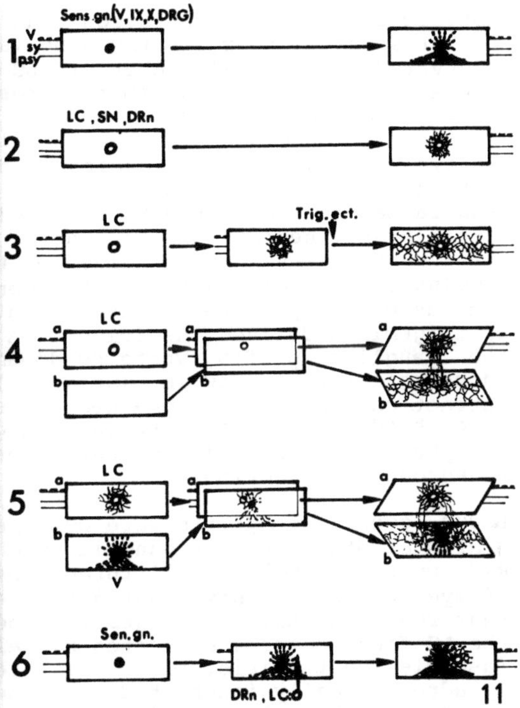

Fig. 10. Intraocularly maturated nodose ganglion graft with a dense, strongly fluorscent innervation from the cografted LC. The noradrenaline-containing fibers are in close apposition both with ganglion perikarya and their myelinated fibers in the periphery of the graft. Falck-Hillarp fluorescence, X 225

Fig. 11. Schematic representation of the various experimental procedures. 1. Single sensory ganglion graft to an intact eye. Outgrowth of axon fascicles of myelinated and unmyelinated fibers, preferentially into the sphincter. 2. Single central monoamine neuron graft to an intact eye. Restricted halo of varicose plexuses intermingled with a few axon bundles. 3. Single LC graft to an eye where, after neuronal maturation, a trigeminal lesion is performed. Reinitiated growth of the LC neurons to cover the entire iris surface, preferentially as axon fascicles. 4. Iris transplant placed on an already maturated LC graft. Directed reinitiated outgrowth only into the iris transplant, with a spread distribution preferentially as axon fascicles. No change of the restricted halo in the host iris (cf. 2 above). 5. Like 4 but with an iris graft already carrying a maturated trigeminal graft with fascicle outgrowth. Reinitiated directed ingrowth into all areas of the iris graft except those covered by the trigeminal graft fascicles. 6. Sequential grafting of central monoamine neurons to an eye with an already maturated sensory ganglion graft. The monoamine fiber outgrowth takes on the course of the sensory axon fascicles wherever they make contact. In areas of the iris where they do not make contact, the monoamine fibers form varicose plexuses (cf. 2 above) erotypic adrenergic innervation was directed and selective, since the iris area to which the ganglia were attached did not become innervated. The total number of LC-derived fibers in the iris was, however, not increased in the presence of a trigeminal ganglion graft as compared to single control LC grafts (Seiger 1980). When serotonergic neurons from the dorsal raphe nucleus were combined in the eye with maturated trigeminal ganglia, a similar guiding of the raphe-derived fiber network in the iris was seen. Similarly, the ganglia became innervated by such yellow fluorescent fibers. Dopaminergic nigral neurons and peripheral sympathetic adrenergic ganglion cells were, however, not guided by the fibers from a simultaneously present trigeminal graft. On the contrary, these monoaminergic neurons seemed to preferentially inervate areas of the host iris devoid of trigeminal-derived fascicles. A similar difference was seen in the ganglion grafts, where no dopaminergic nigral or sympathetic adrenergic fibers grew in (Seiger 1983). Thus, there is a marked attraction of LC and raphe neurons exerted by the cell bodies and fiber bundles of grafted trigeminal ganglia. Nigral dopaminergic and peripheral sympathetic adrenergic neurons are not attracted by such a neuronal interaction.

Locus coeruleus grafts were combined with intraocularly maturated grafts of other sensory ganglia such as superior-jugular, nodose, and petrous ganglia. All three types of ganglia produced fascicles of myelinated and unmyelinated fibers, in patterns rather similar to that found for trigeminal grafts. There was a conspicuous guiding of the LC fiber outgrowth exerted by all three types of sensory fiber fascicles on the iris. Also, all three sensory ganglia became densely adrenergically innervated from the LC graft (Fig. 10.) simultaneous with a lack of ingrowth of adrenergic LC fibers into the iris attachment underneath the ganglion grafts. Obviously, the various sensory ganglia all exert the same kind of guiding influence on the immature LC fibers in oculo.

Conclusions

As summarized in Fig. 11, the eye chamber model can serve as a highly variable and useful means by which central monoamine neurons and peripheral sensory neurons are made to interact. The experiments have scrutinized guiding and growth-stimulating forces acting on grafted neuronal populations. The reticular neurons of LC and dorsal raphe nucleus can be readily influenced by a variety of sensory ganglion grafts. Axon fascicles formed by these grafted sensory neurons on the host iris, as well as the perikarya themselves in the grafts, attract the monoaminergic fibers. Although a similar but less pronounced guidance can be exerted by the normal intrinsic trigeminal fibers of the iris, an acute trigeminal denervation will elicit a strong growth response in the monoaminergic neurons and influence their course of outgrowth to follow the tracks of the degenerated sensory iris fibers. A seemingly contradictory finding is that the monoaminergic fibers will grow and be guided by both the presence and the absence of sensory axon bundles. A tentative explanation is that the <u>maturated</u> monoaminergic reticular neurons will only be initiated to grow and enter an area of the target iris if that area is sensorily denervated. If similar monoaminergic neurons during normal <u>ontogenetic growth</u> appear in the vicinity of sensory axon fascicles, such fascicles and sensory perikarya will attract and guide the central fibers.

Acknowledgments. Supported by the Swedish Medical Research Council 14X-06555, 25P-6326, the Swedish Natural Science Research Council B-BU-1522-102, U-FR 1522-100; C.A.-L. was supported by the French C.N.R.S. We thank Dr. J. Kessler and I. Black, New York, for the generous supply of SP antisera, Ms. Lena Hultgren, Ms. Anna Hultgårdh, and Ms. Barbro Standwerth for technical assistance, and Mrs. Ida Engqvist for expert typing.

References

Ayer-LeLievre C, Granholm A-Ch, Seiger Å (1984) Silver impregnation of intrinsic sensory innervation of rat iris in toto. J Anat 138: 309-321

Coons AH (1958) Fluorescent antibody methods. In: Danielli JF (ed) General cytochemical methods. Academic, New York, pp 399-422

Corrodi H, Jonsson G (1967) The formaldehyde fluorescence method for the histochemical demonstration of biogenic monoamines. A review on the methodology. J Histochem Cytochem 15:65-78

Falck B, Hillarp N-Å, Thieme G, Torp A (1962) Fluorescence of catecholamines and related compounds condensed with formaldehyde. J Histochem Cytochem 10:348-354

Jessell TM, Iversen LL, Cuello AC (1978) Capsaicin-induced depletion of substance P from primary sensory neurones. Brain Res 152:183-188

Linder JE (1978) A simple and reliable method for the silver impregnation of nerves in paraffin sections of soft and mineralized tissues. J Anat 127:543-551

Olson L, Seiger Å, Stroemberg I (1983) Intraocular transplantation in rodents. A detailed account of the procedure and example of its use in neurobiology with special reference to brain tissue grafting. In: Fedoroff S, Hertz L (eds) Advances in cellular neurobiology, vol 4. Academic, New York, pp 407-442

Olson L, Seiger Å, Ålund, M (1978) Locus coeruleus fiber growth in oculo induced by trigeminotomy. Med. Biol 56:23-27

Sartori C (1926) Sugli innesti di tessuti embrionali. Arch Ital Chir 15:3-13

Seiger Å (1980) Growth interaction between locus coeruleus and trigeminal ganglion after intraocular double grafting. Med Biol 58:149-157

Seiger Å (1983) Growth interaction between monoamine neurons and trigeminal ganglion cells grafted to the anterior eye chamber. In: Das G, Wallace R (eds) Neural tissue transplantation research. Springer, Berlin Heidelberg New York, pp 175-190

Seiger Å, Olson L (1977a) Quantitation of fiber growth in transplanted central monoamine neurons. Cell Tissue Res 179:285-316

Seiger Å, Olson L (1977b) Reinitiation of directed nerve fiber growth in central monoamine neurons after intraocular maturation. Exp Brain Res 29:15-44

Seiger Å, Olson L (1977c) Growth of locus coeruleus neurons in oculo independent of simultaneously present adrenergic and cholinergic nerves in the iris. Med Biol 55:209-223

Factors Influencing Axonal Growth: Central vs. Peripheral Nerve Explants as Substrates in Vitro

M. E. Schwab and H. Thoenen

Abteilung für Neurochemie, Max-Planck-Institut für Psychiatrie,
Am Klopferspitz 18A, 8033 Martinsried, FRG

Introduction

The central and the peripheral nervous system of higher vertebrates differ distinctly in their capacity to regenerate fiber connections after a lesion (for review: Kiernan 1979; Lasek et al. 1981). Over the last 5 years it has been shown by a series of elegant experiments that the lack of regenerating capacity in the CNS is not due to a general inability of CNS neurons to produce new neurites, which also elongate over long distances, but is crucially dependent on the environment in which the new axons grow (Richardson et al. 1980, 1984;: Weinberg and Raine 1980). Whereas tissues from the peripheral nervous system, e.g., sciatic nerve transplants, permit or enhance central and peripheral neurite regeneration, the adult CNS environment either is unable to support neurite growth or actively prevents it. Among the possibilities responsible for this important phenomenon are the lack of trophic factors, unfavorable extracellular substrate conditions, mechanical barriers, e.g., by dense astrocytic scars, or as yet undefined inhibitory components in the CNS environment. In the following we will first briefly discuss the present knowledge on trophic factors, starting with the paradigmatic case of nerve growth factor (NGF). We will then describe recent experiments using central and peripheral nervous tissue in culture which demonstrate that, in spite of the probable importance of trophic factors in the CNS, the lack of trophic factors may not be the primary reason why the CNS environment does not permit axonal regeneration.

Trophic Factors, NGF

The general importance of trophic factors becomes apparent mainly from cell culture experiments, where purified factors or specific tissue extracts are often needed to allow survival and appropriate differentiation of nerve cells grown in the absence of non-neuronal cells (Berg 1984; Patterson 1978; Thoenen and Barde 1980). In vivo, the importance of NGF, the only neurotrophic factor which has been extensively characterized so far, can be deduced from the effects of antibodies or the autoimmunisation of animals with the NGF (Harper and Thoenen 1981; Goedert et al. 1984; Ross et al. 1981; Schwartz et al. 1982). In perinatal animals the sympathetic ganglion cells and in embryonic animals also a subpopulation of dorsal root ganglion cells die as a consequence of NGF deprival. Cell death is preceded by a drastic reduction in the activities of tyrosine hydroxylase (TH), dopamine-hydroxylase (DBH) in the sympathetic ganglia, and substance P, and somatostatin in dorsal root ganglia. In

Experimental Brain Research, Suppl. 13

adult animals, antibodies against NGF also lead to a rapid decrease in transmitter production and in sympathetic ganglia to a slow atrophy of the cells which eventually also leads to cell death. NGF is an example of a target-derived trophic factor; it is produced and locally released by, for example the iris or the heart (Korsching and Thoenen 1983a), binds with high affinity to specific receptors on the nerve terminals of sympathetic and sensory ganglion cells, is internalized, and is transported by retrograde axonal transport to the cell bodies (Schwab and Thoenen 1983). The biochemical sequence of intracellular events leading to increased TH and DBH or peptide synthesis, or to metabolic stimulation and the enhanced survival of young nerve cells, is not yet known, except for the fact that second messenger mechanisms have to be involved (Heumann et al. 1981, 1984). On the level of the growing axon NGF also stimulates the growth cone activity and the extent of branching, and growth cones are able to orient themselves in an NGF gradient (Campenot 1982; Gundersen and Barrett 1980; Seeley and Greene 1983). Most interestingly, the effects of blockade of retrograde axonal transport and of axotomy in young or adult sympathetic ganglia closely resemble the effects of NGF antibody administration (Schwab and Thoenen 1983). In fact, they can be overcome by exogenous administration of NGF and may therefore be largely due to the interruption of the continuous supply of NGF from the target area by retrograde axonal transport.

Adult sympathetic ganglion neurons, in contrast to those of neonatal rats, survive axotomy as well as NGF deprival at least for some time and are able to regenerate the lesioned axonal connections (Thoenen and Barde 1980; Hendry 1975). This finding leads to the question of whether NGF is produced in response to a lesion by the denervated Schwann cells or fibroblasts in the peripheral nerve stump, resulting in an enhanced sprouting and elongation and a substitute for the interrupted NGF supply from the target organ. Alternatively, adult neurons may tolerate short-term NGF deprival even under regenerating conditions, or other as yet unknown factors produced by denervated Schwann cells or fibroblasts may be involved. No conclusive data are available yet to distinguish between these possibilities. Although NGF is present in adult rat sciatic nerves, it seems to predominantly result from retrograde axonal transport in sympathetic and sensory axons (Korsching and Thoenen 1983b). Production of NGF by denervated nerves or Schwann cells in culture, e.g. comparable to that of the denervated or cultured iris, has not yet been demonstrated (Ebendal et al. 1980; Richardson and Ebendal 1982). If trophic factors derived from denervated peripheral nerves are important for regeneration, other factors not identical with NGF have also to be postulated for the spinal motoneurons and those dorsal root ganglion neurons which regenerate well but do not respond to NGF and do not have NGF receptors. In fact, fluid collected from silicon tube chambers implanted between the two sectioned ends of rat sciatic nerves contains significant amounts of trophic activities effective on sympathetic, sensory, parasympathetic, and spinal cord neurons of the embryonic chick (Longo et al. 1983), and very specific changes in the pattern of newly synthesized proteins as a consequence of denervation have been found in the rat sciatic but also in the optic nerve (Skene and Shooter 1983).

The requirements and responsiveness of CNS neurons in relation to trophic factors are still very unclear, mainly due to the difficulty of obtaining homogeneous populations of a given type of neuron (Berg 1984). With the exception of basal forebrain cholinergic neurons, which possess NGF receptors and respond to the administration of NGF with an increase in choline acetyltransferase activity; however, NGF does not seem to play a role in the development or maintenance of other mammalian CNS neurons (Gnahn et al. 1983: Schwab et al. 1979; Seiler and Schwab 1984). In goldfish, however, NGF stimulates fiber outgrowth and survival of retinal ganglion cells in vitro and, as in the newt, enhances the regeneration of optic axons in vivo (Turner et al. 1980, 1982; Turner and Delaney 1979). NGF does not act on the retinal ganglion cells of the rat, but preliminary studies indicate a survival effect on ganglion cells in vitro by a new neurotrophic factor derived from pig brain (Johnson, Barde, Schwab, and Thoenen, unpublished observations; Barde et al. 1982). Very defined survival effects are also seen on dorsal root ganglia of embryonic chicken where this factor acts on that neuronal population which is not responsive to NGF (Barde et al. 1982). Peripheral sympathetic and ciliary ganglion neurons are not responsive to this brain-derived factor. The specificity of this new factor for various CNS neurons, its physiological role during development and in the adult nervous system, the range of its effects, and its mechanism of action are not yet known.

The fact that neurotrophic activity very similar to that effected by this brain-derived factor is produced by C6 glioma cells in culture points to a possible glial source for this factor (Barde et al. 1978). Pure cultures of astrocytes have in fact been shown to produce trophic factors which support the survival of peripheral as well as central neurons in culture (Banker 1980; Lindsay et al. 1982). Similar activities have recently also been shown to be present in the wound fluid collected from intracortical wounds in rats (Manthorpe et al. 1983b). These scattered initial findings strongly point to the presence of trophic factors in the CNS or even an induced production at the site of a lesion rather than to a lack of trophic factors. Still, however, the nature, physiological role, and appropriateness of these factors for CNS regeneration remain to be investigated.

Central vs Peripheral Nerve Explants as Substrates for Axonal Growth In Vitro

In vitro systems allow a degree of experimental manipulation which can never be attained in vivo. To evaluate the influence of trophic factors and specific substrates in CNS and PNS environment, we used a chamber culture system whereby a Teflon ring divided the culture dish into a narrow central slit and two adjacent side chambers (Campenot 1979). The ring was tightly sealed onto the culture dish with silicon grease, allowing no penetration of axons into the side chambers. Before sealing the ring to the dish, pieces of rat optic nerve or sciatic nerve were embedded in the silicon grease in a way that the neurons plated into the central slit could use them as bridges or tunnels to the side chambers (Schwab and Thoenen, to be published). As a source of neurons, superior cervical ganglion cells of neonatal rats were used and grown under optimal trophic conditions (5% rat serum, 50 ng/ml NGF) to allow vigorous fiber production and long-term survival. When these neurons were given the choice between optic nerve on one side and sciatic nerve on the other side of the chamber they were frequently seen to grow into the sciatic nerves. After two to three weeks in culture, axons emerged from the 4-5 mm long sciatic nerve segments on the other side of the barrier. Such penetrating axons were never seen in the case of the optic nerves. Electron microscopy confirmed these results; a large number of axons were present in the sciatic nerve explants from adult or 10-day-old rats, but no regenerating axons could ever be

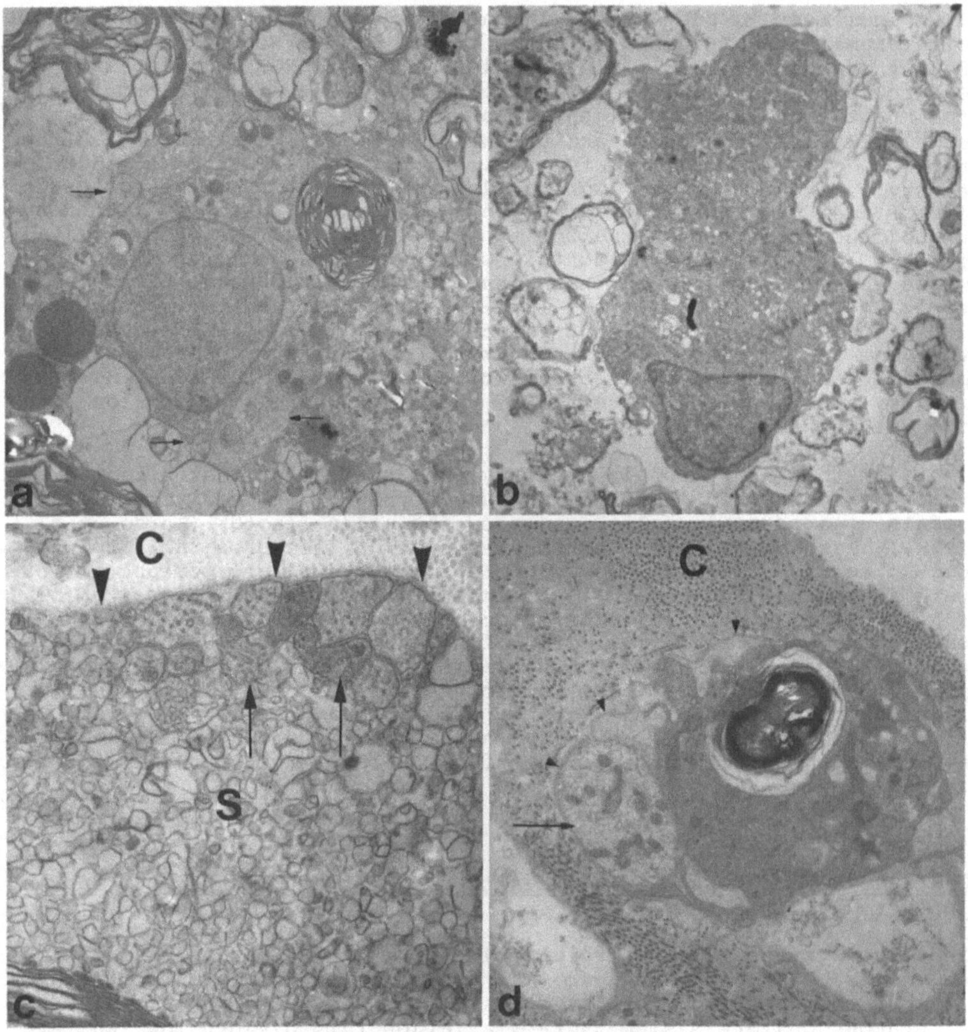

Fig. 1a-d. Electron microscopic structure of rat sciatic and optic nerve explants in chamber cultures (a-c) or regenerating sciatic nerve in vivo (d). a Adult sciatic nerve 23 days in culture; Schwann cells surrounded by debris and axons (arrows); x4500 b adult optic nerve, 31 days in culture: glial cells, no axons; x3600 . c frozen adult sciatic nerve, 17 days in culture, with axons (arrows) growing in bundles along Schwann cell side of basal lamina (arrowheads); x21000 d sciatic nerve in vivo, 3 mm distal to freeze-crush lesion inflicted 4 days earlier. Periphery of nerve bundle. Axon (arrow) in contact with Schwann cell and basal lamina. x6900 S, Schwann cell debris; C, endoneurial collagen seen in the optic nerves (Fig. 1a,b). The density of the tissue and debris was very comparable in both types of nerve explants.

Treatment with cytosine arabinoside prevented the proliferation of Schwann cells and glial cells in the explants and often resulted in a quite loose agglomerate of myelin and cell debris. Again, axons grew exclusively through the sciatic nerves. When the sciatic nerve explants were frozen and thawed three times before implantation the high frequency of axonal ingrowth persisted (Fig. 1c).

These results demonstrate two important points: the ingrowth of axons into sciatic nerve explants and the total absence of axons in the optic nerves in these experiments is not due to a lack of trophic factors in the CNS tissue, since it can also be seen under optimal trophic factor conditions and the selectivity persists in nerves where all endogenous glial cells have been killed. Second, a mechanical barrier cannot be responsible for this selectivity since the density of debris tissue was comparable in sciatic and optic nerves and an astrocytic scar did not form in the optic nerves under the conditions of the present experiments.

Localisation of Regenerating Axons in Sciatic Nerve Explants

Looking at thin sections of sciatic nerve explants containing regenerating axons it became immediately apparent that the axons did not associate with all nerve components at random. When the neighbor relationships of each axon were assessed, the following hierarchy of preferred substrates emerged: the best substrate seems to be other axons, followed by the surface of living Schwann cells and by the Schwann cell side of the basal lamina (BL) (Schwab and Thoenen, to be published). Most notably, this BL appeared to be highly polarized, since axons were never seen on the outer, fibroblast side of the BL (Fig. 1a,c). Collagen seemed to be a much poorer substrate for the axons, which were rarely associated with the endoneurial collagen bundles. This preference for the Schwann cell side of the BL was particularly striking in the case of dead nerve explants, where large empty spaces were often present in the explants, which were, however, not invaded by axons; the axons selectively squeezed themselves between the myelin and the overlying BL (Fig. 1c). Exactly the same behavior of regenerating axons could be seen in the sciatic nerve in vivo, where after an extensive freeze-crush lesion the regenerating axons showed high preference for the Schwann cell surface and the Schwann cell side of the BL as a substrate, besides their tendency for fasciculation (Fig. 1d; also Ide et al. 1983; Nathaniel and Pease 1963). This observation that one side of basement membranes is a very good substrate for axonal growth has also been made in the case of motor reinnervation of frog muscles at the original synaptic sites, and also for muscle basement membranes (Sanes et al. 1978; Keynes et al. 1984). The BL is polarized with regard to anionic sites (Yokota et al. 1983), but the molecular mechanism and the components responsible for this preferred axonal growth have not yet been identified. Laminin is a major component of many basement membranes and has been shown to be produced and secreted by Schwann cells in the sciatic nerve (Cornbrooks et al. 1983; Palm and Furcht 1983). It is also an excellent substrate for growing peripheral and central neurons in culture (Baron van Evercooren et al. 1982; Manthorpe et al. 1983a; Rogers et al. 1983), but preliminary results showed that the presence of antiserum against fibroblast laminin during the entire culture period in our chamber cultures did not inhibit axonal growth in the sciatic nerve explants.

Neuron-Glia Interaction in Dissociated Cultures

Central glial cells differ in many respects from Schwann cells, and, except around blood vessels, basement membranes are not present in the CNS. The possibility therefore exists that the sprouting axons trying to grow into the optic nerves did not find an appropriate substrate for successful growth and elongation. We therefore investigated the neuron-glia interaction on the single cell level by plating dissociated sympathetic neurons onto cultures of dissociated glial cells from 10-day-old rat optic nerves. These cultures contained stellate as well as flat astrocytes (positive for glial fibrillary acidic protein), both of which proved to be a good substrate for sympathetic, sensory and embryonic retina cells (Schwab and Thoenen, in preparation). From the two types of galactocerebroside- and antigen 04 (Sommer and Schachner 1981)-positive cells, one looked like a differentiated oligodendrocyte with a medium sized round cell body and 2-5 thin long fibers. This cell readily associated with the neurons and extended its processes along the neurite bundles. The other cell type had many processes which were highly branched. Very characteristically the neurites actively avoided this cell, so that in very dense cultures typical holes or windows occupied by these cells could be seen in the network of the neurites (Fig. 2). The identity of these cells and the characteristics responsible for this neurite repulsion effect are currently under investigation. It is still unknown how these in vitro findings relate to the in vivo situation, and whether a particular type of glial cell in vivo also exerts such inhibitory effects which would overcome the favorable substrate conditions of, for example, astrocytes.

Whereas the specific association of axons with Schwann cells and basal lamina in the sciatic nerve points to a lack of a corresponding substrate in the optic nerve as a reason for the failure of fiber ingrowth, the findings with dissociated glial cultures point to the additional possibility of the presence of an inhibitory (or extremely poor) substrate influence in the developed central nervous system. As the chamber culture experiments show, lacking appropriate substrates or such inhibitory factors cannot be overcome by trophic factors. Therefore, lack of particular trophic factors may for many neurons not be the primary cause of the absence of regeneration in the CNS or the inability of peripheral neurons to grow past the Schwann cell-CNS barrier.

Acknowledgments. We thank Mr. Ch. Mueller for skillful technical assistance and Mrs. E. Eichler for typing the manuscript.

References

Banker GA (1980) Trophic interactions between astroglial cells and hippocampal neurons in culture. Science 209:809-810

Barde YA, Lindsay RM, Monard D, Thoenen H (1978) New factor released by cultured glioma cells supporting survival and growth of sensory neurons. Nature 274:818

Barde YA, Edgar D, Thoenen H (1982) Purification of a new neurotrophic factor from mammalian brain. EMBO J 1:549-553

Baron van Evercooren A, Kleinman HK, Ohno S, Marangos P, Schwartz JP, Dubois-Dalq ME (1982) Nerve growth factor, laminin and fibronec-

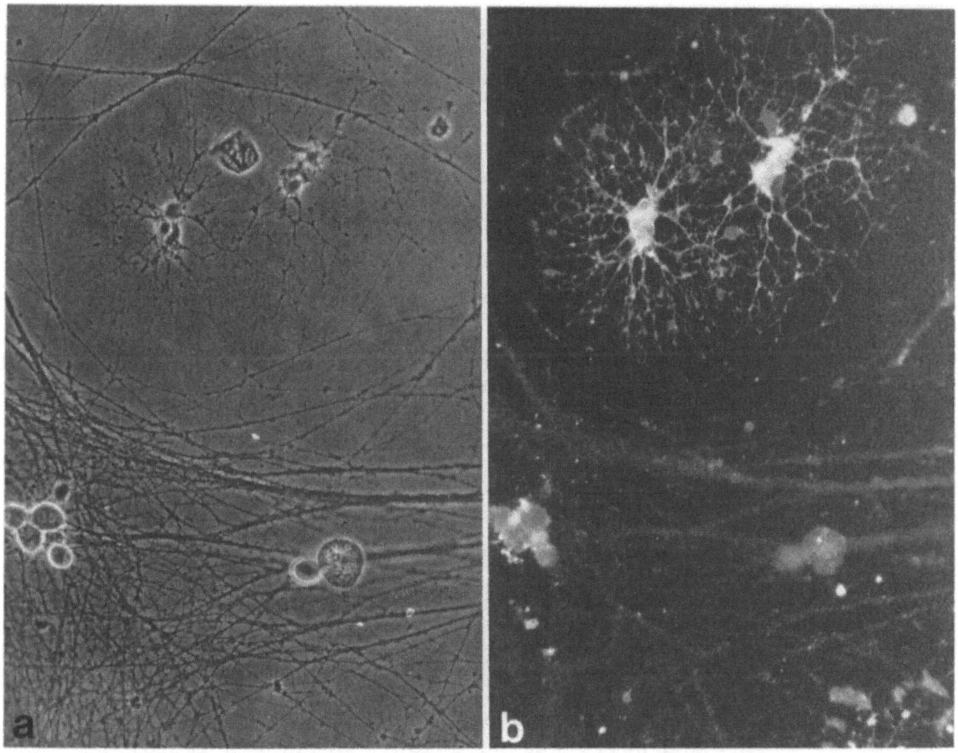

Fig. 2. Dissociated optic nerve glial culture, 23 days in vitro, co-cultured with dissociated sympathetic neurons for 13 days. Dense plexus of neurites is absent around two highly branched glial cells (a) which stain for antigen 04 (b: fluorescent micrograph of a). x 250

tin promote neurite growth in human fetal sensory ganglia cultures. J Neurosci Res 8:179-193

Berg DK (1984) New neuronal growth factors. Annu Rev Neurosci 7:149-170

Campenot RB (1979) Independent control of the local environment of somas and neurites. Methods Enzymol 58:302-307

Campenot RB (1982) Development of sympathetic neurons in compartmentalized cultures. I. Local control of neurite growth by nerve growth factor. Dev Biol 93:1-12

Cornbrooks CJ, Carey DJ, McDonald JA, Timpl R, Bunge RP (1983) In vivo and in vitro observations on laminin production by Schwann cells. Proc Natl Acad Sci USA 80:3850-3854

Ebendal T, Olson L, Seiger A, Hedlund KO (1980) Nerve growth factor in the rat iris. Nature 286:25-28

Gnahn H, Hefti F, Heumann R, Schwab ME, Thoenen H (1983) Selective increase in choline acetyltransferase (ChAT) activity in the forebrain of neonatal rats by NGF. Failure of antibodies against NGF to decrease basal ChAT activity. Dev Brain Res 9:45-52

Goedert M, Otten U, Hunt SP, Bond A, Chapman D, Schlumpf M, Lichtensteiger W (1984) Biochemical and anatomical effects of antibodies against nerve growth factor on developing rat sensory ganglia. Proc Natl Acad Sci USA 81:1580-1584

Gundersen RW, Barrett JN (1980) Characterization of the turning response of dorsal root neurites toward nerve growth factor. J Cell Biol 87:546-554

Harper GP, Thoenen H (1981) Target cells, biological effects, and mechanism of action of nerve growth factor and its antibodies. Annu Rev Pharmacol Toxicol 21:205-229

Hendry IA (1975) The response of adrenergic neurons to axotomy and nerve growth factor. Brain Res 94:87-97

Heumann R, Schwab ME, Thoenen H (1981) A second messenger required for nerve growth factor biological activity? Nature 292:838-840

Heumann R, Schwab ME, Merkl R, Thoenen H (1984) Nerve growth factor (NGF) mediated induction of choline acetyltransferase (ChAT) in PC12 cells: evaluation of the site of action of NGF and the involvement of lysosomal degradation products of NGF. J Neurosci 4:3039-3050

Ide C, Tohyama K, Yokota R, Nitatori T, Onodera S (1983) Schwann cell basal lamina and nerve regeneration. Brain Res 288:61-75

Keynes RJ, Hopkins WG, Huang CLH (1984) Regeneration of mouse peripheral nerves in degenerating skeletal muscle: guidance by residual muscle fiber basement membrane. Brain Res 295:275-281

Kiernan JA (1979) Hypotheses concerned with axonal regeneration in the mammalian nervous system. Biol Rev 54:155-197

Korsching S, Thoenen H (1983a) Nerve growth factor in sympathetic ganglia and corresponding target organs of the rat: correlation with density of sympathetic innervation. Proc Natl Acad Sci USA 80:3513-3516

Korsching S, Thoenen H (1983b) Quantitative demonstration of the retrograde axonal transport of endogenous nerve growth factor. Neurosci Lett 39:1-4

Lasek RJ, McQuarrie IG, Wujek JR (1981) The central nervous system regeneration problem: neuron and environment. In: Gorio A et al. (eds) Posttraumatic peripheral nerve regeneration. Raven, New York, p 59

Lindsay RM, Barber PC, Sherwood MRC, Zimmer J, Raisman G (1982) Astrocyte cultures from adult rat brain. Derivation, characterization and neurotrophic properties of pure astroglial cells from corpus callosum. Brain Res 243:329-343

Longo FM, Manthorpe M, Skaper SD, Lundborg G, Varon S (1983) Neuronotrophic activities accumulate in vivo within silicone nerve regeneration chambers. Brain Res 261:109-117

Manthorpe M, Engvall E, Roslakti E, Longo FM, Davis GE, Varon S (1983a) Laminin promotes neuritic regeneration from cultured peripheral and central neurons. J Cell Biol 97:1882-1890

Manthorpe M, Nieto-Sampedro M, Skaper SD, Lewis ER, Barbin G, Longo FM, Cotman CW, Varon S (1983b) Neuronotrophic activity in brain wounds of the developing rat. Correlation with implant survival in the wound cavity. Brain Res 267:47-56

Nathaniel EJH, Pease DC (1963) Regenerative changes in rat dorsal roots following Wallerian degeneration. J Ultrastruct Res 9:533-549

Palm SL, Furcht LT (1983) Production of laminin and fibronectin by Schwannoma cells: cell-protein interactions in vitro and protein localization in peripheral nerve in vivo. J Cell Biol 96:1218-1226

Patterson PH (1978) Environmental determination of autonomic neurotransmitter functions. Annu Rev Neurosci 1:1-17

Richardson PM, Ebendal T (1982) Nerve growth activities in rat peripheral nerve. Brain Res 246:57-64

Richardson PM, McGuiness UM, Aguayo AJ (1980) Axons from CNS neurons regenerate into PNS grafts. Nature 284:264-265

Richardson PM, Issa VMK, Aguayo AJ (1984) Regeneration of long spinal axons in the rat. J Neurocytol 13:165-182

Rogers AL, Letourneau PC, Palm SL, McCarthy J, Furcht LT (1983) Neurite extension by peripheral and central nervous system neurons in response to substratum-bound fibronectin and laminin. Dev Biol 98:212-220

Ross M, Loefstrandh S, Gorin PD, Johnson EM, Schwartz JP (1981) Use of an experimental autoimmune model to define nerve growth factor dependency of peripheral and central substance P-containing neurons in the rat. J Neurosci 1:1304-1311

Sanes JR, Marshall LM, McMahan UJ (1978) Reinnervation of muscle fiber basal lamina after removal of myofibers. J Cell Biol 78:176-198

Schwab ME, Thoenen H (1983) Retrograde axonal transport. In: Lajtha A (ed) Handbook of Neurochemistry, vol 5. Plenum, New York, p 381

Schwab ME, Otten U, Agid Y, Thoenen H (1979) Nerve growth factor (NGF) in the rat CNS: absence of specific retrograde axonal transport and tyrosine hydroxylase induction in locus coeruleus and substantia nigra. Brain Res 168:473-483

Schwab ME, Thoenen H (1985) Dissociated neurons regenerate into sciatic but not optic nerve explants in culture irrespective of neurotrophic factors. J Neurosci (to be published)

Schwartz JP, Pearson J, Johnson EM (1982) Effect of exposure to anti-NGF on sensory neurons of adult rats and guinea pigs. Brain Res 244:378-381

Seeley PJ, Greene LA (1983) Short-latency local actions of nerve growth factor at the growth cone. Proc Natl Acad Sci USA 80:2789-2793

Seiler M, Schwab ME (1984) Specific retrograde transport of nerve growth factor (NGF) from neocortex to nucleus basalis in the rat. Brain Res 300:33-39

Skene JHP, Shooter EM (1983) Denervated sheath cells secrete a new protein after nerve injury. Proc Natl Acad Sci USA 80:4169-4173

Sommer I, Schachner M (1981) Monoclonal antibodies (O1 to O4) to oligodendrocyte cell surfaces: an immunocytological study in the central nervous system. Dev Biol 83:311-327

Thoenen H, Barde YA (1980) Physiology of nerve growth factor. Physiol Rev 60:1284-1335

Turner JE, Delaney RK (1979) Retinal ganglion cell response to axotomy and nerve growth factor in the regenerating visual system of the newt (notophthalmus viridescens): an ultrastructural morphometric analysis. Brain Res 171:197-212

Turner JE, Delaney RK, Johnson JE (1980) Retinal ganglion cell response to nerve growth factor in the regenerating and intact visual system of the goldfish (Carassius auratus). Brain Res 197:319-330

Turner JE, Schwab ME, Thoenen H (1982) Nerve growth factor stimulates neurite outgrowth from goldfish retinal explants: The influence of a prior lesion. Dev Brain Res 4:59-66

Weinberg EL, Raine CS (1980) Reinnervation of peripheral nerve segments implanted into the rat central nervous system. Brain Res 198:1-11

Yokota R, Tohyama K, Ide C (1983) Distribution of anionic sites on the basal lamina of Schwann cells. Neurosci Lett 41:49-54

Neuronal Cell Cultures and Monosialoganglioside: A Model for Comprehension of Mechanisms Underlying Central Nervous System Repair

A. Leon[1], D. Bevegnù[1], R. Dal Toso[1], O. Giorgi[1], D. Presti[1], G. Tettamanti[2], and G. Toffano[1]

1 Fidia Neurobiological Research Laboratories, Via Ponte della Fabbrica 3/A, 35031 Abano Terme, Italy
2 Institute of Biological Chemistry, University of Milano, Via Saldini 50, 20133 Milano, Italy

Introduction

The response of the adult mammalian central nervous system (CNS) neurons to injury and the potential capability of the latter to recover are at present basic neurobiological problems with vast pharmacological and clinical implications. Critical questions, such as to what extent and by what mechanisms CNS repair processes can occur, are currently being addressed in an attempt to elucidate whether, and to what degree, such processes may be regulated in order to enhance CNS repair after damage.

From lesion and transplantation techniques, it is now clear that not only developing but also adult mammalian CNS has an innate capacity for functional repair after damage. The irreversibly damaged neurons cannot be replaced but may be functionally substituted by alternative pathways following massive sprouting and reactive synaptogenesis of intact undamaged axons in response to adjacent axonal and synaptic degeneration (Björklund and Stenevi 1979; Cotman et al. 1981; Finger and Stein 1982). The intensity of such anatomical and functional synaptic remodeling varies with age, extent, localization, and type of neuronal injury and is presumably determined by appropriate signalling from the adjacent environment and denervated target area. Concepts such as neuronotrophic interactions, neuronal plasticity, and/or activation of novel or usually quiescent processes have been implicated and agents or conditions capable of affecting these processes are currently being examined in an attempt to improve restoration of function of damaged neuronal tissues.

Studies from our own and other laboratories have recently documented the capability of monosialoganglioside GM_1 (nomenclature according to Svennerholm 1963) to enhance functional recovery following specific lesions of adult mammalian brain (Agnati et al. 1983; Toffano et al. 1983; Oderfeld-Nowak et al. 1984; Pedata et al. 1984; Sabel et al 1984; Toffano et al 1984a; Toffano et al 1984b). Approaches for investigating such an effect of chronic ganglioside administration essentially derive from evidence indicating that such molecules may play a prominent regulatory role in neuronal development and regeneration (for a recent review, see Ando 1983). However, the molecular mechanisms by which GM_1 enhances CNS recovery are unknown. As a first insight, one can state that should the ganglioside stably associate with neuronal cell surfaces, then it could be expected that GM_1 effects may be mediated through the plasma membrane and, as such, may affect the fine interplay occurring at this level between the cellular intrinsic

Experimental Brain Research, Suppl. 13

programs and the triggering factors of the extracellular environ ment. A role for gangliosides as mediator molecules of neuronotrophic influences, known to be induced by injury and necessary for CNS recovery (Nieto-Sampedro et al. 1983), has already been suggested by the many studies documenting pronounced changes in concentration and pattern of ganglioside which accompany neuronal developmental and adult pathological processes (Ledeen et al. 1983).

In this paper we report our approach to comprehending GM_1 effects on CNS repair processes in vitro by study and analysis of its effects in various neuronal culture systems. In brief, the possibility of GM_1 association to neuronal cells and its relationship to neuritogenesis was verified in neuroblastoma cells. Verification of the occurrence of GM_1 effects on CNS neurons and its relationship to neuronotrophic factors was conducted in primary fetal sensory and mesencephalic dopaminergic neurons in culture respectively.

GM_1 and Neuroblastoma Cells in Culture

Neuroblastoma cells have been extensively utilized as a model system for study of the crucial role of plasma membrane constituents and their activity in the regulation of neurite outgrowth and cellular differentiation. Although they are malignant-transformed cells, under suitable experimental conditions neuroblastoma cells may be induced to form neurites and to express biochemical properties characteristic of mature neurons (de Laat and van der Saag 1982). Such differentiation is usually accompanied by profound changes of membrane architecture and properties, including membrane fluidity (de Laat and van der Saag 1982) and ganglioside content and complexity (Dawson 1979; Rebel et al. 1980).

The addition of exogenous gangliosides, including GM_1, but not neutral glycolipids, sulfatides, or free sialic acid, has been shown to promote neurite formation of N_2A murine neuroblastoma cells (Dimpfel et al. 1981; Roisen et al. 1981; Leon et al. 1982; Byrne et al. 1983; Spoerri 1983), BIO4 rat brain tumor cells (Seifert 1981), and SB21BI somatic neurohybrid clonal cell line (Rybak et al. 1983). Biochemical correlates of this morphological differentiation include increased ornithine decarboxylase activity, cAMP content and expression of mRNA sequences for tubulin, decreased thymidine incorporation into DNA, and elongation of the G_1 cell cycle phase.

The mechanisms by which gangliosides cause the morphological and biochemical changes referred to above are still unknown. Although exogenous gangliosides have been reported to be incorporated as integral components of the cell membrane bilayers (Toffano et al. 1980; Kanda et al. 1982; Radsak et al. 1982), it has not been demonstrated that this process occurs in neuroblastoma cells and whether it is related to the neuritogenic effect referred to above. Analysis of the process of GM_1 binding revealed three forms of GM_1 association to neuroblastoma N_2A cells (Table 1) (Facci et al. 1984): one removable by treatment with serum-containing solutions and thus superficially associated to the cell surface, presumably as micelles; one serum resistant and labile to trypsin treatment, corresponding to GM_1 single molecules or micelles interacting with polypeptide domains on the cell surface; and one resistant to serum and trypsin treatment, probably constituted by ganglioside molecules (single monomers?) inserted into the mem-

Table 1. Proportions of the different forms of associated GM_1[a]

Form of GM_1 association	Cell-associated GM_1 %
Total associated	100 ± 0.04
Serum labile	67 ± 0.02
Serum stable:	33 ± 0.01
trypsin labile	17 ± 0.01
trypsin stable	16 ± 0.008

brane bilayer. The proportions of the three forms of associated GM_1 varied in relation to the duration of initial ganglioside exposure and ganglioside concentration in medium. The serum-removable form of associated GM_1 was predominant during the initial stages of association and at the highest ganglioside concentrations (over 10^{-6} M), while the trypsin-labile and -stable forms tended to increase with decreasing ganglioside concentration. The recorded effect on neuritogenesis was almost completely retained by the cells bearing the trypsin-labile and -stable forms of associated GM_1. Instead the serum-removable form appeared not to be involved in the neuritogenic effect. All this seems to suggest that GM_1 molecules, once stably associated, induce modifications of the cell membrane properties and that these modifications are directly involved in the process of stimulation of neuritogenesis. Similarly, the stable insertion of small amounts of GM_1 into neuronal membranes has been shown to enhance the (Na^+, K^+) ATPase activity (Leon et al. 1981). Whether these membrane functional changes are the same as those elicited by endogenous gangliosides in the physiological milieu has to be investigated.

Neuroblastoma N_2A cells were exposed to serum-free culture medium containing ^3H-GM_1 (10 M). Following incubation for 2 h at 37°C in a humidified incubator (5% CO_2/95% air), the cells were submitted to successive treatments so as to quantify the different forms of associated GM_1. Total associated GM_1 is that aliquot of ^3H-GM_1 associated to cells following washings with phosphate buffer saline. For further details concerning washings with serum-containing solutions and trypsin treatment, see Materials and Methods reported in Facci et al. (1984). The data reported are the mean values ± SD of five experiments

GM_1 and Dorsal Root Ganglionic Neurons in Culture

To date, nerve growth factor (NGF) is the only macromolecular agent directed to neurons that has been fully purified and characterized. NGF is known to be essential for the survival, neurite outgrowth, proper development, and presumably also maintenance and regeneration of its target sympathetic and sensory ganglionic neurons in vivo and in vitro (Levi-Montalcini 1966; Thoenen and Barde 1980; Varon and Adler 1980; Varon and Adler 1981). Consequently, a molecular approach to the problem of the possible ganglioside mediation of neuronotrophic activity was attempted by analyzing the monosialoganglioside potentiation of the NGF-induced neurite outgrowth from fetal dorsal root ganglionic (DRG) cells in culture (Leon et al. 1984).

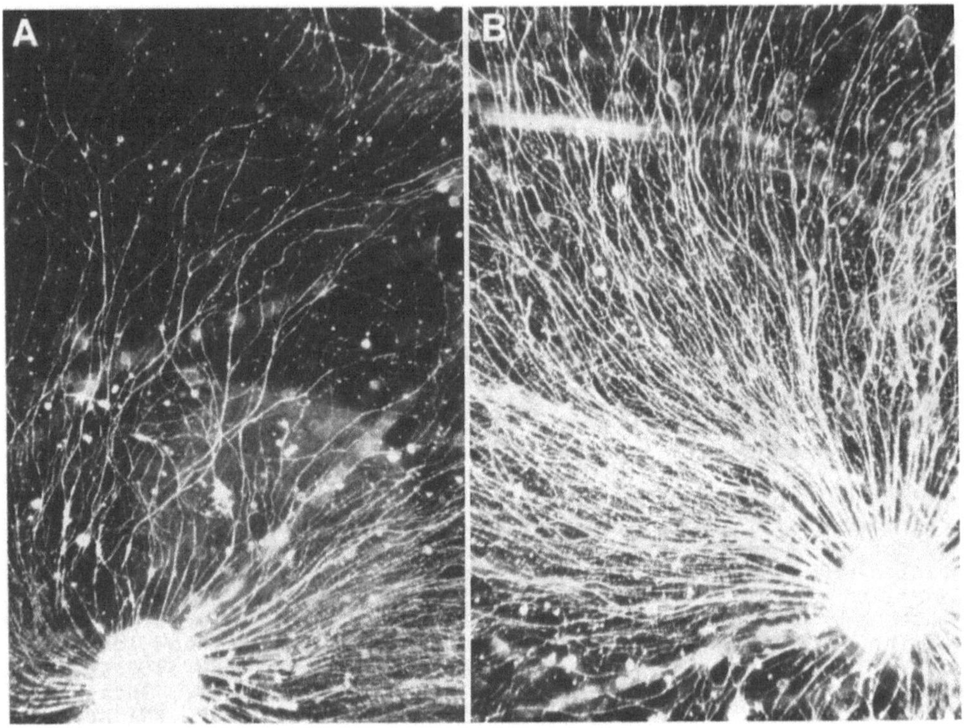

Fig. 1 A,B. Phase contrast photomicrographs of NGF-induced neurite outgrowth from embryonic day 8 chick DRG explants after 2 days in culture: effect of GM_1. DRG explants were plated on collagen-coated Falcon tissue culture dishes in the presence of culture medium [Dulbecco's Modified Eagle's Medium + medium 199 (3:1 v/v) supplemented with 15 mM $NaHCO_3$, 15 mM Hepes, 33.3 mM glucose, 2 mM glutamine, 50 units/ml penicillin, 50 µg/ml streptomycin, and 10% heat inactivated fetal calf serum] containing 2.5S NGF or NGF + GM_1. A: NGF 10 ng/ml; B: GM_1 (10^{-6} M) + NGF 10 ng/ml

As shown in Fig. 1, the addition of GM_1 to culture medium of DRG explants facilitates the NGF-induced neuritic outgrowth. Appreciable GM_1 effects occurred after 48 h in vitro and at concentrations of GM_1 ranging from 10^{-6} to 10^{-7} M. Similar effects were also observed in dissociated neuronal cell cultures (>95% pure), indicating that GM_1 can affect directly DRG neurons and that its effects can take place independently of the presence of non-neuronal cells or of phenomena such as neurite fasciculation. Analogously, GM_1 has been reported to potentiate NGF-induced outgrowth of pheochromocytoma PC12 cells (Ferrari et al. 1983), whereas addition of ganglioside mixtures (Hauw et al. 1981; Roisen et al. 1981) or GM_1 antibodies has been shown, respectively, to stimulate or block NGF-induced sprouting from DRG explants (Schwartz and Spirman 1982).

Hypothetically, the above-reported GM_1 facilitation of NGF trophism may be due to a neuronal growth-promoting activity of GM_1 or to a modulation of the NGF-dependent neuronal cell behavior. However, without NGF, GM_1 was totally ineffective in promoting neurite outgrowth or neuronal survival. Therefore in cells with an absolute NGF requirement, GM_1 cannot substitute for the neuronotrophic agent, the necessary trigger for specific transciptional events. It is likely that GM_1 effects occur via modification of cell surface properties consequent upon its stable insertion. NGF association with the cells is, however, apparently unaffected by GM_1 (Ferrari et al. 1983). Possible explanations could involve modulation of cell-surface transduction mechanisms or enhancement of specific NGF-induced post-translational events.

GM_1 and Primary Neuronal Central Nervous System Cultures

Significant progress has been made recently in using CNS tissue culture systems for identification and study of the physiological role of cellular and environmental influences which affect neuronal survival, neurite outgrowth, and specificity of neuronal connections during CNS development and regeneration (Toffano and Leon 1985). Although heterogeneous, this choice is largely supported by the opportunity offered by the in vitro system to work with "normal" living cells in a much less complex and more accessible structure than the intact brain. The extracellular environment can now be efficiently manipulated, enabling the investigator a degree of control over the composition and organization of the cell populations under study. Among the major criticisms raised against the use of these cells in culture is that once neuronal cells are transferrred out of their natural environment they may lose some of their inherent properties or do not fully develop them as in situ. From extensive studies of neuronal cells in culture to date, it would appear that these fears are unwarranted. Many of the fundamental properties of neurons described in vivo have frequently been observed in vitro indicating that many traits under genetic control are expressed or can be induced in vitro.

We will now briefly discuss some of the characteristics of GM_1 effects on the development of mesencephalic dopaminergic neurons in vitro. These cells, when plated in serum-free hormone-supplemented media, which impede glial cell and fibroblast proliferation, have been reported to survive and to express common features with in vivo lesioned dopaminergic neurons (Berger et al. 1982). As such, this model may prove to be particularly relevant for the comprehension of GM_1 effects in vivo, especially when referred to the GM_1 facilitation of functional dopaminergic reinnervation of the striatum in unilaterally hemitransected rats (Agnati et al. 1983; Toffano et al. 1983, 1984a,b). When GM_1 is added to culture medium of dissociated fetal mesencephalic neurons ($>98\%$ pure), it enhances neurite outgrowth and biochemical maturation of dopaminergic neurons (Dal Toso et al. submitted). In particular, GM_1 produced in time a concentration-dependent increase of specific ^3H-dopamine (DA) uptake (Table 2) associated with an increase of apparent Vmax. The stimulatory effect of GM_1 occurred at various cell densities and was coincident with the occurrence of adequate titers in culture medium of growth-promoting influences. Indeed, in time GM_1 also enhanced DA cell survival in culture, a phenomenon known to be dependent on the occurrence of permissive trophic interactions. Similarly, GM_1 increased the specific ^{14}C-γ-aminobutyric acid (GABA) uptake in both mes-

Table 2. Effect of the addition of GM on specific ^{3}H-DA and ^{14}C-GABA uptake of dissociated fetal mouse mesencephalic cells in culture[a]

Addtion	^{3}H-DA uptake (fmol/plate)	^{14}C-GABA uptake (pmol/plate)
None	44.18 ± 72.67	0.91 ± 0.3
GM_1 10^{-8}	80.87 ± 16.3**	5.49 ± 1.2*
GM_1 5 x 10^{-8}	107.48 ± 16.6*	5.96 ± 2.7*
GM_1 10^{-7}	112.59 ± 23.38*	6.41 ± 1.7*

*$P < 0.01$, **$P < 0.05$ vs control (Student's t-test)

[a] GM_1 at different concentrations was added to serum-free hormone-supplemented media on plating day (10^6 cells/35 mm Falcon dishes), and benztrophine-sensitive ^{3}H-DA uptake and L-2-4-diaminobutyric acid-sensitive ^{14}C-GABA uptake were assayed on day 4 in vitro, as previously described (Prochiantz et al. 1981). Values are the mean of triplicate analyses ±SEM

encephalic (Table 2) and striatal cell cultures. This indicates that GM_1 is capable of affecting various neuronal cell populations. In similar experimental conditions, asialo GM_1, sialic acid, and the oligosaccharide portion of GM_1 were without effect. Furthermore, GM_1 antiserum addition produced a marked decrease of both neurite extension and ^{3}H-DA uptake in mesencephalic cells whilst preabsorption of the antiserum with GM_1 totally or partially neutralized these effects.

In the above model system, exogenous GM_1 was observed to associate stably with mesencephalic cells in a manner similar to that reported for neuroblastoma cells. Thus it appears that the GM_1 molecules, either endogenous or following stable insertion of exogenous molecules, affect CNS neuronal cell behavior in vitro. Most likely this occurs through modulation of as yet unknown neuronotrophic influences present in the neuronal cell culture system. It is noteworthy that NGF in this latter system is ineffective in promoting neuronal survival and maturation, indicating that GM_1 may affect the efficacy of different neuronotrophic agents acting on specific neuronal subpopulations.

Future Prospects

The use of neuronal tissue cultures as model systems for the study of the potential role of gangliosides in regulation of neuronal behavior during development and regeneration may provide useful information. From the work reviewed, it seems safe to conclude that exogenous monosialoganglioside can affect morphological and biochemical characteristics of various neuronal cell types in vitro and that such effects essentially occur via stable association of the ganglioside molecule with the plasma membrane. In addition, it may be concluded that the obtainment of GM_1 effects on neuronal behavior in vitro necessitates the occurrence of adequate extracellular triggers, i.e., neuronotrophic agents, essential for the expression of characteristic phenotypic traits. Data obtained in primary neuronal cell cultures, especially DRG, support such a conclusion and furthermore suggest that GM_1 facilitates the acquisition or the maintenance of the induced differentiative state independently of the nature of the neuronotrophic influence or of the neuronal type. The molecular mechanisms underlying such effects are still obscure. An important step will be to identify the molecular reactions affected by monosialogangliosides and how such events are related to its morphological and biochemical effects not only in vitro but also in vivo. This latter work will inevitably involve purification of injury-induced neuronotrophic molecules which may regulate expression of key phenotypic features of CNS neurons.

Concluding, the present data obtained are consistent with the hypothesis that the presence of GM_1 molecules on the neuronal cell surface is of major importance in determining the properties of the plasma membrane with respect to its capacity to transfer, modulate, and/or respond to signals from the extra- to the intracellular spaces and vice versa. In particular, several lines of evidence support the possibility that GM_1 can modulate functional neuronal cell behavior in response to neuronotrophic signals. This may be relevant for the comprehension of GM_1 facilitation of CNS repair process.

References

Agnati LF, Fuxe K, Calzà L, Benfenati F, Cavicchioli L, Toffano G, Goldstein M (1983) Gangliosides increase in the survival of lesioned nigral dopamine neurons and favour the recovery of dopaminergic synaptic function in the striatum of rats by collateral sprouting. Acta Physiol Scand 119:347-363

Ando S (1983) Review gangliosides in the nervous system. Neurochem Int 5:507-537

Berger B, Di Porzio U, Daguet MC, Fay M, Vigny A, Glowinski J, Prochiantz A (1982) Long term development of mesencephalic dopaminergic neurons of mouse embryos in dissociated primary cultures: morphological and histochemical characteristics. Neuroscience 7:193-205

Björklund A, Stenevi U (1979) Regeneration of monoaminergic and cholinergic neurons in the mammalian central nervous system. Physiol Rev 59:62-100

Byrne MC, Ledeen RW, Roisen FJ, Yorke G, Scalfani JR (1983) Ganglioside-induced neuritogenesis: verification that gangliosides are active agents, and comparison of molecular species. J Neurochem 1:1214-1222

Cotman CW, Nieto-Sampedro M, Harris EW (1981) Synapse replacement in the nervous system of adult vertebrates. Physiol Rev 61: 684-784

Dal Toso R, Leon A, Benvegnù D, Presti D, Facci L, Giorgi O, Toffano G (submitted) Effect of GM_1 ganglioside on the expression of dopaminergic characteristics of embryonic mesencephalic cells in culture

Dawson G (1979) Complex carbohydrates of cultured neuronal and glial cell lines. In: Margolis RU, Margolis RK (eds) Complex carbohydrates of nervous tissue. Plenum, New York, p 29 de Laat SW, van der Saag PT (1982) The plasma membrane as a regulatory site in growth and differentiation of neuroblastoma cells. Int Rev Cytol 74:1-54

Dimpfel W, Moller W, Mengs U (1981) Ganglioside-induced neurite formation in cultured neuroblastoma cells. In: Rapport MM, Gorio A (eds) Gangliosides in neurological and neuromuscular function, development and repair. Raven, New York, p 119

Facci L, Leon A, Toffano G, Sonnino S, Ghidoni R, Tettamanti G (1984) Promotion of neuritogenesis in mouse neuroblastoma cells by exogenous ganglioside. Relationship between the effect and the cell association of ganglioside GM_1. J Neurochem 42:299-305

Ferrari G, Fabris M, Gorio A (1983) Gangliosides enhance neurite outgrowth in PC12 cells. Dev Brain Res 8:215-222

Finger S, Stein DG (eds) (1982) Brain damage and recovery. Academic, New York

Hauw JJ, Fenelon S, Boutry JM, Nagai Y, Escourolle R (1981) Effect of brain gangliosides on neurite outgrowth in guinea pig spinal ganglia tissue cultures and on fibroblast cell cultures. In: Rapport MM, Gorio A (eds) Gangliosides in neurological and neuromuscular function, development and repair. Raven, New York, p 171

Kanda S, Inoune K, Noijma S, Utsumi H, Wiegandt H (1982) Incorporation of spin-labelled ganglioside analogues into cell and liposomal membranes. J Biochem 91:1709-1718

Ledeen RW, Byrne MC, Roisen FJ, Yorke G, Scalfani JR (1983) Gangliosides of the neuronal membrane and the neuritogenic phenomenon. In: Sun GY, Porcellati G, Bazan N, Wu JY, Sun AY (eds) Neural membranes. Humana, Clifton, p 459

Leon A, Facci L, Toffano G, Sonnino S, Tettamanti G (1981) Activation of (Na+, K+)ATPase by nanomolar concentration of GM_1 ganglioside. J Neurochem 37:350-357

Leon A, Facci L, Benvegnù D, Toffano G (1982) Morphological and biochemical effects of gangliosides in neuroblastoma cells. Dev Neurosci 5:108-114

Leon A, Benvegnù D, Dal Toso R, Presti D, Facci L, Giorgi O, Toffano G (1984) Dorsal root ganglia and nerve growth factor: a model for understanding the mechanism of GM_1 effect on neuronal repair. J Neurosci Res 12:277-287

Levi-Montalcini R (1966) The nerve growth factor: its mode of action on sensory and sympathetic nerve cells. Harvey Lect 60: 217-259

Nieto-Sampedro M, Manthorpe M, Barbin G, Varon S, Cotman CW (1983) Injury-induced neuronotrophic activity in adult rat brain: correlation with survival of delayed implants in the wound cavity. J Neurosci 3:2219-2229

Oderfeld-Nowak B, Ulas J, Jezierska M, Skup M, Wojcik M, Domanska-Janik K (1985) Role of GM_1 ganglioside in repair process after hippocampal deafferentiation in rats. In: Bignani A, Bloom FE, Bolis CL, Adeloye A (eds) Central nervous system plasticity and repair. Raven, New York, p 85

Pedata F, Giovanelli L, Pepeu G (1984) GM_1 ganglioside facilitates the recovery of high-affinity choline uptake in the cerebral cortex of rats with a lesion of the nucleus basalis magnocellularis. J Neurosci Res 12:421-427

Prochiantz A, Daguet MC, Herbert A, Glowinski J (1981) Specific stimulation of in vitro maturation of mesencephalic dopaminergic neurons by striatal membranes. Nature 293:570-572

Radsak K, Schwarzmann G, Wiegandt H (1982) Studies on the cell association of exogenously added sialo-glycolipids. Hoppe Seylers Z Physiol Chem 363:263-272

Rebel G, Robert J, Mandel P (1980) Glycolipids and cell differentiation. In: Svennerholm L, Mandel P, Dreyfus H, Urban J (eds) Structure and function of gangliosides. Plenum, New York, p 159

Roisen FJ, Bartfeld H, Nagele R, Yorke G (1981) Ganglioside stimulation of axonal sprouting in vivo. Science 214:577-578

Rybak S, Ginzburg I, Yavin E (1983) Gangliosides stimulate neurite outgrowth and induce tubulin mRNA accumulation in neural cells. Biochem Biophys Res Commun 116:980-994

Sabel BA, Slavin MD, Stein DG (1984) GM_1 ganglioside treatment facilitates behavioral recovery from bilateral brain damage. Science 225:340-342

Schwartz M, Spirman N (1982) Sprouting from chicken embryo dorsal root ganglia induced by nerve growth factor is specifically inhibited by affinity-purified antiganglioside antibodies. Proc Natl Acad Sci USA 79:6080-6083

Seifert W (1981) Gangliosides in nerve cell cultures. In: Rapport MM, Gorio A (eds) Gangliosides in neurological and neuromuscular function, development and repair. Raven, New York, p 99

Spoerri PE (1983) Effects of ganglioside on the in vitro development of neuroblastoma cells: an ultrastructural study. Int J Dev Neurosci 1:383-391

Svennerholm L (1963) Chromatographic separation of human brain gangliosides. J Neurochem 10:613

Thoenen H, Barde Y-A (1980) The physiology of nerve growth factor. Physiol Rev 60:1284-1335

Toffano G, Leon A (1985) Central nervous system repair and nerve cell cultures. Trends in Pharmacological Sciences (submitted)

Toffano G, Benvegnù D, Bonetti AC, Facci L, Leon A, Orlando P, Ghidoni R, Tettamanti G (1980) Interaction of GM_1 ganglioside with crude rat brain membranes. J Neurochem 35:861-866

Toffano G, Savoini G, Moroni F, Lombardi G, Calzà L, Agnati LF (1983) GM_1 ganglioside stimulates the regeneration of dopaminergic neurons in the central nervous system. Brain Res 261:163-166

Toffano G, Savoini G, Moroni F, Lombardi G, Calzà L, Agnati LF (1984a) Chronic GM_1 ganglioside treatment reduces dopamine cell body degeneration in the substantia nigra after unilateral hemitransection in rat. Brain Res 296:233-239

Toffano G, Agnati LF, Fuxe K, Aldinio C, Consolazione A, Valenti G, Savoini G (1984b) Effect of GM_1 ganglioside treatment on the recovery of dopaminergic nigro-striatal neurons after different types of lesion. Acta Physiol Scand (to be published)

Varon S, Adler R (1980) Nerve growth factor and control of nerve growth. In: Kevin Hunt R (ed) Current topics in developmental biology, vol 16, part II. Academic, New York, p 207

Varon S, Adler R (1981) Trophic and specifying factors directed to neuronal cells. In: Federhoff S, Hertz L (eds) Advances in cellular neurobiology, vol 2. Academic, New York, p 115

Regeneration on Collagen Gels of Central and Peripheral Neurons Dissociated by a New Enzyme, RDB

A. Shahar[1], A. Amir[1], D. Ben-Nathan[1], R. Barzilai[1], H. H. Althaus[2], and V. Neuhoff[2]

1 Israel Institute for Biological Research, P.O. Box 19, Ness Ziona 70450, Israel
2 Max-Planck-Institut für Experimentelle Medizin, Herrman-Rein-Strasse 3, 3400 Göttingen, FRG

Introduction

Two main techniques are commonly in use for separation of central and peripheral neurons of fetal origin: mechanical dissociation by trituration (Lasher and Zagon 1972) or by passage through a series of meshes of graded pore sizes (Sensenbrenner 1977), and enzymatic digestion of cellular network by trypsin (Varon 1975). Following dissociation most of the neurons loose their processes and become rounded. There is some evidence that the enzymatic procedure causes damage to the neuronal membrane (Lodin et al. 1970). Nevertheless, with the use of the above methods, sprouting of neuronal processes is usually achieved in many of the cultured cells within 24-72 h after seeding.

While searching for a mild way to dissociate cells from cerebral hemispheres we came across a new enzyme, code named RDB, derived from a plant. It is very effective in releasing dividing cells from monolayers in culture flasks (Ben-Nathan et al., to be published). We describe here a practical use of RDB in the preparation of dispersed cell cultures derived from elements of fetal central (CNS) and peripheral (PNS) nervous system.

In addition, the present report describes an attempt to cultivate isolated nerve cells, as well as neuronal explants, in collagen gels.

Materials and Methods

Tissue Dissection

Dorsal root ganglia (DRG) were dissected from 12-day-old chick embryos as previously described (Shahar 1983). Brains were excised from 15 day rat or mouse fetuses and freed from their meninges under a stereomicroscope (Yavin and Yavin 1974). The striatum and midbrain were separated according to the anatomical descriptions of fetal rat brain (Olson and Seiger 1972). Spinal cords from 15-day-old mouse and rat fetuses were removed as whole mounts and the meninges, containing most of the DRG, were peeled off (Shahar et al. 1982).

Experimental Brain Research, Suppl. 13

Cell Dissociation

About 20-30 DRG (6-8 lumbar from each embryo) were washed twice in Eagle's and MEM transferred into a 15 ml conical tube containing 5 ml of RDB diluted in Eagle's MEM, 1:50, from the commercially available sample. The tube was then placed for 10 min in a 37°C water bath and shaken every 3-4 min. Final dissociation was achieved by a gentle trituration. Cells were centrifuged at 200 g, resuspended in nutrient medium, and seeded.

Brain parts and spinal cords were finely chopped with a cataract knife and exposed to RDB as described for DRG cells. After 10 min incubation at 37°C, Fetal Calf Serum (FCS)(15%) was added, to attenuate the action of RDB. At this stage many single dissociated cells were already present in the medium. The remaining tissue was resuspended in a medium containing FCS, and a mild trituration yielded more single cells.

Culture Conditions

DRG cells were cultured on collagen-coated dishes (Bornstein 1958) and CNS cultures were made on polyornithine-coated dishes (Prochiantz et al. 1979). Both tissues were also cultured in collagen gels (Michalopoulos and Pitot 1975) prepared in the following way: a mixture of 1 part of 5 times concentrated Dulbecco's MEM adjusted to pH 8.5 with 75 mM Hepes and 4 parts of undialyzed rat tail collagen was prepared. The pH of the mixture becomes 7.4. 0.1 ml of a concentrated cell suspension (10^7 cells/ml) is quickly added per 1 ml mixture. The cell-collagen preparation is quickly poured into 32-mm plastic dishes (1 ml per dish), and allowed to gel for 20 min in a CO_2 incubator. The gelled collagen containing the cells or explants is then covered with 1 ml of nutrient medium.

The media were prepared as previously described for DRG (Shahar 1983) and for CNS cells (Porzio et al. 1980). Medium was changed 2 h after seeding and every 4 days. Cultures were observed daily with a phase contrast microscope.

Results

About 90% of the DRG cells which had been dissociated with RDB exhibited under phase contrast a distinct glow indicative of cell viability. A large number of cells retained one or more residual processes after dissociation (Fig. 1A, B). In contrast, DRG cells dissociated with the conventional methods (free hand dissociation or trypsin digestion) usually yielded about 50% viable cells and only very few cells retained residual processes. It can be assumed, therefore, that the RDB dissociation is a mild procedure in comparison with the conventional procedures being used. Axonal sprouting of the residual fibers in the RDB dissociated cells started already during the first 5 h after seeding. Sprouting of new fibers from the perikarya occurred mainly after 24-72 h in culture. In collagen gels the regenerated fibers ramified in a tridimensional pattern showing active growth cones (Fig. 1C).

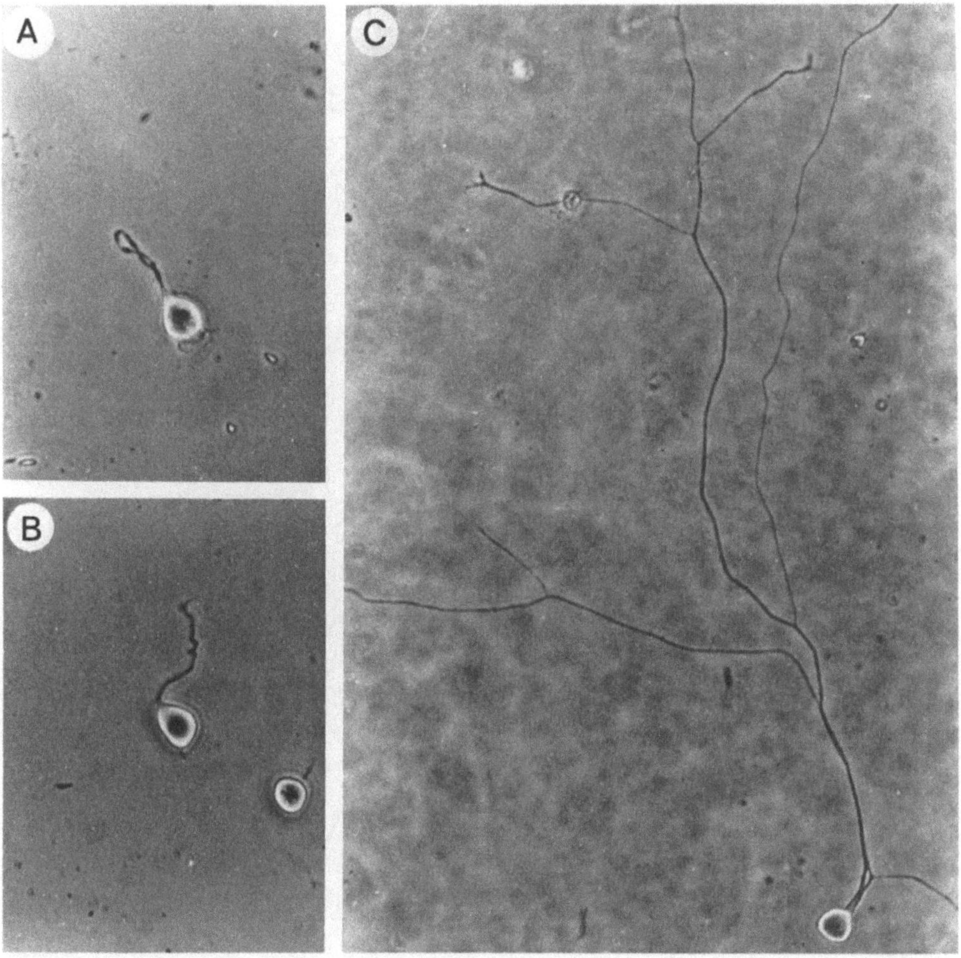

Fig. 1A-C. DRG cells dissociated with RDB. A, B: Glowing cells exhibiting residual processes. C: A single DRG neuron cultured for 4 days in collagen gel. Phase contrast: x 200

The incubation of fetal CNS tissue in the presence of RDB causes the release of single dissociated cells into the medium without the need for mechanical action of repetitive triturations. In the remaining tissue a loosening of cellular network can be observed, such that by further relatively gentle trituration it can be dissociated into single cells. Many of the separated neurons still possess apical dendrites and often also other residual processes. Here too, as in the case of DRG dissociation, the presence of these processes in many cells after dissociation is indicative of the mild action of RDB. Sprouting of residual processes commences already 1-2 h after seeding (Fig. 2A). The most active fiber regeneration takes place after 24-72 h. The fibers interconnected and developed during the first week into an intensive network (Fig. 2B-D). When labeled with ^{3}H-dopamine (Prochiantz et al. 1979; Porzio et al. 1980; Berger et al. 1982), RDB-dissociated midbrain cultures showed specific dop-

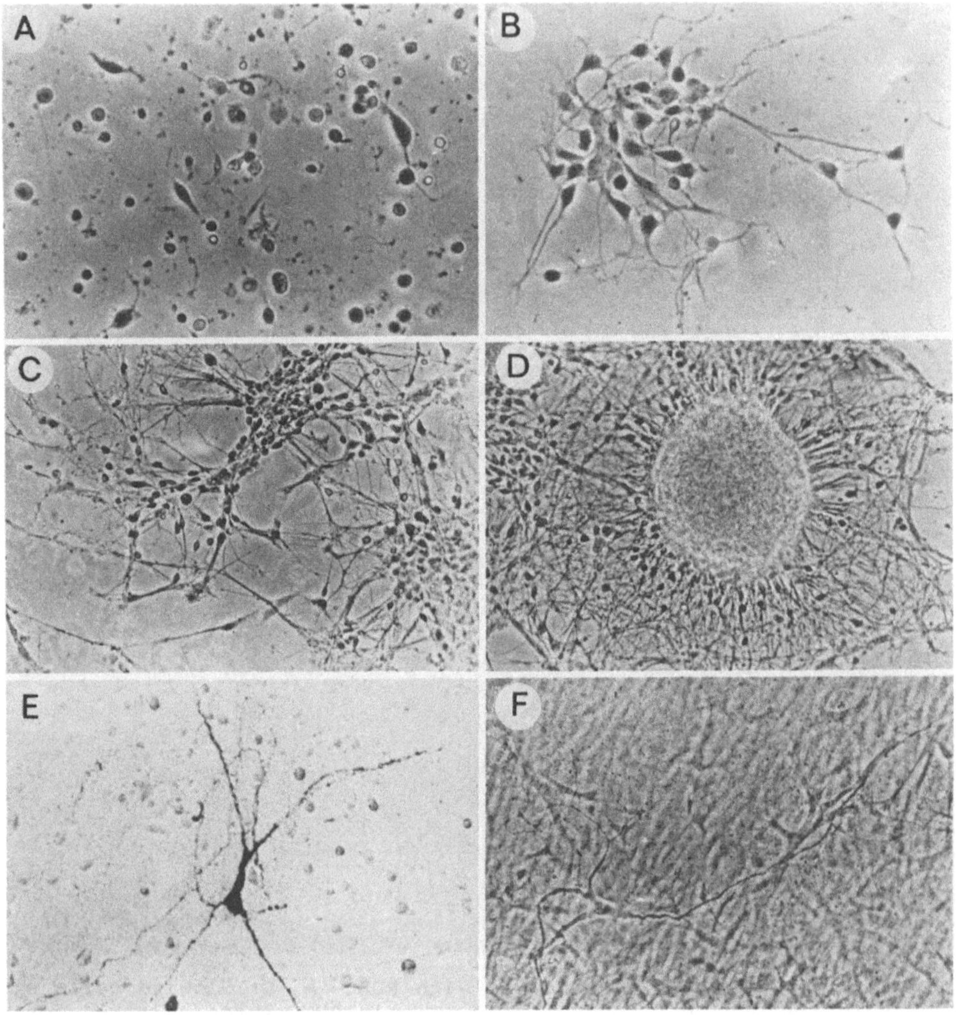

Fig. 2A-E. Rat brain cells dissociated with RDB, cultured on polyornithine: 1 h in culture. B-D: 5-day cultures of (B) striatum and (C,D) midbrain. E: Autoradiography of ^{3}H-dopamine labeled midbrain neuron after 9 days in culture. F: A motor neuron from chick enbryo spinal cord after 3 weeks in a collagen gel culture. Phase contrast: A, B, E x 240; C, D, F x 120 amine uptake by single neurons, as visualized by autoradiography (Fig. 2E). When cultured in collagen gels, RDB-dissociated neurons showed an active tridimensional sprouting of neuronal fibers as well as growth of non-neuronal dividing elements which took place during the first week in culture (Fig. 2F).

Discussion

In this study we describe the successful use of a new product, RDB, of plant origin, in dissociating peripheral and central neurons of fetal origin. RDB causes a mild loosening of the neuronal tissue and as a result facilitates the release of single neurons. The major number of PNS and CNS neurons separated by RDB show good morphological features (like a distinct glow and conserved processes) and have an early start of fiber regeneration in culture. During the next 2-3 weeks the RDB cultures have a similar growth pattern to mechanically dissociated cultures. We are now completing our studies on long-term differentiation of RDB cultures as well as on the ability of RDB to dissociate adult cerebral neurons.

Neurons, cells, and explants begin to sprout when cultured in plasma clots, or on reconstituted rat tail collagen (Murray 1965). This may not only be due to the nutritional value of the plasma and the collagen but also to the creation of a semisolid environment which allows a tridimensional neuronal sprouting. We here took advantage of a collagen gel technique described for cultivation of liver cells (Michalopoulos and Pitot 1975) and modified it to permit growth of neuronal cells. This method, which is simple and easy to apply has another advantage. It enables the localization of cells or explants from different origins in the same culture dish, by gelling each type in a separate collagen spot. These spots can be connected by collagen bridges. This will hopefully provide a system suitable for the study of taxis and trophic interactions.

Acknowledgments. Supported by DFG grant AL 184/2-2.

References

Ben-Nathan D, Barzilai R, Lazar A, Shahar A (to be published) RDB - a new product of plant origin for cell culture dispersion. In: Developments in biological standardization. 6th general meeting of European society of animal cell technology. Karger, Basel

Berger B, Porzio U di-, Daguet MC, Vigny A, Glowinski J, Prochiantz A (1982) Long-term development of mesencephalic dopaminergic neurons of mouse embryos in dissociated primary cultures: morphological and histochemical characteristic. Neuroscience 7:193-205

Bornstein MB (1958) Reconstituted rat tail collagen used as a substrate for tissue culture on coverslips. Lab. Invest 7:134-137

Lasher R, Zagon IS (1972) The effect of potassium on neuronal differentiation in cultures of dissociated newborn rat cerebellum. Brain Res 41:482-488

Lodin Z, Booher J, Kasten FH (1970) Long-term cultivation of dissociated neurons from embryonic chick dorsal root ganglia in the Rose chamber. Exp Cell Res 59:291-298

Michalopoulos G, Pitot HC (1975) Primary culture of parenchymal liver cells on collagen membranes. Exp Cell Res 94:70-78

Murray MR (1965) Nervous tissues in vitro. In: Willmer EN (ed) Cells and tissues in culture, vol 2. Academic, London, p 373

Olson L, Seiger A (1972) Early ontogeny of central monoamine neurons in the rat: fluorescence histochemical observations. Z Anat Entwickl Gesch 137:301-316

Porzio U di-, Daguet MC, Glowinski J, Prochiantz A (1980) Effect of striatal cells on in vitro maturation of mesencephalic dopaminergic neurons. Nature: 288:370-373

Prochiantz A, Porzio U di-, Kato A, Berger B, Glowinski J (1979) In vitro maturation of mesencephalic dopaminergic neurons from mouse embryos is enhanced in presence of striatal target cells. Proc Natl Acad Sci USA 76:5387-5391

Sensenbrenner M (1977) Dissociated brain cells in primary cultures. In: Fedoroff S, Hertz L (eds) Cell, tissue and organ cultures in neurobiology. Academic, New York, pp 191-213

Shahar A (1983) Isolation and culture of cells of the dorsal root ganglia. In: Fedoroff S, Hertz L (eds) Advances in cellular neurobiology. Academic, New York, p 311

Shahar A, Schupper H, Althaus HH, Neuhoff V (1982) "Aging" in cerebral and motor neurons of fetal and adult origin in long term cultures. In: Giacobini E, Filogamo G, Giacobini G, Vernadakis A (eds) Aging 20: the aging brain: cellular and molecular mechanisms of aging in the nervous system. Raven, New York, p 35

Varon S (1975) Neurons and glia in neural cultures. Exp Neurol 48:93-134

Yavin E, Yavin Z (1974) Attachment and culture of dissociated cells from rat embryo cerebral hemispheres on polylysine-coated surface. J Cell Biol 62:540-546

Environmental Factors that Influence the Differentiation and the Development of Voltage-Dependent Sodium Channel in Cultured Dorsal Root Ganglion Cells of Newborn Rats

H. Meiri, G. Omri, I. Zeitoun, and N. Savion

Rappaport Family Institute for Research in the Medical Sciences,
Efron St., P. O. Box 9697, Haifa 31096, Israel

Introduction

Nerve cells are distinguished from many other cells by their ability to generate and conduct electrical impulses (Hodgkin and Huxley 1952). This excitability provides the neurons their ability to conduct sensory information from the periphery to the CNS, to process the information along the CNS, and to send the motor drive for muscle contractions.

When a neuron recovers from a lesion and starts to regenerate and send out neurites, the newly formed membrane has to acquire its excitability (Fukuda and Kamayama 1979; Spitzer 1979; Fugii et al. 1982). It is interesting, therefore, to study how this process occurs.

Excitability is dependent upon the presence and activity of specific membrane macromolecules: the ion channels (Hodgkin and Huxley 1952; Catterall 1984). These channels are gated by a transmembrane depolarization to allow a rapid and transient movement of ions through the membrane (Hodgkin and Huxley 1952). Thus, one way to look at the development of excitability in neurons is to monitor channel appearance in the newly formed membrane.

Among the channels that participate in the nerve impulse, the voltage-dependent sodium channel plays a major role (Ritchie and Rogart 1977; Catterall 1984). This channel, which is repsonsible for the early depolarization of the action potential (Hodgkin and Huxley 1952; Catterall 1984), consists of three important functional entities: a voltage-dependent activation gating mechanism ("m" entity), a voltage-dependent inactivation gating mechanism ("h" entity), and a selectivity filter (Hodgkin and Huxley 1952; Ritchie and Rogart 1977; Catterall 1984). These functional entities correspond with toxin-binding sites. Tetrodotoxin (TTX) and saxitoxin bind to a selectivity filter site, thereby blocking the channel and arresting nerve conduction (Ritchie and Rogart 1977). α-Scorpion toxins and sea anemone toxin "knock out" channel inactivation by binding to a second site (Catterall 1984). β-Scorpion toxins bind to channel-activation sites, thereby producing repetitive firing (Barhanin et al. 1983). Finally, the lipid-soluble alkaloids, such as veratroidine and batrachotoxins, interact with both channel-activation and -inactivation sites, leading to permanent channel activation at resting potential (Catterall 1984).

The development of new immunological techniques has increased the possibility of producing a virtually unlimited repertoire of mono-

Experimental Brain Research, Suppl. 13

specific antibodies against any membrane macromolecule, including ion channels (Koehler and Milstein 1976). These techniques allow one to select probes of desired properties for monitoring channel development during the formation of new membrane. Monospecific antibodies produced by the hybridization technique (Koehler and Milstein 1976; Goding 1980; Galfree and Milstein 1981) share the properties of their parents: like the spleen cells, they are capable of generating a single type of antibody molecule against a single antigenic determinant; and, like a myeloma cell, they are capable of growing in culture, proliferating, and producing an immmortalized cell line, thus allowing the obtaining of their monospecific antibodies in high quantities. Using a specific screening procedure, one can then select, among the various hybrid lines, those which are appropriate to a goal.

Our strategy for making antibodies against the sodium channel, for their future use as a probe to detect channel appearance during neuronal outgrowth, consisted of two steps. First, we generated sodium channel monospecific antibodies against active sites of the channel which also stain the channel in the natural membrane. These antibodies were then employed to analyze the factors involved in the control of channel appearance along cultured neurons growing in defined conditions.

The "Making" of the Antibodies

Our screening procedure, which revealed at least 12 monospecific clones, was based upon the following three assumptions:

1. The channels as they are embedded in the natural membrane are immunogenic.

2. Active sites of the intact channels can serve as an antigenic determinant.

3. Active sites of the channel were conserved through evolution and thus are similar among different species.

All these assumptions, which were totally speculative when made, were proved to be correct and thus shed new light on the immunogenic properties of the sodium channel.

As an antigen we used membrane fragments of the eel electroplax (Grunhagen et al. 1981) from Electrophorus electricus (EMF). This antigen was chosen since it is the richest source of sodium channels. We used membrane fractions, anticipating that the channel in its natural configuration allows the immune system to interact with the active sites of the channels. The preliminary screening procedure (Meiri et al. 1984) was based upon four steps (Fig. 1):

1. Antibody recognition of the original antigen (EMF) in a solid--phase radioimmunoassay (SP-RIA).

2. Measuremet of the ability of sodium channel-specific neurotoxins to modulate the binding of the antibodies to the original antigen in a second SP-RIA. This step implies a possible association of antibody-binding sites with neurotoxin-binding sites.

3. Same measurement as in step 2, but using rat brain synaptosomal membrane (RMV) as another species of membrane preparation which is rich in sodium channels (Meiri et al. 1984). This step demonstrates a

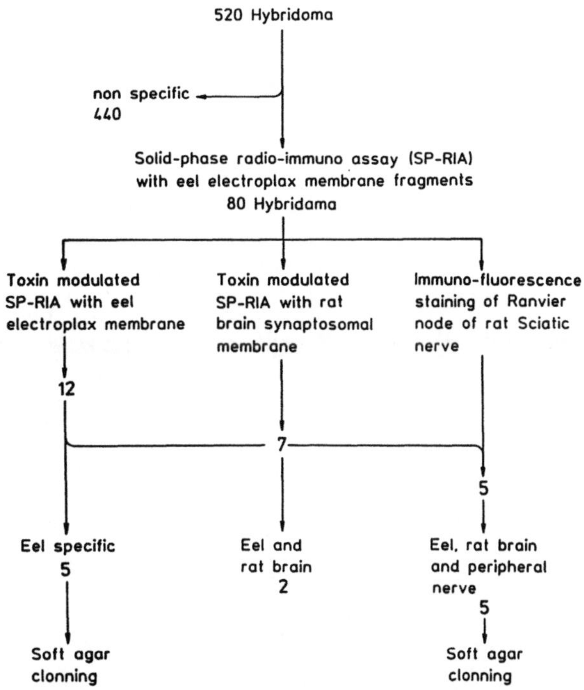

Fig. 1. Diagrammatic description of the screening procedure for selecting sodium channel-specific mAbs. Of 520 hybridomas received after fusion, 80 were selected according to their potent binding to the original antigen (EMF). A solid-phase radioimmunoassay (SP-RIA) revealed that Na channel-specific neurotoxins can modulate the binding of 12 hybridomas to EMF and of 7 hybridomas to rat brain synaptosomal membrane (RMV). Among these, five were able to stain the node of Ranvier of rat peripheral nerve when tested using indirect immunofluorescence. Those were then cloned and recloned on soft agar and taken for electrophysiological studies

cross-reactivity of antibodies with eel electroplax and mammalian CNS in association with Na channel-specific neurotoxin-binding sites.

4. The immunofluorescent staining of the nodes of Ranvier, where Na channels have been found (Ritchie and Rogart 1977; Waxman and Foster 1980) to be of high density ($\sim 2000/\mu m^2$). This step implies the location of the site for binding of monoclonal antibodies (mABs) at the external surface of the membrane and their cross-reactivity with mammalian peripheral nerve.

Hybrids which overcame these four steps and others were cloned and recloned on soft agar to ensure their monospecificity, and their electrophysiological properties were analyzed.

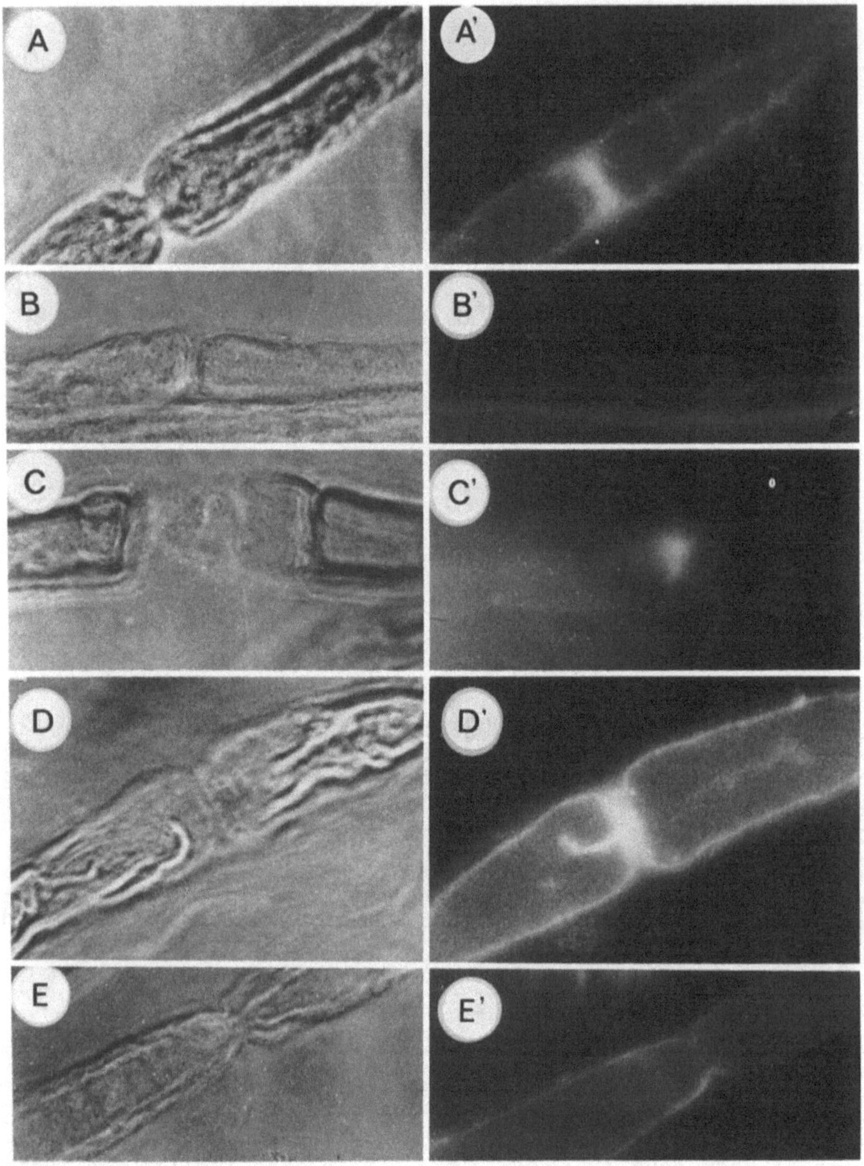

Fig. 2 A-E. Immunofluorescence staining of the node of Ranvier of a single fiber of rat sciatic nerve. A,A' Staining with mAbs of clone SC-66-5. B,B' The same, following preincubation with veratridine. C,C' The same, after acute demyelination. D,D' Staining with clone SC-72-14. E,E' The same, following veratridine preincubation (following Meiri et al. 1984).

Properties of Monoclonal Antibodies

SC-72-14: A Channel-Blocker Clone

The mAbs produced by this clone bind to EMF and RMV in the same potency (Table 1 in Meiri et al. 1984). Their binding is decreased by veratridine, is enhanced by α-scorpion toxin (Leirus), but is unaffected by TTX, which implies their potential interaction with antigenic determinants associated with the gating mechanisms of the channel (see Ritchie and Rogart 1977; Catterall 1984). These mAbs stained the nodes of Ranvier (Fig. 2D), and the staining was abolished following preincubation with veratridine (Fig. 2E). Finally, these mAbs block the compound action potential as recorded in either optic or sciatic nerve of rat (Fig. 1 in Meiri et al. 1984b). This action-potential blockage is due to attenuation of the inward (sodium) current as measured by voltage clamping single nodes of Ranvier of rat sciatic nerve (Fig. 3A). This effect is specific, as neither outward nor leakage current are affected (Meiri et al. 1985). The antibodies attenuate the current without shifting the membrane-potential values where the current normally "peaks" and without affecting E_{Na}. The attenuation is dose dependent, and the curve is very steep (Fig. 3B). This steepness is probably due to multiple interactions of the mAbs with several similar antigenic determinants of the sodium channel. These antigenic determinants are associated with the voltage-dependent channel-inactivation mechanism as h_∞ (Fig. 3C) is shifted along the voltage axis by 10-15 mV. In addition, they are associated with the voltage-dependent activation mechanism of the sodium channel as m_∞ is reduced in a voltage and time-independent manner (Fig. 4B in Meiri et al. 1985). Thus, this clone, which recognizes channel-gating entities and is capable of staining them in the membrane, can serve as a good probe for the localization of active entities of the channels in various preparations.

SC-72-38: A Channel-Activator Clone

These mAbs, when added either to neuroblastoma cells received from mice (NE-155) or to myoballs received from rat hind limb (Barhanin et al. 1984), produce repetitive firing (Fig. 4A and B). This effect results from the activation of the channel at resting membrane potential value, where the channel is normally closed (Fig. 4C). These antibodies mimic the electrophysical effect of the β-scorpion toxin of *Tityus serrulatus* (Barhanin et al. 1983) and also compete with the toxin for the binding to the channel (Figs. 1 and 2 in Barhanin et al. 1985). This way, these mAbs, which can be regarded as immunotoxins (Barhanin et al. 1985), can serve for monitoring the appearance of another antigenic determinant associated with the channel-activation gating mechanism.

Two other clones are in a preliminary stage of characterization. Monoclonal antibodies of clone SC-66-5 block action potential, stain specifically the node of Ranvier (Fig. 2A-C), and intensify the veratridine-stimulated Na influx into rat myoblasts or rat brain synaptosomes.

Another clone, SC-79-17, augments the *Anemone*-toxin enhancement of veratridine-stimulated ^{22}Na flux. It also increases and then blocks the propagation of the action potential and stains both nodes and

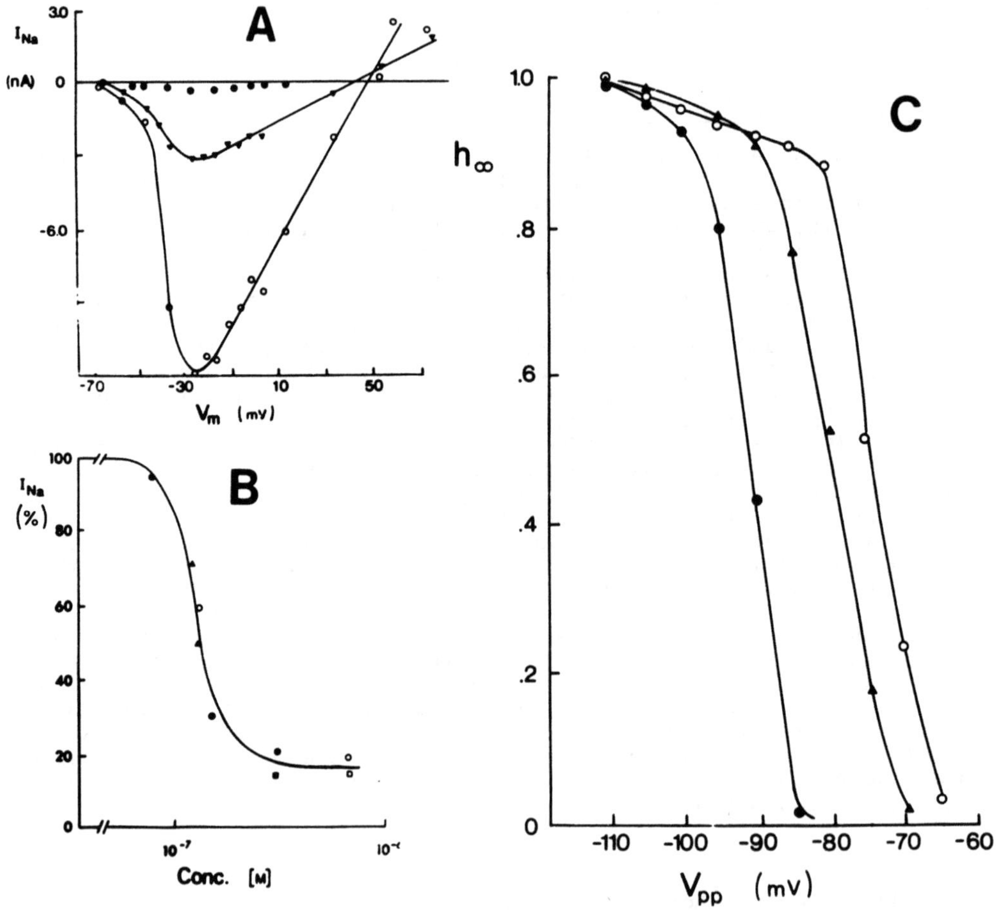

Fig. 3. A Inward sodium current (I/V curve) recorded from a voltage-clamped node of Ranvier of a single rat sciatic nerve. o, control: Δ, in presence of 100 µg/ml of mAbs SC-72-14. B Dose-response curve of inward current at various mAb concentrations. Various symbols represent measurements from various fibers. C The steady-state inactivation parameter (h_∞) is shifted along the voltage axis by 10 mV toward hyperpolarization

internodes. These findings imply the possibility that this mAb interacts with an antigenic determinant which is common to both Na^+ and K^+ channels.

These four clones and others which are in the process of characterization will now be used to detect channel development on outgrowing neurons.

Differentiation and Outgrowth of Cultured Dorsal Root Ganglion Cells

In this study, dorsal root ganglion (DRG) cells of newborn rats were chosen, due to the ease of culturing these cells (Fenton 1970; Varon

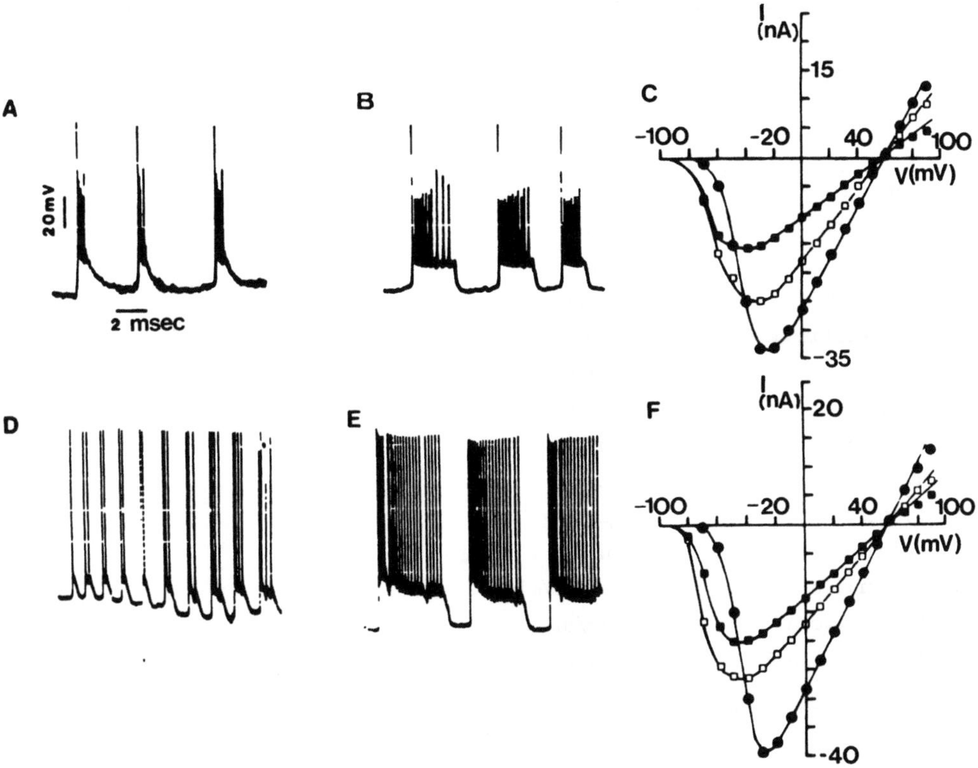

Fig. 4 A-F. Repetitive firing induced in mouse myoblast cell following incubation with 5 µg /ml mAbs of clone SC-72-38. A,D Control. B Presence of mAbs. C I/V curve of the sodium current recorded in control conditions (●) or in presence of mAbs (□,■). Note that Tityus-γ toxin produces similar effects (E,F)

et al. 1973), their high number, and relative homogeneity. Moreover, during the outgrowth of these cells, the action potential in the cell bodies depends on activations of Na^{+} channel, while the action potential at the growth cone depends on activation of Ca^{2+} channels (Matsuda et al. 1978; Yoshida et al. 1978; Fukuda and Kamayama 1979, 1980). Thus, this preparation can serve as an appropriate model to follow the interaction between channel development, membrane outgrowth, and functional excitability.

DRG cells were obtained from newborn rats after mechanical and enzymatic treatment (Varon et al. 1983). They were then seeded on a coverglass coated with either poly-L-lysine (PLL) or extracellular matrix (ECM) produced by bovine corneal endothelial cells (Kander et al. 1982; Voldavski et al. 1982).

We grew these cells in serum-free defined medium (SFM) where all medium components (Honegger and Lenoir 1982; Honegger et al. 1979) were carefully controlled (Table 1). Medium was replaced after 24 h with fresh SFM or with SFM containing various humoral factors, and at 5 days the cells were taken for visualization of growth properties and for their channel distribution.

Table 1. Composition of our serum-free medium (SFM)

Serum Free Medium (SFM)	
Basal Medium Eagle's (BME) supplemented with	
1. Glucose	1-2 mg/ml
2. Glutamine	2 mM
3. Insulin	25 μg/ml
4. Transferrin	25 μg/ml
5. Hydrocortisone	100 nM
6. Biotin	0,07 μg/ml
7. Vit. B_{12}	1,36 μg/ml
8. Carnitine	10 μM
9. T_3	30 nM
10. Linoleic acid	10 μM
11. Selenium	30 nM
12. Penicillin	50 units/ml
13. Streptomycin	50 μg/ml
14. Amphotericin B	125 ng/ml

These cells can grow and send neurites in SFM when cultured on PLL (Fig. 5). Addition of 1% fetal calf serum (FCS) enhanced neuronal survival and outgrowth (Fig. 5). The nerve growth factor (NGF) supported nerve survival (Fig. 5), enhanced outgrowth, and induced neuronal aggregation, as had previously been found (Varon et al. 1973; Dichter et al. 1977; Schwartz and Spirman 1982). Interestingly, the high-density lipoprotein (HDL), which is a plasma factor that supports the growth and survival of other cell types (Gospodarowicz and Tauber 1980), was found to induce nerve cell aggregation (Fig. 5D). When the matrix was made by bovine corneal endothelial cells (Gospodarowicz et al. 1980), nerve outgrowth and survival were strongly enhanced (Fig. 5E), as previously found for chick DRG (Fugii et al. 1982; Voldavski et al. 1982). Under these conditions, neither NGF nor HDL caused aggregation. Thus, our studies provided another example that the morphometric differentiation and the survival of the neurons are controlled by environmental factors.

Channel Appearance in DRG Cells is Regulated by Environmental Factors

Our sodium channel-specific mAbs are now being used to monitor channel distribution along the neurolemma of the DRG.

All studies were made in the following manner. Cells were washed well in 10 mM phosphate buffer containing 150 mM NaCl (PBS), 1.5 mM $CaCl_2$, 0.5 mM $MgCl_2$, 3.1 mM KCl, and 10 mM glucose, pH 7.4 (PBS*). They were first incubated for 30 min with 10% horse serum in PBS*, then with mAbs (10-50 μg/ml) for 2-3 h at room temperature. After being washed

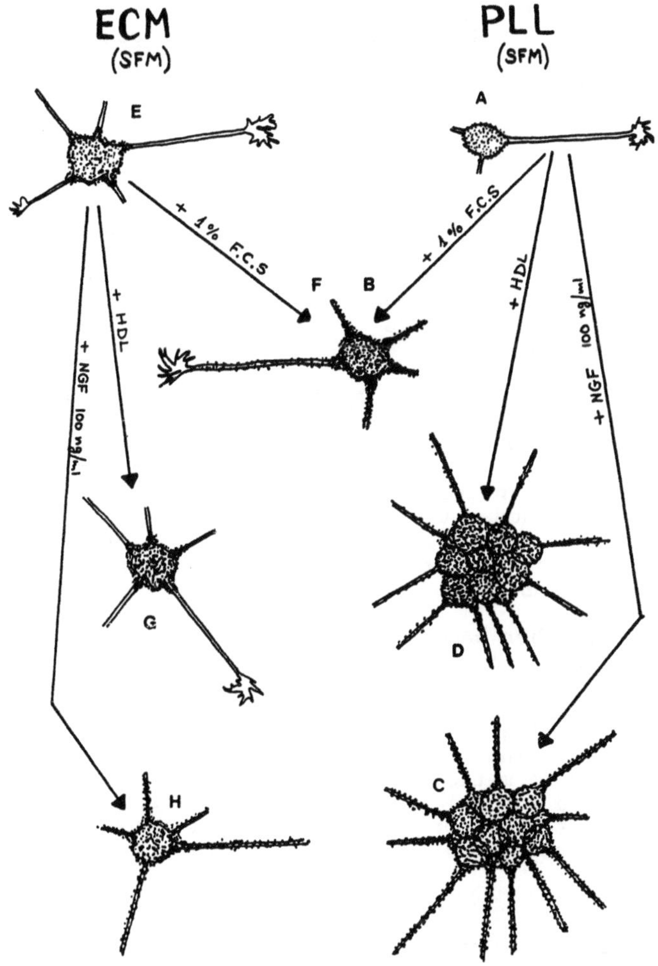

Fig. 5. Diagrammatic description of the influence of environmental factors on channel appearance in relation to differentiation of cultured rat DRG neurons. A-D Cultures grown on PLL-coated dishes. E-H Cultures grown on ECM/coated dishes. A, E SFM. B, F SFM + 1% FCS. H, C SFM + 100 µg/ml NGF. D, G SFM + 10 µg/ml HDL

well, the cells were exposed to a 30-min fixation with 3% paraformaldehyde in PBS* without Ca and Mg. Then they were washed well and exposed to rhodamine-labeled goat or rabbit anti-mouse IgG (50 µg/ml) for 1-2 h. Then the slides were washed well and mounted in 90% glycerol in PBS* and observed at 63 x 10 magnification in a Zeiss fluorescent microscope. Staining specificity was determined, using veratridine preincubation to prevent staining of Sc-72-14 mAbs or using Tityus-γ toxin preincubation to prevent staining of SC-72-38 mAbs.

The results were as follows (Fig. 5): When the cells were cultured in SFM on PLL only cell bodies were stained. FCS induced channel appearance along the neurites toward the growth cone. Since FCS is a "black box" containing various constituents, we analyzed several serum factors for their ability to affect the channel formation.

NGF was found in this study, as in many previous ones, to support nerve survival and morphometric differentiation. Previous findings showed that NGF also enhances the excitability of neurons (Burnstein and Green 1978). When we exposed the neuron to NGF, it induced channel appearance along the neurites. The possible association between the effect of NGF's on Na-channel appearance and its known effect on hyperpolarizing membrane potential by activating the Na/K-ATPase is being carefully analyzed by measuring staining at potassium-depolarized conditions. This acute depolarization was not sufficient to prevent staining. However, it would be of interest to identify a possible link between NGFs effect on Na/K-ATPase, resting potential, and Na channels.

HDL induced channel appearance along the neurites (Fig. 5). This effect, like NGF's was dose dependent and also correlated with the degree of aggregation produced by these factors. It is important to note that other lipoproteins have been found to influence channel appearance in heart tissue (Renaud et al. 1982). Combining NGF and HDL or those two with FCs led to intense staining of the entire length of neuronal sprouts (Fig. 6). These results show that humoral factors are involved in regulating channel appearance along the newly formed membrane.

On the other hand, when DRG were grown on ECM which support nerve survival and outgrowth, channel staining was found on cell bodies only (Fig. 5F). Addition of our humoral factors (NGF, HDL) or serum led to channel appearance on initial zones only, but not along the sprouts (Fig. 5G). Thus, it seems that the matrix is involved mostly in supporting morphometric differentiation and later plays a role in synapse formation, while humoral factors support both channel appearance and morphological differentiation. In addition, these results imply that membrane formation and channel appearance are not necessarily regulated by the same factors.

Summary and Conslusions

This study revealed the potential use of immunological probes to localize channel formation along the thin neuronal processes during neuronal outgrowth and for analyzing the factors that are involved in regulating this process. Humoral factors as well as various matrices were analyzed for their potential influence on channel appearance. A possible role of calcium ions in this process (Hinnen and Monard 1980; Anglister et al. 1982; Koike 1983; Sherman and Catterall 1984) is yet to be examined.

It is obvious that this study just opens the gate for extensive research. Only early stages of outgrowth were examined. The mechanisms by which environmental factors influence channel appearance have yet to be worked out. The factors involved in the process which leads to nerve maturation and to the development of the node of Ranvier (where in the mature neuron the channels are found exclusively) have to be worked out (Ritchie and Rogart 1977; Waxman and Foster 1980; Ritchie and Rang 1983; Rieger et al. 1984). In addition, we would like to know if the determinants recognized by the antibodies in the developing neurons are associated with Na-channel activity and the actual excitability of the developing neurons (Yoshida et al. 1978; Spitzer 1979; Lombet et al. 1983). It is also important to analyze if devel-

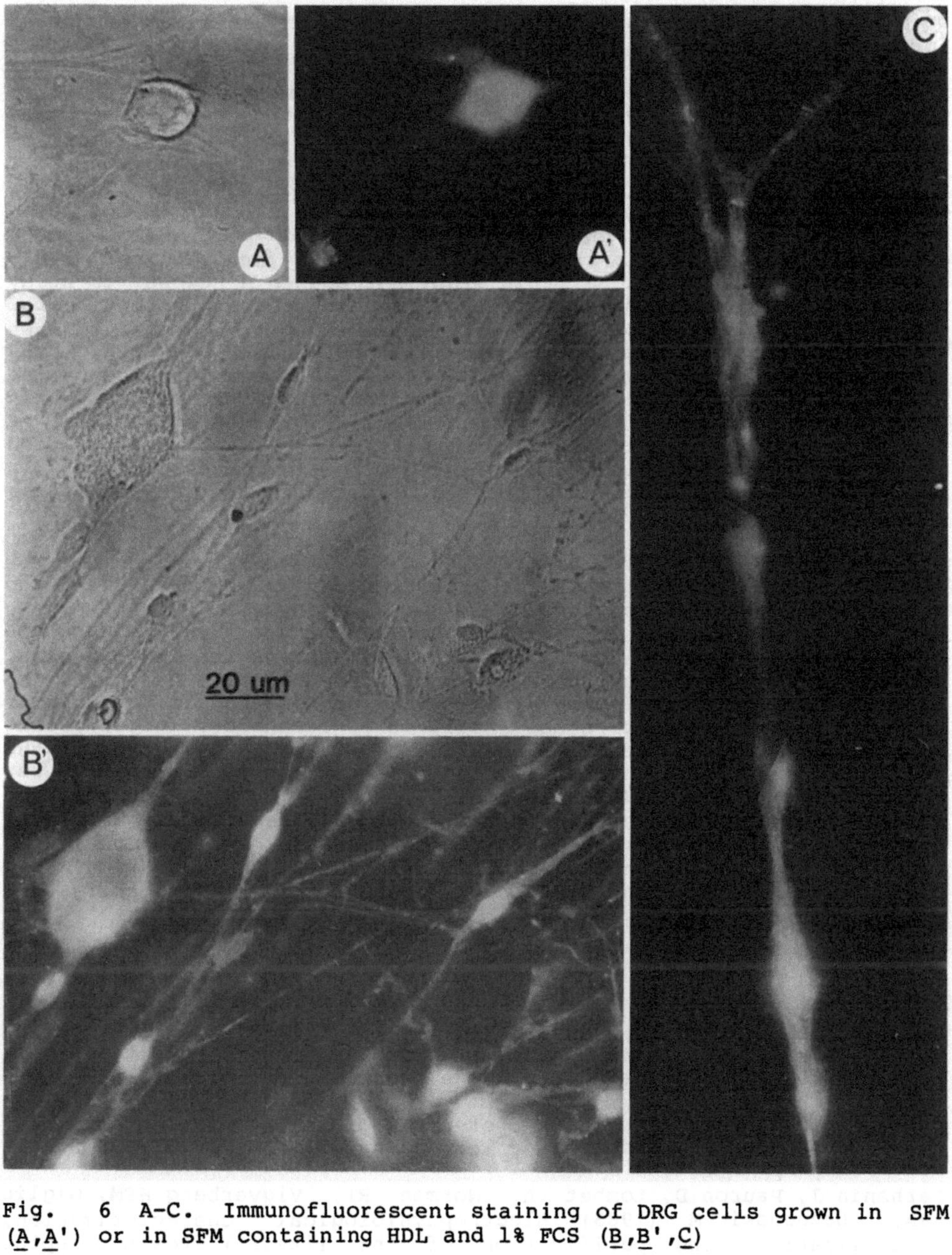

Fig. 6 A-C. Immunofluorescent staining of DRG cells grown in SFM (A,A') or in SFM containing HDL and 1% FCS (B,B',C)

oping channels are affected by the antibodies in a similar manner as in the mature nerve.

Finally, we would like to know how the environmental factors which were studied in this work produced their effect on channel appearance. A number of mechanisms are of particular interest: 1) activation of a previously existing channel to function (Renaud et al. 1981, 1982), 2) support of channel assembly in the membrane (Takenaka et al. 1983; Strichartz et al. 1984), and 3) induction of gene expression to support channel synthesis and membrane assembly (Baccaglini 1978; Blair and Dionne 1985; Bixby and Spitzer 1984; O'Dowd 1983). Nevertheless, this study has opened a new way to study these questions on the molecular level. An interesting potential of our method involves comparison of the staining pattern received when different clones against various channel determinants are used. Previous studies revealed a discrepancy between the appearance of the binding sites to TTX and that of the binding sites to scorpion toxin (Baumgold et al. 1983; Frelin et al. 1981). Our antibodies gave us the opportunity to analyze the molelcular exposure and organization of the antigenic determinants associated with the channel activity in the membrane during the development of excitability (Berwald-Netter et al. 1981).

Acknowledgments. We are grateful to Dr. Y. Schlessinger of the Weizmann Institute of Science for his continuous encouragement and helpful discussions throughout this work, to Dr. Z. Eshhar of the Weizmann Institute for immunological advice, and to H.H. Grunhagen from Germany for providing eel electroplax membrane for immunization. We appreciate very much the contribution of Drs. Y. Palti and E. Goren of the Rappaport Family Institute for Research in the Medical Sciences at the Technion-Israel Institute of Technology in conducting the voltage-clamp studies. We thank V. Lev-Ram and I.R. Cohen of the Weizmann Institute, who participated in the pilot recording from rat optic nerves, and Drs. J. Barhanin, G. Romey, and M. Lazdunski, who conducted the analysis of neurotoxin-antibody interactions. This work was supported by the Israel Center for Psychobiology, the Charles E. Smith Family Foundation (Grant 41/81), the Shauder Foundation, and the Schrieber Foundation. H.M. and N.S. are Bat-Sheva Scholars.

References

Anglister L, Farber IC, Shahar A, Grinvald A (1982) Localization of voltage-sensitive calcium channels along developing neurites: their possible role in regulating neurite elongation. Dev Biol 94:351-365

Baccaglini PI (1978) Action potential of embryonic dorsal root ganglion neurons in Xenopus tadpoles. J Physiol (Lond) 283:585-604

Barhanin J, Pauron D, Lombet A, Norman RI, Vigverberg HPM, Giglio JRA, Lazdunski M (1983) Electrophysiological characterization, solubilization, and purification of the Tityus-γ -toxin receptor associated with the gating component of the Na^+ channel from rat brain. EMBO J 2:915-920

Barhanin J, Romey G, Lazdunski M, Meiri H (1985) A monoclonal immunotoxin acting on the Na^+ channel with properties similar to those of a scorpion toxin. Proc Natl Acad Sci USA 82:1842-1846

Baumgold J, Zimmerman I, Bambrick L (1983) Appearance of [^{3}H] saxitoxin binding sites in developing rat brain. Dev Brain Res 9: 405-407

Berwald-Netter Y, Martin-Moutot M, Koulakoff A, Couraud F (1981) Na-channel-associated scorpion-toxin receptor sites as probes for neuronal evolution in vivo and in vitro. Proc Natl Acad Sci USA 78: 1245-1249

Bixby JL, Spitzer NC (1984) Early differentiation of vertebrate spinal neurons in the absence of voltage-dependent Ca and Na influx. Dev Biol 106:89-96

Blair LAC, Dionne VE (1985) Developmental and acquisition of Ca^{2+}-sensitivity of K^+ channels in spinal neurons. Nature 315:329-330

Burnstein DE, Green LA (1978) Evidence for RNA synthesis-dependent and independent pathways in stimulation of neurite outgrowth by nerve growth factor. Proc Natl Acad Sci USA 75:6059-6063

Catterall WA (1984) The molecular basis of neuronal excitability. Science 223:653-661

Dichter MA, Tischler AS, Green CA (1977) Nerve growth factor-induced increase in electrical excitability and acetylcholine sensitivity of a rat pheochromocytoma cell line. Nature 268:505-505

Fenton EL (1970) Tissue culture assay of nerve growth factor and of the specific antiserum. Exp Cell Res 59:383-392

Frelin C, Lombet A, Vigne P, Romey G, Lazdunski M (1981) The appearance of voltage-sensitive Na^+ channels during the in-vitro differentiation of embryonic chick skeletal muscle cells. J Biol Chem 256:12355-12361

Fugii DK, Massoglia SL, Savion N, Gospodarowicz D (1982) Neurite outgrowth and protein synthesis by PC_{12} cells as a function of substratum and nerve growth factor. J Neurosci 2:1157-1175

Fukuda J, Kamayama M (1979) Enhancement of calcium spikes in nerve cells of adult mammals during neurite outgrowth in tissue culture. Nature 279:546-548

Fukuda J, Kamayama M (1980) Tetrodotoxin-sensitive and tetrodotoxin--resistant sodium channels in tissue-cultured spinal ganglion neurons from adult mammals. Brain Res 182:191-197

Galfree G, Milstein C (1981) Preparation of monoclonal antibodies: strategies and procedures. In: Van Vunakis H, Langore JJ (eds) Methods in enzymology, vol 73b. Academic, New York, p 1

Goding JW (1980) Antibodies production by hybridomas: a review article. J Immunol Methods 39:285-308

Gospodarowicz D, Tauber J-P (1980) Growth factors and the extracellular matrix. Endocr Rev 3:201-227

Gospodarowicz D, Delgado D, Vlodavsky I (1980) Permissive effect of the extracellular matrix on cell proliferation in vitro. Proc Natl Acad Sci USA 77:4094-4098

Grunhagen HH, Dahl G, Reiter P (1981) Tetrodotoxin receptors in membrane fragments: purification from Electrophorus electricus electroplax and binding properties. Biochim Biophys Acta 642:267-285

Hinnen R, Monard D (1980) Involvement of calcium ions in neuroblastoma neurite extension. In: Jimenez de Asua LD (ed) Control mechanisms in animal cells. Raven, New York, p 315

Hodgkin AL, Huxley AF (1952) A quantitative description of membrane current and its application to conduction and excitation in nerve. J Physiol (Lond) 117:500-544

Honegger P, Lenoir D (1982) Nerve growth factor (NGF) stimulation of cholinergic telecephalic neurons in aggregating cell cultures. Dev Brain Res 3:229-238

Honegger P, Lenoir D, Faurod P (1979) Growth and differentiation of aggregating fetal brain cells in a serum-free defined medium. Nature 282:305-307

Kander AD, Fujii DK, Gospodarowicz D, Reichardt IF (1982) Characterization of a factor that promotes neurite outgrowth: evidence linking activity to heparan sulfate proteoglycan. J Cell Biol 94:574-585

Koehler G, Milstein C (1976) Deviation of specific antibodies producing tissue culture and tumor lines by cell fusion. Eur J Immunol 6: 511-519

Koike T (1983) Nerve growth factor-induced neurite outgrowth of rat pheochromocytoma PC_{12} cells: dependence on extracellular Mg^{2+} and Ca^{2+} . Brain Res 289:292-303

Lombet A, Kazazoglou T, Delpont E, Renaud JF, Lazdunski M (1983) Ontogenic appearance of Na^+ channels characterized as high-affinity binding sites for tetrodotoxin during development of the rat nervous system and skeletal muscle system. Biochem Biophys Res Commun 110: 894-901

Matsuda Y, Yoshida S, Yonezawa T (1978) Tetrodotoxin sensitivity and calcium component of action potentials of mouse dorsal root ganglion cells cultured in vitro. Brain Res 154:69-83

Meiri H, Zeitoun I, Grunhagen HH, Lev-Ram V, Eshhar Z, Schlessinger J (1984) Monoclonal antibodies associated wtih sodium channel block nerve impulse and stain nodes of Ranvier. Brain Res 310:168-173

Meiri H, Goren E, Bergmann H, Zeitoun I, Palti Y (1985) Specific modulation of sodium channels in mammalian nerve by monoclonal antibodies. Proc Natl Acad Sci USA (to be published)

O'Dowd DK (1983) RNA synthesis-dependence of action potential development in spinal cord neurones. Nature 303:619-621

Renaud JF, Romey G, Lombet A, Lazdunski M (1981) Differentiation of fast Na channel in embryonic heart cells: interaction of the channel with neurotoxins. Proc Natl Acad Sci USA 78:5348-5352

Renaud JF, Scanu AM, Kazazoglou T, Lombet A, Romey G, Lazdunski M (1982) Normal serum and lipoprotein-deficient serum give different stages of differentiation in chicken cardiac cells in culture. Proc Natl Acad Sci USA 79:7768-7772

Rieger F, Pincon-Raymond M, Lombet A, Ponsio G, Lazdunski M, Sidman RL (1984) Paranodal dysmyelination and increase in tetrodotoxin-binding sites in the sciatic nerve of the motor end-plate disease (med/med) mouse during postnatal development. Dev Biol 101:401-409

Ritchie JM, Rang HP (1983) Extraneuronal saxitoxin-binding sites in rabbit myelinated nerve. Proc Natl Acad Sci USA 83:2803-2805

Ritchie JM, Rogart RB (1977) The binding of saxitoxin and tetrodotoxin to excitable tissue. Rev Physiol Biochem Pharmacol 79:1-50

Schwartz M, Spirman N (1982) Sprouting from chicken embryo dorsal root ganglia induced by nerve growth factor is specifically inhibited by affinity-purified gangliosides antibodies. Proc Natl Acad Sci USA 79:6080-6083

Sherman SJ, Catterall WA (1984) Electrical activity and cystolic calcium regulate levels of tetrodotoxin-sensitive sodium channels in cultured rat muscle cells. Proc Natl Acad Sci USA 81:262-266

Spitzer NC (1979) Ion channels in development. Annu Rev Neurosci 2: 363-397

Strichartz GR, Small RK, Pfenninger KH (1984) Components of the plasma membrane of growing axons. III. Saxitoxin binding to sodium channels. J Cell Biol 98:1444-1452

Takenaka T, Horie H, Kawasaki Y (1983) Effects of fatty acids on the membrane fluidity of cultured chick dorsal root ganglion measured by fluorescence photobleaching recovery. J Neurobiol 14:457-461

Varon S, Raiborn C, Tyska E (1973) In-vitro studies of dissociated cells from newborn mouse dorsal root ganglia. Brain Res 54:651-673

Voldavski I, Levi A, Lux I, Fuks Z, Schlessinger J (1982) Induction of cell attachment and morphological differentiation in pheochromocytoma cell line and embryonal sensory cells by the extracellular matrix. Dev Biol 93:285-300

Waxman SG, Foster RE (1980) Ionic channel distribution and heterogeneity of the axon membrane in myelinated fibers. Brain Res Rev 2: 205-234

Yoshida S, Matsuda Y, Samajina A (1978) Tetrodotoxin-resistant and calcium component of action potentials in dorsal root ganglion cells of adult mouse. J Neurophysiol 41:1096-1106

Behavioral Recovery After Trauma

Recovery from Biochemical and Behavioral Deficits in Adult Rats with Partial Lesions of the Fimbria

A. R. Dravid[1], A. L. Jaton[1], and E. B. Van Deusen[2]

1 Klinische Forschung, Sandoz Ltd., 4002 Basel, Switzerland
2 Department of Development and Cell Biology, University of California, Irvine, Ca 92717, USA

Introduction

The suggestion that re-establishment of connections damaged by neural trauma could lead to functional recovery (Raisman 1969) has been supported by recent evidence for postlesion recovery of function subsequent to reinnervation of denervated targets by appropriate neural transplants (Dunnett et al. 1982). The ability of intact central axons present in the vicinity of a lesion to sprout and replace synapses lost to deafferentation (Cotman et al. 1981) led us to postulate that homotypic and therefore functionally relevant reinnervation might also occur if homologous efferents innervating regions near to the denervated ones were spared by partial lesions of a neuronal pathway.

In support of this possibility we found that partial lesions of septal cholinergic efferents induced by transection of only a part of the fimbrial bundle resulted in an initial decline in enzyme markers for hippocampal cholinergic terminals (choline acetyltransferase, ChAT; acetylcholinesterase, AChE) followed by a substantial recovery during the 8-week postlesion interval (Dravid and Van Deusen 1983). In subsequent studies in which the topography of cholinergic innervation of the hippocampus by dorsal and ventral routes (Meibach and Siegel 1977; Gage et al. 1983a) was taken into consideration, we have further found that selective (lateral, medial, or complete) fimbrial transections resulted in region-specific enzyme reductions along the septotemporal axis of the hippocampus, followed by significant recovery 8 weeks after lesioning (Dravid and Van Deusen 1984).

In the present communication we present evidence indicating the involvement of septal cholinergic efferents spared by partial fimbrial lesions in the recovery of hippocampal ChAT and AChE activities. In a second part of this study we show that recovery from behavioral deficits induced by such lesions also occurs.

Methods

Female Wistar rats weighing 180-200g at the time of surgery were used. Under hexobarbital anesthesia, the lateral half, the medial half, or the entire fimbrial bundle was unilaterally transected with a 1-mm microknife using stereotaxic coordinates previously described in detail (Dravid and Van Deusen 1984). Male rats with bilateral transection of the lateral half of the fimbria were used for behavioral studies in which passive and active avoidance conditioned

Experimental Brain Research, Suppl. 13

responses were investigated according to the methods described by Ader et al. (1972) and Lo Conte et al. (1982), respectively. In order to avoid damage to the striatum, fimbrial transections were performed at an angle 40° anterior to the vertical plane through the interaural line.

The animals were housed in standard animal quarters maintained on a 12-h light-dark cycle and had free access to food and water. The procedures for processing the hippocampus and for the determination of cholinergic enzyme activities were as described previously (Dravid and Van Deusen 1984).

Results and Discussions

In order to provide the background necessary for appreciation of the present results, our recently published findings (Dravid and Van Deusen 1984) are briefly summarized as follows:

1. In unlesioned control animals ChAT and AChE activities were distributed in a highly significant ascending gradient along the septotemporal axis of the hippocampus, indicating a relatively higher density of cholinergic terminals in temporal regions, in agreement with recent anatomical findings (Milner et al. 1983). This structural heterogeneity within the hippocampus extends earlier findings of higher ChAT activity in ventral relative to dorsal hippocampus (Fonnum and Walaas 1978; Hoover et al. 1978), uneven distribution of pyramidal and granule cells along the septotemporal axis in the rat (Gaarskjaer 1978) and human (Ball 1977) hippocampus, and selective vulnerability of the posterior hippocampus to neuronal degeneration in senile dementia of Alzheimer's type (Ball 1977).

2. Enzyme depletions in hippocampal regions following lateral, medial, or complete transection of the fimbrial bundle were in accordance with the topography of cholinergic innervation of this structure by the dorsal and ventral projections (Meibach and Siegel 1977; Gage et al. 1983a).

Table 1. Statistically significant recovery of cholinergic enzymes in septotemporal hippocampal regions at 8 weeks relative to maximal depletions at 1 week after lesioning: Ratio 8 wk / 1 wk

Hippocampal Regions		SP 1	2	3	4	5 TP
CFT	ChAT	-	-	-	2.00	1.93
	AChE	-	-	-	-	1.33
MFT	ChAT	-	-	-	1.76	1.47
	AChE	-	-	-	1.32	1.20
LFT	ChAT	1.14	1.21	1.55	2.23	1.68
	AChE	1.09	1.17	1.35	1.68	1.49

SP and TP represent septal and temoporal poles of the hippocampus, respectively. CFT, MFT, and LFT represent complete, medial, and lateral fimbrial transection, respectively. p values: <0.05 to <0.001. For details see Dravid and Van Deusen (1984)

3. The statistically significant recovery of enzyme markers after the three types of fimbrial lesions, previously described in detail (Dravid and Van Deusen 1984), are summarized in Table 1.

It can be seen that following medial or complete lesions enzyme recoveries were confined to temporal regions whereas after lateral fimbrial lesions all hippocampal regions exhibited significant recovery of ChAT and AChE activities. We interpret these findings to reflect reinnvervation of denervated regions following sprouting of undamaged cholinergic fibers, particularly fimbrial fibers spared by lateral transection of the fimbria. Further studies to demonstrate the role of these fibers in the recovery process, and to investigate lesion-induced behavioral consequences, were carried out in animals with lateral fimbrial lesions.

Animals with standardized, unilateral transection of the lateral half of the fimbrial bundle were distributed at random into three groups. Group I was killed 1 week after lesioning and the resultant maximal enzyme deficits in the five hippocampal regions were determined. Spontaneous recovery of enzyme activities was estimated in group II, killed 8 weeks after lesioning. In group III, the remaining medial fimbrial fibers were transected 8 weeks after the first lesion, the animals killed after another week of survival, and hippocampal enzyme activities determined. These results are shown in Fig. 1.

It can be seen in Fig. 1 that 1 week after lesioning (open columns) the significant loss of ChAT and AChE activities was progressively more severe towards the temporal pole of the hippocampus. Eight weeks after lesioning (hatched columns) a substantial recovery of both enzymes was clearly apparent in all regions, confirming our earlier findings summarized in Table 1. The magnitude of initial degeneration seems to determine the extent of enzyme recovery, which probably represents reinnervation of denervated regions due to local proliferation of cholinergic fibers spared by the first lesion. The complete restitution of enzyme activities in the septalmost region may then reflect a ceiling effect. However, the total loss of recovery after transection (by second lesion) of the remaining medial fimbrial fibers (stippled columns) clearly demonstrates the essential role of these fibers in the recovery of cholinergic enzymes in all regions of the hippocampus. It appears, therefore, that intact fibers terminating at a distance may also contribute to the recovery, since medial fibers which mainly innervate septal hippocampal regions (Meibach and Siegel 1977; Dravid and Van Deusen 1984) effect recovery in temporal regions. The same topographic feature can also account for the relatively marked enzyme depletions in septal regions resulting from the second lesion.

Several studies have shown a correlation between behavior and the hippocampal cholinergic system (Grossman 1976). For example, rats with hippocampal lesions exhibit deficits in retention of passive avoidance conditioned response (Glick et al. 1973). Such animals are also reported to be hypersensitive to external stimuli and to respond repetitively in a stereotypic manner (Kimble and Kimble 1965). It was of interest, therefore, to investigate if partial fimbrial lesions which induced depletion and subsequent recovery of hippocampal cholinergic enzymes also result in behavioral impairments which improve with time. In our behavioral studies passive and active avoidance conditioned responses of male rats with partial lesions of the fimbria were determined. Animals with bilateral partial fimbrial lesions were used since unilateral lesions did not induce marked behavioral alterations (see also Gage et al. 1983b).

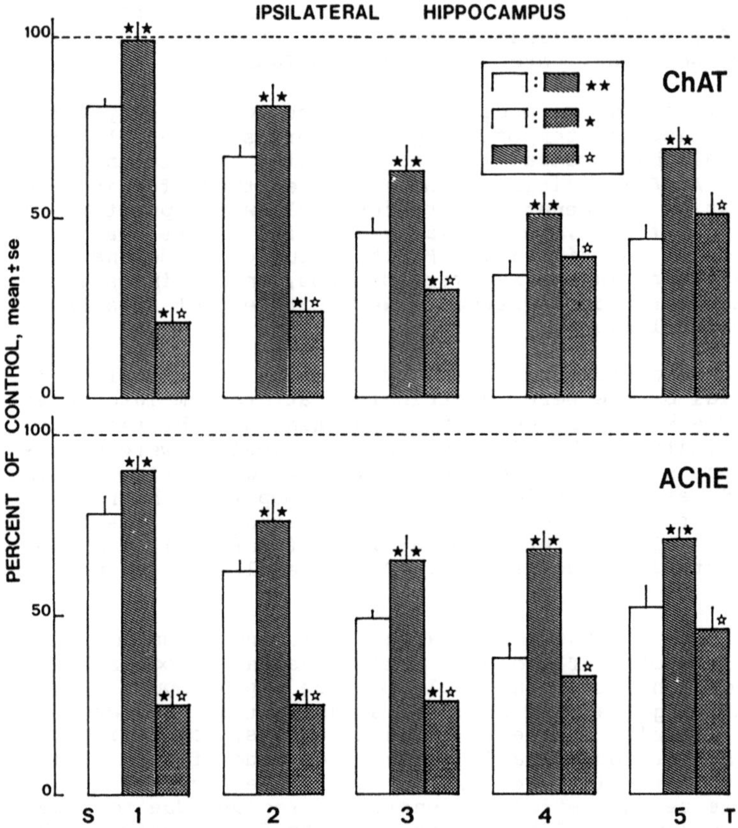

Fig.1. Hippocampal ChAT and AChE activities 1 week (open columns, group I, n=10) and 8 weeks (hatched columns, group II, n=8) after partial (lateral) fimbrial lesions, and 1 week after the second (medial) fimbrial transection at 8 weeks after the first lesion (stippled columns, group III, n=9). The numbers below denote the five septo(S)-temporal(T) regions. Enzyme activities are expressed as percent of those in corresponding regions from control rats. Stars denote statistically significant changes following comparisons among the three groups (see insert); $p < 0.05$

In the first experiment, 24-h retention following one trial learning of a passive avoidance task was investigated in control and lesioned animals. Animals were tested 2 and 10 weeks after lesioning. These times were chosen since lesion-induced depletion of hippocampal cholinergic enzymes was still near maximal at 2 weeks, and substantial recovery of enzyme activities was observed prior to the 10-week postlesion interval (Dravid and Van Deusen 1984).

The results of this experiment, presented in Fig. 2, indicate that lesioned animals exhibit significant impairment in the retention of passive avoidance response 24 h after the initial trial (50% shorter latency as compared with controls). These findings are in agreement with earlier studies (Glick et al. 1973). Subsequent testing of animals 10 weeks after lesioning, however, revealed that the performance of lesioned animals was identical to that of the controls

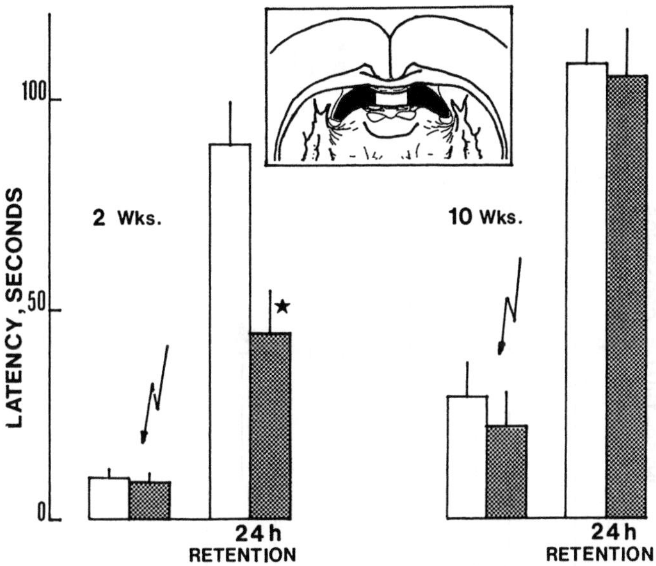

Fig. 2. Twenty-four hour retention in the passive avoidance task was determined as follows: Testing at 2 weeks postlesion was performed in a two-chamber system separated by a guillotine door. During the first session, control and lesioned animals were trained keeping the door to the dark chamber open. Once the rat was inside the second chamber, the partition was lowered and an electric shock (0.3 mA) was applied for 3 s. After this training trial the animals were returned to their home cages to await testing the retention 24 h later; this was accomplished as before, and the latency (in seconds) to enter the second chamber was recorded. Testing of animals 10 weeks after lesioning was performed as above, except that the animals were required to enter the chamber from an attached platform. The darkened area in the insert shows the approximate extent of the bilateral fimbrial lesion. Open columns: control animals, n=10. Stippled columns: lesioned animals, n=10. Star represents statistically significant reduction in the latency, p = <0.05, Student's t-test

(Fig. 2). Thus the time course of recovery in passive avoidance agrees well with the recovery of hippocampal cholinergic enzyme activities.

In the second experiment, rats with bilateral fimbrial lesions were trained to perform an active avoidance task in an automated two-way shuttle box. The first training session was begun 7 days after lesioning and lasted for 4 days. Thereafter, the animals were tested at several postlesion intervals. The results shown in Fig. 3 indicate that relative to controls, lesioned animals exhibit a marked elevation in the number of intertrial crossings which gradually diminishes with time, becoming significantly lower 64 days after lesioning.

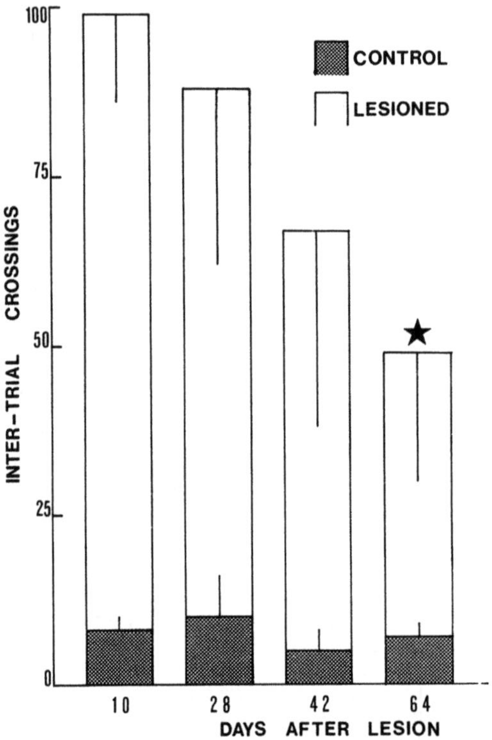

Fig. 3. Intertrial crossings in an automated shuttle box. Rats were trained to avoid an electric shock (0.4 mA, 5 s duration). The animals were placed in the shuttle box and were subjected to the following schedule: 30 s without stimulus, 5 s tone, and 5 s tone + shock. Sixty such trials were given per day for 4 days beginning 7 days after lesioning. Subsequent trials were carried out at the postlesion days shown below each column. Intertrial crossings = (Total crossings) - (Crossings during tone + crossings during tone and shock). The 10-day score represents the mean of the first 4 days during which time there was no significant day to day variation. The star denotes a significant reduction in intertrial crossings as compared with the score at 10 days postlesion, p = <0.05 (Student's t-test). Control animals, n=8; lesioned animals, n=8

The score for shock avoidance during the tone period was, however, identical in control and lesioned animals (results not shown), which probably reflects masking of avoidance deficit in lesioned animals due to the large increase in intertrial crossings. We therefore calculated the number of correct avoidance responses, i.e. avoidance during tone as a percent of total number of crossings as shown in Fig. 4. These results reveal that unlesioned control rats reach an 85% correct response level within the first 4 days of testing and maintain this level of performance throughout the experimental period. Lesioned animals, on the other hand, exhibit a significant deficit of about 50% during the acquisition period (first 4 days of testing), remain at this performance level up to 42 days, and thereafter show significant improvement 64 days after lesioning.

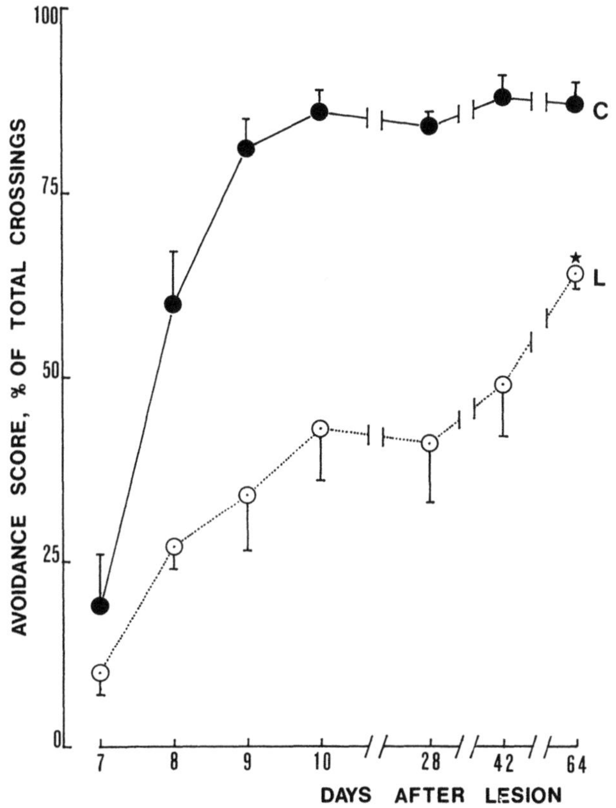

Fig. 4. Avoidance of shock in an automated shuttle box. The avoidance score is expressed as avoidances during tone as a percent of the total number of crossings. C = control group, n = 8. L = lesioned group, n = 8. Star represents a significant improvement as compared with the performance level of lesioned animals between 10 and 42 days postlesion, p = <0.05. For other details, see legend to Fig. 3

In summary, the present studies indicate that a substantial recovery of enzyme markers for hippocampal cholinergic terminals occurs within 8 weeks of partial lesions of the septo-hippocampal cholinergic projection (fimbria). Since homologous fimbrial fibers spared by the partial lesions constitute the source of this recovery, it seems reasonable to consider such reconstruction to be homotypic and therefore functionally meaningful. Accordingly, our studies also demonstrate that animals with partial fimbrial lesions exhibit significant recovery from lesion-induced behavioral impairments. Although the time course of behavioral recovery agrees well with that of hippocampal cholinergic parameters, these results by themselves are insufficient to unequivocally establish a correlation between these two consequences of partial fimbrial lesions. Nevertheless,the present findings suggest that occurrence of compensatory reinnervation may, at least partly, underlie recovery from neural trauma. Furthermore, the partial lesion paradigm could constitute a useful model system to identify possible means for promoting compensatory collateral sprouting in the central nervous system.

Acknowledgments: We are grateful to Adrian Haerri, Bernadette Wimmer, Ulrich Furrer, and Michel Patrick for expert technical assistance, and to J.M. Vigouret and Dr. P.H. Kelly for rewarding discussions and suggestions during the course of this study.

References

Ader R, Weijnen JAWM, Moleman P (1972) Retention of a passive avoidance response as a function of the intensity and duration of electric shock. Psychol Sci 26:125-128

Ball MJ (1977) Neuronal loss, neurofibrillary tangles and granulovacuolar degeneration in the hippocampus with aging and dementia. Acta Neuropathol 37:11-118

Cotman CW, Nieto-Sampedro M, Harris EW (1981) Synapse replacement in the nervous system of adult vertebrates. Physiol Rev 61:684-784

Dravid AR, Van Deusen EB (1983) Recovery choline acetyltransferase and acetylcholinesterase activities in the ipsilateral hippocampus following partial transection of the fimbria in rats. Brain Res 277: 169-174

Dravid AR, Van Deusen EB (1984) Recovery of enzyme markers for cholinergic terminals in septo-temporal regions of the hippocampus following selective fimbrial lesions in adult rats. Brain Res 324:119-128

Dunnett SB, Björklund A, Stenevi U, Iversen SD (1982) CNS transplantation: Structural and functional recovery from brain damage. In: Bujis RM, Pevet P, Swaab DF (eds) Chemical transmission in the brain: the role of amines, amino acids and peptides. Progress in Brain Research, vol. 55. Elsevier, Amsterdam, p 431

Fonnum F, Walaas I (1978) The effect of intrahippocampal kainic acid injections and surgical lesions on neurotransmitters in hippocampus and septum. J Neurochem 31:1173-1181

Gaarskjaer FB (1978) Organization of the mossy fibre system in the rat studied in extended hippocampi. I. Terminal area related to number of granule and pyramidal cells. J Comp Neurol 178:48-72

Gage FH, Björklund A, Stenevi U (1983a) Reinnervation of the partially deafferented hippocampus by compensatory collateral sprouting from spared cholinergic and noradrenergic afferents. Brain Res 268: 27-37

Gage FH, Björklund A, Stenevi U, Dunnett SB (1983b) Functional correlates of compensatory collateral sprouting by aminergic and cholinergic afferents in the hippocampal formation. Brain Res 268: 39-47

Glick SD, Mittag TW, Green JP (1973) Central cholinergic correlates of impaired learning. Neuropharmacology 12:291-296

Grossman SP (1976) Behavioral function of the septum: a reanalysis. In: DeFrance JF (ed) The septal nuclei. Plenum, New York, p 361

Hoover DB, Muth EA, Jacobowitz DM (1978) A mapping of the distribution of acetylcholine, choline acetyltransferase and acetylcholinesterase in discrete areas of rat brain. Brain Res 153:295-306

Kimble DP, Kimble RJ (1965) Hippocampectomy and response preservation in the rat. J Comp Physiol Psychol 60:472-476

Lo Conte G, Bartolini L, Casamenti F, Marconcini-Pepeu I, Pepeu G (1982) Lesions of cholinergic forebrain nuclei: changes in avoidance behavior and scopolamine actions. Pharmacol Biochem Behav 17: 933-937

Meibach RC, Siegel A (1977) Efferent connections of the septal area in the rat: an analysis utilizing retrograde and anterograde transport methods. Brain Res 119:1-20

Milner TA, Loy R, Amaral DC (1983) An anatomical study of the development of the septo-hippocampal projection in the rat. Dev Brain Res 8:343-371

Raisman G (1969) Neuronal plasticity in the septal nuclei of the adult rat. Brain Res 14:25-48

The Effects of Surgical Manipulations of the Optic Tectum

S. C. Sharma

Department of Ophthalmology, New York Medical College, Valhalla, New York, NY 10595, USA

Introduction

The formation of precise retinotopic projections during ontogeny or regeneration depends on the emergence of certain interactions between presynaptic optic axons and postsynaptic tectal cells. Within this commonly accepted framework, the theory of chemoselective recognition, derived largely from the work of Sperry (1945, 1963) postulated that position-dependent chemoselective labels develop in the retina and in the tectum and that connections are formed by the markers carried by the two populations.

This hypothesis has come to seem inadequate to account for the experimental results. If one considers the lower vertebrate visual systems on which much of the work has been done, there are aspects of both development and regeneration that seem to require something other than just target affinity. In particular, the phenomenon of compression and expansion of the retinotectal projection indicates that localized target affinity can only partially explain the results (Gaze and Sharma 1970; Yoon 1971, 1972a,b; Sharma 1972a,b; Horder 1971; Schmidt 1978; Sharma and Tung 1979). And in development, similar conclusions follow from the phenomenon associated with the "compound" eye projection in Xenopus. In both situations the final pattern of connections formed between the eye and the brain is other than what would have been predicted on a simple target affinity basis.

From the above, it may be inferred that much of the discussion in recent years about the validity of Sperry's hypothesis is misconstrued. This is perhaps because the theory was not articulated in sufficient detail and different workers have interpreted it in different ways. As a result there has occurrred, over the past several years, a series of attempts to account for the formation of ordered connections by mechanisms other than target affinity. These have included the arrow model of Hope et al. (1976), the tea trade model of Wilshaw and Von der Malsburg (1979), and the differential adhesion model of Fraser (1970). Most of these models remain inadequate in explaining all available results in the retinotectal system, and these models remain to be tested. Furthermore, the complexities of results obtained from regeneration studies seem to obscure rather than to clarify the rules governing neural development.

As an alternative to chemoaffinity, Horder and Martin (1978) emphasized the role of fiber ordering in their "morphogenetic" theory of retinotectal specification. In their scheme, the array of axons arriving at the tectum is retinotopically ordered, and the pattern of

Experimental Brain Research, Suppl. 13

terminal contacts is a direct result of this ordering. One strategy for testing this hypothesis involves determining whether sufficient order is present in the optic tract to account for the topography displayed by the mature projection. Recent studies have concluded that some degree of ordering is present throughout the optic nerve and tract in several species (amphibians and fishes: Bunt 1982; Easter et al. 1981) and have shown that the axons of adjacent ganglion cells travel together throughout their trajectory to the tectum. In contrast, Fawcett (1980) found that retinotopic order in Xenopus decays rapidly past the optic nerve head.

In this chapter I shall first describe how regenerating optic axons innervate an isolated portion of tectal tissue. I will then discuss induced changes in isthmic connections following half tectal ablation. Since recovery of function ensues following traumatic ablation of tectal tissue, the information presented below should provide some clues to the mechanisms responsible for such recovery.

Topographic Ordering in Tectal "Island"

The lower vertebrate visual system has provided a useful model for investigating the anatomical sequence of events during the course of optic nerve regeneration and topographic ordering of optic axons in the tectum. The precise topographic ordering follows from an initial diffuse pattern of innervation (Meyer 1980; Fujisawa et al. 1982; Murray and Edward 1982). In addition, Edwards and Jacobson (1985) and Schwartz et al. (1982) have postulated that initial regenerating optic axonal growth within the tectum is promoted by diffusable factors produced by target tissue. In order to evaluate the mechanisms responsible for the initially imprecise projections and the subsequent reorganization, Drs. Marion Murray, Michael Edwards, and I developed a preparation in the visual system of adult goldfish in which regenerating optic fibers had to pass through foreign terrain in order to reach the proper target field (Edwards et al. 1985). In this situation, a large piece of the mediolateral tectum was removed in adult goldfish and a small piece of isolated caudal tectum was separated from the intact rostral tectum. This tectal "island" preparation was used to assess the role of deafferented tissue in eliciting ingrowth by regenerating optic axons.

The sequence of innervation of the tectal island involves three distinct steps. In the first phase, large regenerating optic axonal bundles bridge the gap by passing over the surface of exposed subtectal structures (valvula cerebelli and the lateral tegmental structures). These fibers innervate the tectal island 2-3 months postoperatively. When axonal bundles are labeled retrogradely with HRP, a large portion of the nasodorsal retina is labeled, suggesting that the tectal island receives convergent input of retinal origin. In the second phase, selective loss of axonal inputs from less appropriate areas of the retina occurs. Only 20-30% of the retina is involved at this stage. This is roughly four to five times the amount of retina involved in normal control animals. By the third phase (6-7 months postoperatively), only a restricted nasal peripheral retina innervates the island. There is about a 60% reduction in the total area of the retina involved as compared to the first phase. Although this is a still larger retinal area than the normal retinal area innervating the caudal tectum, there is a definite tendency towards the control retinal values in these cases during regenerative phases.

Electrophysiological mappings of the retinotopic projections to the caudal tectal island were appropriate only for the caudal tectum, and although the larger retinal area innervated the island in the mid phases, no greater degree of representation of the retina was observed. This apparent contradiction could be due to two factors.

Either only appropriate fibers innervated the tectal island or, perhaps, an insufficient number of synapses were formed by the anomalous fibers which could be recorded electrophysiologically. The lack of larger retinal representation in the island also reflected maturational events which are normal during innervation of the normal tectum by regenerating optic axons (Jacobson and Gaze 1965; Cook et al. 1983).

Other points relative to the optic axonal innervation to the isolated tectal island (Edwards and Murray 1984) include: (a) During early phases of regeneration, when axons form a bridge between the rostral normal tectum and the caudal tectal island, no direct relationship exists between the extracellular matrix formed by meningial elements and glia. Only in the later stages are the elements involved in the fasciculation of the regenerating axons. (b) During regeneration, optic axons transiently innervate wrong pathways (Springer 1981), suggesting a trial-and-error process in successful reinnervation of the proper target. Since a large number of early exploratory axons do not associate with the degenerating axonal or glia cells, which may produce growth-promoting effects (Adler 1982; Schwartz et al. 1982), it is highly likely that the first phase of regeneration is that of the diffuse exploratory outgrowth leading to the innervation of the tectal island. (c) In the absence of the tectal island, initial outgrowth exhibits very few exploratory axons. However, in the presence of the tectal island, the number of axons or their branches increased 16-fold. Hence, the role of target tissue appears to be in the maintenance of early fascicles and the promotion of subsequent growth. Once the initial axons have formed a fascicle, the other regenerating axons perhaps use it as a substrate to innervate the island. (d) The erroneous optic fibers from the "island" are displaced and retracted to the intact tectum where they are again responsible for the rearrangement of order in the rostral half tectum. The appropriate terminals in the caudal tectum and the "compressed" visual projection in the rostral half tectum together form a complete normal retinotopically organized map. A reduction in the number of ganglion cells in the contralateral retina was not observed.

These results provide evidence that during regeneration of the optic nerve, the initial establishment of a diffuse projection occurs, but the topography of the projection is rather poor. This is followed by maturation of the appropriately projecting axonal branches which would lead to an appropriate projection map as defined electrophysiologically. Finally, inappropriately projecting axonal branches degenerate or withdraw. As discussed earlier, these observations would place doubt on the role of order in the optic pathways on the final refinement of projection onto the tectum, since in all submammalian species studied so far the regenerating optic nerve and tract have a rather diffused topography (Meyer 1980; Cook et al. 1983; Fujisawa 1981).

The results of most studies on regenerating optic axons do not unequivocally support the notion that recognition of the target region is involved in axonal modelling for the generation of a topographically organized projection. It has been implied by most workers that optic axonal arbors organize into continuously ordered arrays with the available target field, and evidence is available that regenerating optic axonal interactions can operate independent of tectal polarity cues (Sharma 1975; Rho and Hunt 1980; Horder and Martin 1982). The evidence cited above that only appropriate synapses for the isolated tectal island prevailed and other axons probably withdrew supports the earlier conception of Sperry's original hypothesis (1963). Our study also provided evidence that wide ranging exploratory phases of the regenerating axons are restricted and only appropriate synapses prevail, perhaps having a competitive advantage over the nonappropriate axons. Through this mechanism specificity for the connections can be expressed.

Rearrangements of Isthmic Connections

Following half tectal ablation in adult goldfish, the visuotopic projection returns to normal, with the projection visuotopically organized but compressed on the remaining half tectum (Gaze and Sharma 1970). In the ensuing 15 years since this observation, a number of workers have confirmed our results and have extended the observations further to the anatomical (Murray et al. 1982) and behavioral (Yager et al.1977; Northmore and Sharma 1979; Northmore 1981) studies of the "compression". The behavioral observations of normal animals have suggested that localization or detection of an object in space depends upon the visuotopic map to the tectum. The orientation response is fast and is perhaps controlled by a motor map or motor center which is in register with the visuotopic map. Northmore (1981) provided some experimental support for the notion that the "motor map" to the tectum is not fixed and that it must alter following half tectal ablation. In order to explore the consequences of compression, it was essentially first to know the intrinsic and extrinsic tectal connections (Grover and Sharma 1981). Most diencephalic and tegmental centers projecting to the tectum in goldfish are rather small in size, except the torus longitudinalis and the nucleus isthmi. In goldfish, both of these centers are connected to the ipsilateral optic tectum. In collaboration with Dr. Dunn-Meynell I have studied the topography of the isthmus to the tectum and vice versa and have explored the changes which the half tectal ablation brings to the isthmal projection (Dunn-Meynell and Sharma 1984).

In goldfish there is a topographically ordered projection from the nucleus isthmi onto the ipsilateral optic tectum (Grover and Sharma 1981; Dunn-Meynell and Sharma 1984).The cells in the isthmi which are efferent to the tectum are located on the periphery of the nucleus. This nucleus consists of a cell-sparse medulla covered by a cellular cortex which is thick on the rostral, medial, and dorsal surfaces of the nucleus. All cells projecting to the ipsilateral tectum are located in the thick cortical cap. In normal goldfish, HRP injected into the rostral tectum labeled cells in the rostroventral portion of the thick cortex; a midtectal injection labeled cells located more caudodorsally; and caudal tectal injections labeled cells more caudally in the extreme dorsal cortex. Hence, the rostroventral to caudodorsal isthmic axis projects to the rostrocaudal axis of the ipsilateral optic tectum. The isthmal projection to the ipsilateral tectum in amphibia is slightly different. In amphibia, the rostrocaudal isthmic axis projects to the rostrocaudal optic tectum (Grobstein and Comer 1983). The major distinction between the two animals is that dorsoventral cells in the isthmus project to the mid and caudal tectum in goldfish. This normal isthmotectal projection was compared with the projection obtained in animals in which the visuotopic projection had "compressed" following caudal half tectal ablation and crush of the contralateral optic nerve.

Following complete removal of one tectum, the ipsilateral nucleus isthmi degenerates. However, when the caudal half tectum was removed, the isthmic cells deprived of their normal terminal zones did not degenerate. Instead, the isthmotectal projection changed in response to tectal surgery. Since the caudodorsal portion of the nucleus projected to the caudal half of the tectum, the fibers from these tectal cells terminated in the caudalmost portion of the remaining tectum - actually in the midtectum of the original intact tectum. In so doing, these fibers apparently displaced the fibers normally terminating in the midtectum. The fibers which originally terminated in the midtectum now terminated in the rostral tectum. Finally, the cells projecting to the rostral tectum did not change their terminals. These results demonstrate that topographic order is retained in the isthmotectal projection following caudal half tectal ablation and subsequent compression of the visuotopic projection to the remaining rostral half tectum.

In the developing visual system of Xenopus, rotation of one eye prior to metamorphosis leads to the rotation of both the contralateral and indirect ipsilateral retinotectal projections (Gaze et al. 1970). Udin and Keating (1979, 1981) have shown that the contralateral isthmotectal projections are similarly rotated. These altered isthmotectal projections represent a developmental consequence of eye rotation, which in adult amphibia does not lead to a rotated ipsilateral projection (Keating 1975). From the studies of Gaze et al. (1970), Keating (1974), and Grobstein and Comer (1983), it is safe to conclude that tectoisthmic fibers relay a pattern of neural activity to the contralaterally projecting isthmic cells which seek the complementary pattern of neural activity in the contralateral tectum. The recent studies of Constantine-Paton and Ferrari-Eastman (1981) and Udin (1983) have shown that the isthmotectal fibers in developing amphibia project initially to their normal termination zones in the tectum and subsequently abandon these sites for an area with complementary neural activity. The ipsilateral isthmotectal fibers are initially guided by some factor intrinsic to this system. Although direct evidence is lacking regarding the role of neural activity in patterning the isthmotectal connections in the amphibian brain, the changes in the contralateral isthmic projection following eye rotation seem to suggest such a role.

The role of neural activity in patterning the connections of the isthmic projection to the "compressed" half ipsilateral tectum in fish is, at best, difficult to discern. However, it should be emphasized that neural circuitry necessary for the effect of neural activity in this system does exist. In addition to the isthmotectal projection, the reciprocal tectoisthmal projection also exists in goldfish (Dunn-Meynell and Sharma 1984). Following HRP injection in various locations in the tectum, the axonal terminations in the isthmic medulla suggest that the rostral tectum projects to relatively ventral nucleus isthmi and the caudal tectum projects to the dorsal portion of the nucleus. The midtectum projects in the middle of these two positions. Predictable changes occurred in the tectoisthmal projection following half tectal ablation and subsequent visuotopic "compression". In these cases the tectal projection to the isthmic nucleus conformed to their normal values.

The factors which control the tectoisthmic projection are not well understood. It is conceivable that tectoisthmic projections have a direct effect upon the isthmotectal projection involving neural activity. Some credence to this notion is provided by another series of experiments (Dunn-Meynell and Sharma 1984). In some fish where the caudal half tectal removal without contralateral optic nerve crush did not lead to "compression" of the projection, it was noted that no change had occurred in the tectoisthmic or isthmotectal visuotopic projection. Assuming that the retinal projections affect the tectoisthmic projection, one could conclude that the isthmotectal projection was in turn affected, and hence no change was noted in the topography. Although speculative, it is tempting to propose that neural activity (which has been implicated in patterning retinotectal connections - Meyer 1982) from the tectum controls the projections to the isthmic nucleus by initially inducing the positional marker on the isthmic nucleus. The isthmic nucleus then sends the appropriate axons to the reciprocal sites in the tectum.

The exact role of the nucleus isthmi in goldfish is not understood. The patterns of isthmotectal and tectoisthmic connectivity are now known. Recent electrophysiological findings of Williams et al. (1983) have shown that cells of the nucleus isthmi are electronically coupled. Such coupling would cause isthmic cells to fire synchronously. In addition, Luiten (1981) has shown in the carp the existence of an isthmocerebellar pathway in which isthmic cells project to the corpus cerebelli. It is appropriate to assume that some topography between the two centers might exist. In the absence of any detailed understanding of such connectivity, the precise function of the nucleus isthmi cannot be ascertained.

The mechanism of transformation of sensory signals into commands or signals for the control of movement (i.e., a motor map or a motor center) is poorly understood. Although the optic tectum in mammals has been implicated as an integration center for sensory inputs, in lower vertebrates it has not been easy to assign such a role for the tectum, primarily because connectivity patterns to various centers are not fully understood. Aside from the nucleus isthmi, which has been discussed above, the torus longitudinalis is another nucleus which receives tecto-topic projections in goldfish. This nucleus is only found in teleosts and its homologous nuclei in higher vertebrates are not known. The torus longitudinalis receives its input from the valvula cerebelli and from the ipsilateral optic tectum (Grover and Sharma 1981; Ito and Kishida 1978). Recent studies by Northmore et al. (1983) and Northmore (1984) suggest that photic responses in the torus longitudinalis are derived from ipsilateral optic tectum, whereas saccadic eye responses originate in the cerebellum. The saccadic bursts and the eye movements can be recorded in the torus longitudinalis. The torus longitundinalis is a superficial structure lying in between the two tecta. Since it appears to be a less complicated and easily approachable nucleus, perhaps the transformation of sensory signals to motor command centers could be analysed in this nucleus. With this hope in mind, experiments are in progress to analyse the effects of half tectal ablation on the tecto-topic organization in the ipsilateral torus longitudinalis and subsequent recovery of function.

References

Adler R (1982) Regulation of neurite outgrowth in purified retina neural cultures: effects of PNPF, a stratum bound, neurite promoting factor. J Neurosci Res 8:165-177

Bunt SM (1982) Retinotopic and temporal organization of the optic nerve and tracts in the adult goldfish. J Comp Neurol 206:209-226

Constantine-Paton M, Ferrari-Eastman P (1981) Topographic and morphometric effects of bilateral embryonic eye removal in the optic tectum and nucleus isthmus of leopard frog. J Comp Neurol 196:645-661

Cook JE, Pilgrim AJ, Horder TJ (1983) Consequences of misrouting goldfish optic axons. Exp Neurol 79:830-844

Dunn-Meynell AA, Sharma SC (1984) Changes in the topographically organized connections between the nucleus isthmi and the optic tectum after partial tectal ablation in adult goldfish. J Comp Neurol (to be published)

Easter SS, Rusoff AC, Kish PE (1981) The growth and organization of the optic nerve and tract in juvenile and adult goldfish. J Neurosci 1:793-811

Edwards MA, Jacobson M (1985) Effects of permeability of mid-tectal barrier in goldfish on compression of the visuotectal projection rostrally and regenerative escape caudally. J Comp Neurol (to be published)

Edwards MA, Murray M (1985) Selective retinal reinnervation of a surgically created tectal island in goldfish. II. Electron microscopic analysis. J Comp Neurol (to be published)

Edwards MA, Sharma SC, Murray M (1985) Selective retinal reinnervation of a surgically created tectal island in goldfish. I. Light microscopic analysis. J Comp Neurol (to be published)

Fawcett JW (1981) How optic axons grow down Xenopus optic nerve. J Embryol Exp Morphol 65:219-233

Fraser S (1980) A differential adhesion approach to the patterning of nerve connections. Dev Biol 79:453-464

Fujisawa H (1981) Retinotopic analysis of fiber pathways in the regenerating retinotectal system of adult newt Cynops pyrrhogaster. Brain Res 206:27-37

Fujisawa H, Tani N, Watanabe K, Ibata Y (1982) Branching of regenerating retinal axons and preferential selection of appropriate branches for specific neuronal connections in the newt. Dev Biol 90:43-57

Gaze RM, Sharma SC (1970) Axial differences in the reinnervation of goldfish optic tectum by regenerating optic fibers. Exp Brain Res 10:171-181

Gaze RM, Keating MJ, Szekely G, Beazley L (1970) Binocular interaction in the formation of specific intertectal neuronal connexions. Proc R Soc Lond (Biol) 175:107-147

Grobstein P, Comer C (1983) The nucleus isthmi as an intertectal relay for the ipsilateral oculotectal projection in the frog Rana pipiens. J Comp Neurol 217:54-74

Grover BG, Sharma SC (1981) Organization of extrinsic tectal connections in goldfish (Carassius auratus) J Comp Neurol 196:471-488

Hope RA, Mommond BJ, Gaze RM (1976) The arrow model: retinotectal specificity and map formation in the goldfish visual system. Proc R Soc Lond (Biol) 194:447-466

Horder TJ (1971) The course of recovery of retinotectal projection during regeneration of the fish optic nerve. J Physiol (Lond) 217:53-54P

Horder TJ, Martin KAC (1978) Morphogenetics as an alternative to chemospecificity in the formation of nerve connections. In: Curtis ASG (ed) Cell-cell recognition. Soc exp biol symposium 32. Cambridge University Press Cambridge, pp 275-358

Horder TJ, Martin KAC (1982) Some determinants of optic terminal localization and retinotopic polarity within fiber populations in the tectum of goldfish. J Physiol (Lond) 333:481-509

Ito H, Kishida R (1978) Afferent and efferent fiber connections of the carp torus longitudinalis. J Comp Neurol 181:465-476

Jacobson M, Gaze RM (1965) Selection of appropriate tectal connections by regenerating optic nerve fibers in adult goldfish. Exp Neurol 13:418-430

Keating MJ (1974) The role of visual function in the patterning of binocular visual connexions. Br Med Bull 30:145-151

Keating MJ (1975) The time course of experience-dependent synaptic switching of visual connections in Xenopus laevis. Proc R Soc Lond (Biol) 189:603-610

Luitgen PGM (1981) Afferent and efferent connections of the optic tectum in the carp (Cyprinus carpio L.). Brain Res 220:51-65

Meyer RL (1980) Mapping the normal and regenerating retinotectal projection of goldfish with autoradiographic methods. J Comp Neurol 189:175-190

Meyer RL (1982) Ordering of retinotectal connections: a multivariante operational analysis. Curr Top Dev Biol 15: 101-145

Murray M, Edwards MA (1982) A quantitative study of the reinnervation of the goldfish optic tectum following optic nerve crush. J Comp Neurol 209:363-373

Murray M, Sharma SC, Edwards MA (1982) Target regulation of synaptic number in the compressed retinotectal projection of goldfish. J Comp Neurol 209:374-385

Northmore DPM (1981) Visual localization after rearrangement of retinotectal maps in fish. Nature (Lond) 293:142-144

Northmore DPM (1984) Visual and saccadic responses of the goldfish torus longitudinalis. J Comp Physiol (to be published)

Northmore DPM, Sharma SC (1979) Visual sensitivity and spatial summation in goldfish with compressed retinotectal projections or total tectal ablation. Brain Res 180:344-348

Northmore DPM, Skeen LC, Pindzola (1981) Visuomotor perimetry in fish: A new approach to the functional analysis of altered visual pathways. Vision Res 21:843-853

Northmore DPM, Williams B, Vanegas H (1983) The teleostean torus longitudinalis: response related to eye movements, visuotopic mapping, and fundamental relations with the optic tectum. J Comp Physiol 150:39-50

Rho JH, Hunt RK (1980) Retinal projection to a surgically prepared "double lateral" tectum in Xenopus: electrophysiological studies. Dev Biol 80:436-453

Schmidt J (1978) Retinal fibers alter tectal positional markers during the expansion of the half retinal projection in goldfish. J Comp Neurol 177:279-300

Schwartz M, Mizrachi Y, Eschar N (1982) Factor(s) from goldfish brain induce neuritic outgrowth from explanted regenerating retinas. Dev Brain Res 3:29-35

Sharma SC (1972a) The retinal projections in the goldfish: An experimental study. Brain Res 39:213-223

Sharma SC (1972b) Reformation of retinotectal projections after various tectal ablations in adult goldfish. Exp Neurol 34:171-181

Sharma SC (1975) Visual projection in surgically created "compound" tectum in adult goldfish. Brain Res 93:497-501

Sharma SC, Tung YL (1979) Interactions between nasal and temporal hemiretinal fibers in adult goldfish tectum. Neuroscience 4:113-119

Springer AD (1981) Normal and abnormal retinal projections following the crush of one optic nerve in goldfish (Carassius auratus). J Comp Neurol 199:87-95

Sperry RW (1945) Restoration of vision after crossing of optic nerve and after contralateral transplantation of eye. J Neurophysiol 8:57-69

Sperry RW (1963) Chemoaffinity in the orderly growth of nerve fiber patterns and connections. Proc Natl Acad Sci USA 50:703-710

Udin SB (1983) Abnormal visual input leads to development of abnormal axon trajectories in frogs. Nature 301:336-338

Udin SB, Keating MJ (1979) Altered isthmo-tectal topography after eye rotation in Xenopus tadpoles. Soc Neurosci Absts 5: 183

Udin SB, Keating MJ (1981) Plasticity in a central nervous pathway in Xenopus: Anatomical changes in isthmotectal projection after larval eye rotation. J Comp Neurol 203:575-594

Williams B, Hernandez N and Vanegas H (1983) Electrophysiological analysis of the teleostean nucleus isthmi and its relationship with the optic tectum. J Comp Physiol (A) 152:545

Willshaw DJ and Von der Malsburg C (1979) A marker induction mechanism for the establishment of ordered neural mappings; its application to the retinotectal problem. Philos Trans R Soc Lond (Biol) 287:203-243

Yager D, Sharma SC, Grover BG (1977) Visual function in goldfish with unilateral and bilateral tectal ablation. Brain Res 137:267-275

Yoon M (1971) Reorganization of retinotectal projection following surgical operations on the optic tectum in goldfish. Exp Neurol 33:395-411

Yoon M (1972a) Reversibility of the reorganization of retinotectal projection in goldfish. Exp Neurol 35:565-577

Yoon M (1972b) Transposition of the visual projection from the nasal hemiretina on to the foreign rostral zone of the optic tectum in goldfish. Exp Neurol 37:451-462

Pharmacological Approaches to Enhance Regeneration

Histological Factors Influencing the Growth of Axons into Lesions of the Mammalian Spinal Cord

L. Guth, Ch. P. Barrett, and E. J. Donati

University of Maryland, School of Medicine, 655 West Baltimore Street, Baltimore, MD 21201, USA

Introduction

The failure of axonal regeneration in the injured mammalian spinal cord contrasts markedly with the extensive regenerative capacity of both the mammalian peripheral nervous system and the spinal cord of certain inframammalian vertebrates. The fundamental issue is whether neurons of the mammalian CNS are unable to regrow their severed processes or whether some property of the non-neuronal cells of the CNS environment prevents them from expressing an innate regenerative capacity. The phenomenon of abortive regeneration, originally described by Ramon y Cajal (1928) and verified many times subsequently (Windle 1956), suggested that mature intrinsic neurons of the CNS have the ability to grow, and the more recent demonstrations of growth of CNS axons into peripheral nerve grafts (Benfey and Aguayo 1982; David and Aguayo 1981; Richardson et al. 1980, 1982) substantiate the view that mature CNS neurons retain considerable potential for elongation. Consequently attention must now be given to the mechanisms by which the environment modulates this regenerative capacity. One possibility is that the non-neuronal cells of the mammalian CNS lack the trophic ability to maintain the regenerative efforts of neurons. However, this seems unlikely since collateral sprouting of mammalian CNS neurons occurs in both central and peripheral nervous systems qfter suhtotal tissue damage, and since CNS glial cells support neuronal regeneration in tissue culture. An alternative explanation for the failure of regeneration is that the pathological alterations at the site of spinal cord injury are incompatible with neuritic growth; this possibility will be addressed in the present communication.

Comparative Histopathology of Spinal Injury in Vertebrates

Light and electron microscopical observations on the traumatized spinal cord of those vertebrate species which exhibit axonal regeneration have defined to a considerable degree the cell interactions that contribute to the growth process (Michel and Reier 1979; Selzer 1978; Singer et al. 1979; Stensaas 1983; Wood and Cohen 1981). In such inframammalian vertebrates, the repair process is generally initiated by proliferation of cells in and adjacent to the lesion. Some of the dividing cells give rise to an embryonal mesenchyme which rapidly fills the gap between the stumps of the transected spinal cord. Meanwhile, the ependymal cells lining the central canal of both stumps of the cord also proliferate and migrate into the mesenchyme where they meet and form a continuous bridge uniting the rostral and caudal stumps. Only then is axonal elongation initiated; most of the axons grow in proper longitudinal orientation, apparently guided by

Experimental Brain Research, Suppl. 13

the surface membranes of the ependymal/glial cells. In fact, in some instances the ependymal elements elaborate radial cytoplasmic processes that delineate longitudinal channels prior to the ingrowth of axons, and the regenerating neurites appear to become organized into fascicles as a result of their growth into these compartments (Singer et al. 1979).

These fundamental and sequential processes of mesodermal proliferation, ependymal migration, and neuritic regeneration are characteristic of the injured mammalian spinal cord as well, the most crucial difference being that the mesodermal cells which invade the site of a mammalian spinal injury are fibroblasts rather than mesenchymal cells (Windle 1956). As a result of the secretion of collagen by these cells, the matrix into which the ependymal and neuritic elements must grow is densely fibrous rather than loosely cellular (Matthews et al. 1978). Although ependymal cells proliferate and migrate into the lesion (Gilmore and Leiting 1980; Guth et al. 1980; Matthews et al. 1979), they fail to restore continuity to the central canal and instead become aggregated into isolated clusters (Guth et al. 1980). Some neurites invade the lesion despite its dense fibrous constitution (Matthews et al. 1978); many of these derive from dorsal roots, but some may arise from intrinsic spinal elements. In any event, their number is small, their growth is disoriented, and the period of their regenerative activity is brief (Ramon y Cajal 1928; Windle 1956). In addition to the cellular disorganization within the lesion, pathological alterations occur in the adjacent tissue that has not been directly subjected to trauma. In this region the astrocytes hypertrophy, proliferate, and elaborate an extensive network of thick processes which form a "fibrous glial scar" covering both stumps of the spinal cord (Guth et al. 1980; Matthews et al. 1978; Windle 1956). For reasons that are still unclear, but which are generally considered to be related to alterations of blood flow, a progressive necrosis ensues and extends both rostrally and caudally from the original site of injury (Kao 1974). This pathological process is so severe that by 4 months after spinal transection in the rat, the lesion is two to four vertebral segments long and is composed of dense accu mulations of collagen, ependymal-lined cysts, and large cavities that are partially lined by astrocytic processes (Guth et al. 1980).

Histogenesis of Progressive Necrosis and Its Relation to Axonal Regeneration

Twenty-four hours after spinal transection, the lesion is narrow and the ends of the spinal cord are separated by a space filled with extravasated blood. However, the site of injury soon becomes invaded by macrophages which phagocytize the red blood cells and the myelin and axonal debris of degenerating nerve fibers. As these tissue elements are removed, large cavities and cysts are left behind which become partially walled off by the enlarged processes of reactive astrocytes. At the same time, fibroblasts begin to migrate into the site of injury where they give rise to a transverse plate of dense connective tissue which demarcates the original site of transection (Fig. 1). These fibroblasts derive primarily from the connective tissue components of the bone and muscle and surrounding the cord rather than from the dura mater, since the invasion of these cells and the subsequent formation of the collagenous scar can be prevented to a great degree by placing a sheet of synthetic dura extradurally over the transected spinal cord (Krikorian et al. 1981). As a result of these histopathological reactions, the lesion, which was initially quite narrow, gradually and progressively enlarges, and by 4 months postoperatively it can be as many as four vertebral segments long. This huge lesion is composed of dense collagenous connective tissue, macrophages, large cysts and smaller cavities, clusters of migrating ependymal cells, and reactive astrocytes whose thick processes partially wall off the cysts and form a dense cap of glial fibers over the widely separated stumps of the spinal cord (Figs. 1,2).

This necrotizing reaction is frequently designated progressive ischemic necrosis, and studies on spinal concussion indicate that hemodynamic alterations are indeed responsible (Balentine 1978). Immediately after the trauma, there is a fall in blood pressure and a loss of autoregulation; presumably the resultant reduction of blood flow and edema gives rise to frank tissue necrosis (De La Torre 1981). The pathophysiological importance of blood flow is supported by recent experiments showing that drugs which reduce hypotension reduce the severity of paraplegia after spinal concussion (Faden et al. 1983), but it is by no means certain that this is indeed the mechanism (Faden et al. 1983; Hall and Braughler 1982). Although we do not know why tissue necrosis continues to extend centrifugally from the lesion for weeks and months after injury, Kao (1974) has shown that the process can be aborted within 1-2 weeks by surgically excising the necrotic tissues, a result which suggests that the progressive aspect of the reaction results from the presence of necrotic debris. However, because of these uncertainties, it would be most judicious at this juncture to designate this reaction simply as a necrotizing response or as progressive necrosis so that we do not imply that we understand its underlying cause.

The significance of these observations to the process of axonal regeneration is twofold. First of all, it is clear that the cellular organization of the tissues composing the lesion constitutes a formidable barrier to the oriented ingrowth of regenerating axons. Although some nerve fibers are found within the lesion, these are primarily peripheral nerve fibers originating from the dorsal roots of spinal ganglia. Secondly, even if fibers of spinal origin have considerable intrinsic regenerative capacity (as is now generally suspected), they are unable to invade the lesion site in large numbers because the growing neurites are destroyed by the progressive and gradual necrosis of the intact stumps of the cord. It should also be emphasized that although this sequence of events is described in relation to lesions made by spinal transection, similar changes have been reported after concussive and compressive types of injury as well.

Non-Necrotizing Spinal Lesions

Since progressive necrosis destroys both the neuritic sprouts as well as the tissue environment into which the axons are regenerating, one can evaluate the in vivo regenerative capacity of the spinal cord only by using an experimental animal model in which the necrotizing response is minimized. For this reason we initiated a study on spinal cord injury in the hibernating ground squirrel. During hibernation, the squirrel's body temperature falls to 8°C, the heart and respiratory rates fall to 1/minute, and the overall metabolic rate is greatly depressed. Consequently, not only are the body tissues less sensitive to the effects of ischemia, but synthesis of collagenous scar tissue is greatly reduced. We therefore transected the spinal cord of hibernating ground squirrels and studied them histologically at intervals up to 4 months postoperatively (Guth et al. 1981). The reaction to injury in the hibernating squirrels (Figs. 3, 4) was in

Figs. 1-6 ►
Longitudinal sections through the spinal cord of rats and hibernating ground squirrels 3 months after a complete transection of the spinal cord, illustrating the difference in response to transection of the spinal cord in rat and hibernating ground squirrel. Three months postoperatively in the rat the site of injury is marked by a transverse plate of connective tissue with large cysts and cavities on either side which effectively separate the stumps of the spinal cord (Fig. 1; H-OG-E, x 45). Ingrowth of ependymal cells in an acinarlike arrangement is characteristic in this species (Fig. 2, arrows; trichrome x 75). In contrast, 3 months after spinal cord transection in the hibernating squirrel, the lesion remains narrow (Fig. 3; H-OG-E, x 30). There is no connective tissue scar, cavitation, cyst formation, or ependymal proliferation, and the lesion site is characterized by extensive vascularity, macrophages, and other mononuclear cells (Fig. 4; trichrome, x 75). Silver impregnation of the same animal reveals regenerating nerve fibers which, upon reaching the margin of the lesion (arrows), make a right angle turn and run perpendicularly to the long axis of the spinal cord (Fig. 5; Protargol, x 75). Higher magnification shows that many of the axons present in the zone adjacent to the lesion are varicose and possess terminal swellings (Fig. 6; Protargol, x 300)

every respect different from that seen in nonhibernating squirrels or in rats (Figs. 1, 2): (i) The lesion remained narrow (1-2 mm in width), (ii) there was no collagenous scarring, (iii) there was minimal cavitation and cyst formation, (iv) there was no reactive astrocytosis, and (v) the cells within the lesions were largely mononuclear elements of hematogenous origin. Nerve fibers in the spinal cord of hibernating quirrels grew in great profusion to the interface between the cut end of the cord and the lesion and, at this point, they turned at right angles and remained within the spinal tissue rather than enter the lesion (Fig. 5). These fibers exhibited the varicosities, end-swellings, and undulations (Fig. 6) that are characteristic of regenerating nerve fibers in the peripheral nervous system.

These observations indicate (i) that neurons of the mammalian spinal cord have considerable intrinsic growth capacity: (ii) that this capacity can be manifested in vivo provided the necrotizing response to injury is reduced or obviated; and (iii) that an appropriate and possibly highly specific cellular terrain must be established within the lesion site in order to encourage the nerve fibers to leave the neural tissue environment provided by the spinal cord and grow into the injured tissue.

Consequently, the next goal was to develop a non-necrotizing model of injury in the rat (rather than to examine further the rather unique hibernation model). In this way we hoped to be able to study the relationship between the intrinsic regenerative capabilities and extrinsic non-neuronal factors under physiological conditions more relevant to those found in the majority of mammals. We initiated this approach with an investigation into the effects of injury to the optic nerve, the histological simplicity of which renders it particularly useful for such an initial experimental analysis of a complex problem. We found that transection of the rat's optic nerve at most levels resulted in a significant tissue necrosis at the site of injury. On the other hand, a brief crush of moderate intensity (50 g) and brief duration (1 s) produced complete axotomy without concomitant necrosis. The presence or absence of necrosis was correlated

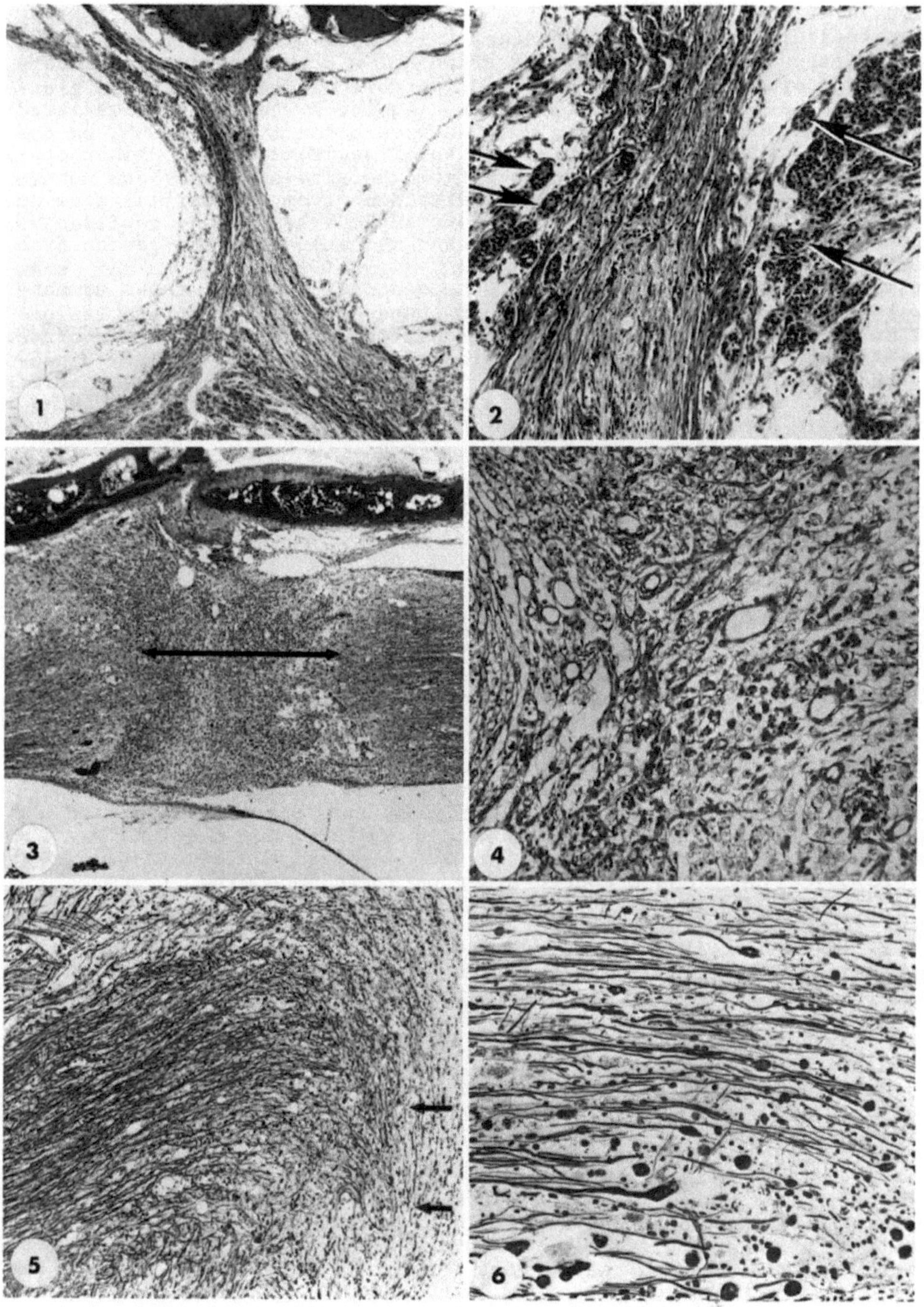
1
2
3
4
5
6

Figs. 7-12
Longitudinal sections illustrating the effect of crushing the spinal cord of the rat: Fig 7. One week after a 1-s crush the lesion is narrow and cellular, resembling to considerable degree that of the transected spinal cord of the hibernating squirrel which was illustrated in Fig. 1. HOGE, x 75 Fig. 8. One week after a 10-s crush, on the other hand, the lesion is three vertebral segments long. This photomicrograph was taken two segments from the site of crush and shows cystic degeneration even at this distance from the initial site of injury. HOGE, x 75 Fig. 9. Two weeks after a 1-s crush, capillaries and cordons of astrocytes (arrows) have migrated into the lesion from both margins of the spinal cord. HOGE, x 150 Fig. 10. In the same animal, a section at the level of the central canal (*) shows comparable proliferation and migration of ependymal cells into the lesion. HOGE, x 150 Fig. 11. Silver impregnation 1 week after a 1-s crush shows that essentially all nerve fibers within the lesion have degenerated. Protargol, x 150 Fig. 12. Silver impregnation 2 weeks after 1-s crush shows nerve fibers to have grown into the lesion in an appropriate longitudinal orientation. Protargol, x 150

with the degree of preservation of blood flwo (as monitored directly under the operating microscope). If blood flow was not rapidly restored (as after a nerve transection or a severe crush) the optic nerve on either side of the lesion underwent necrosis (Chaban et al. to be published).

These findings led us to explore the possibility that a similarly brief crushing injury of the spinal cord might produce a non-necrotizing lesion which would permit investigation of intrinsic neuronal growth capacity. We therefore performed a laminectomy at T5 on anesthetized rats and crushed the spinal cord for various durations (1--10 s) with a Dumont 5 forceps while leaving the dura mater intact to prevent the ingrowth of connective tissue (Krikorian et al. 1981). After intervals of 1-3 weeks, the animals were again anesthetized and killed by intracardiac perfusion with a modified Bouin solution. The vertebral column with spinal cord was removed and decalcified, and histological sections prepared after paraffin embedding.

The results are illustrated in Figs. 7-12. We found that a compression injury of 10 s (Fig. 8) produced a necrotizing lesion with extensive cavitation, cyst formation, and reactive gliosis which was similar to that seen after spinal transection (except for the absence of connective tissue scarring). However, a brief crush of 1 s produced an entirely different reaction: At 1 week the lesion was narrow and filled with mononuclear cells (Fig. 7), degenerating nerve fibers, and capillaries oriented parallel to the long axis of the spinal cord (Fig. 9). No intact (surviving) axons were found within the lesion (Fig. 11). By 2 weeks the mononuclear cells had assumed phagocytic activity and were engorged with myelin and other cellular debris. Longitudinally oriented cords of ependymal cells (Fig. 10) and astrocytes (Fig. 9) had now migrated into the lesion from the adjacent spinal cord, and similarly oriented nerve fibers had begun to regenerate into the lesion along the surface of these capillaries, ependymal cells, and astrocytes (Fig. 12). After 3 weeks, the astrocytes had formed thick cellular bundles within which glial fibers could be discerned immunocytochemically using antibodies to glial fibrillary acidic protein (GFAP). The regenerated nerve fibers were still oriented longitudinally but had increased in number and had formed small fascicles. For descriptive purposes we will designate this type of lesion as a "non-necrotizing lesion".

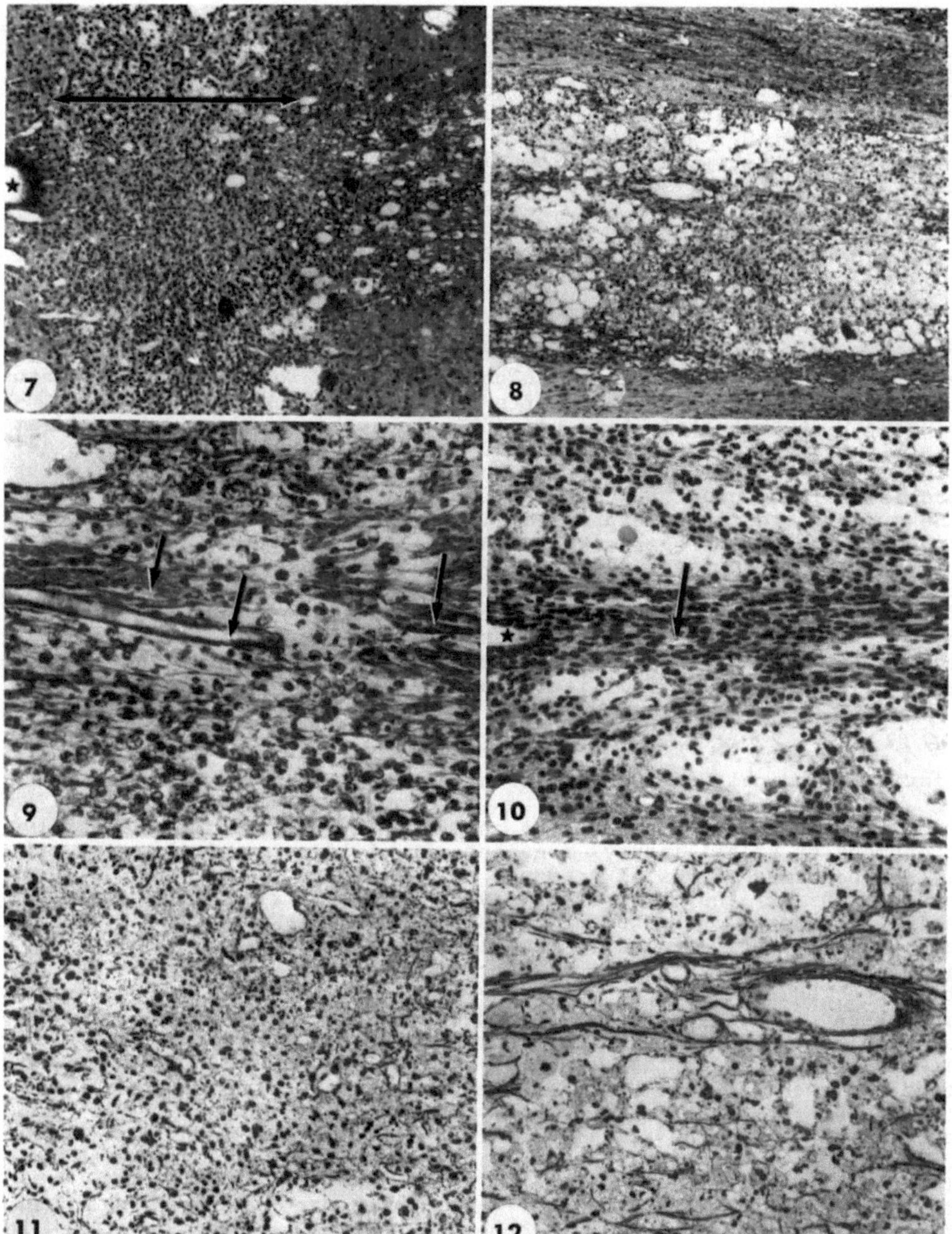
7
8
9
10
11
12

Figs. 13-18
Silver impregnation of rat spinal cords, 2 weeks after a 1-s crush, showing effect of treatment with regeneration-promoting drugs: Fig. 13, 15, and 17. Vehicle-treated (control) rats showing narrow lesion without cavitation (Fig. 13, x 30). Note the paucity of fibers within the lesion (Figs. 15 and 17, x 75 and x 150) Figs 14, 16, and 18, Drug-treated rats, showing extensive ingrowth of nerve fibers into the lesion in comparison with the control rats. Fig. 14, x 75; Fig. 16, x 150; Fig. 18, x 300. Note that the fibers do not have any consistent orientation and that at high magnification (Fig. 18) the fine fibers are seen to have formed fascicles

The significance of these observations to the subject of CNS regeneration is twofold. First, although growth of spinal axons into intraspinal or extraspinal grafts of peripheral nerves has been conclusively demonstrated, the present observations provide the first histological evidence of sustained growth of nerve fibers into a lesion of the rat spinal cord. They therefore enable us to draw the provisional conclusion that the intrinsic regenerative capacity of the spinal cord can be expressed (i) if progressive necrosis and collagenous scarring are prevented and (ii) if the injured tissue is reconstructed in an appropriate way by the vascular, ependymal, and astrocytic components of the spinal cord. Secondly, the availability for the first time of a procedure that results in modest, though significant, axonal regeneration makes possible the utilization of this model for evaluating the efficacy of drug and other therapeutic modalities that might faciliate the regenerative process.

Histopathological Considerations in the Treatment of Spinal Trauma

From the foregoing histological analysis, it follows that effective treatment of spinal injury requires a twofold approach: we must utilize procedures that will reduce or eliminate tissue necrosis and cavitation at the site of injury, and we must also use drugs to stimulate growth of axons into this now-favorable environment. Consequently, a rational and systematic approach to the development of improved treatments for spinal cord injury requires that the therapy be subjected to a thorough and objective histopathological evaluation in which its effectiveness in reducing necrosis is evaluated separately from its effectiveness in promoting axonal growth. We examined the validity of this approach by testing two synthetic compounds that had been shown to stimulate axonal outgrowth in tissue cultures of dorsal root ganglion neurons. A non-necrotizing lesion was produced by crushing the spinal cord of rats for 1 s while leaving the dura intact. Two days later, the dura was opened and a polyethylene tube was sutured to the vertebral spines and soft tissues such that one opening of the tube lay over the injured part of the spinal cord. The other end was brought through a subcutaneous tunnel so that it emerged at the base of the skull. A syringe adapter was attached to the external opening by means of which the drugs were injected. It was found to be essential to wait 2 days between operations; if the dura was opened within 2 days of the crush injury, the edematous spinal cord herniated through the opening and was severely damaged. The drugs and vehicle control solutions were administered four times a day in volumes of 0.5 ml. Alzet osmotic pumps were not used because the total volume delivered would be insufficient to fully bathe the injured tissue with drug. All animals were treated for 14 days before autopsy. Histological slides were prepared and evaluated indepen-

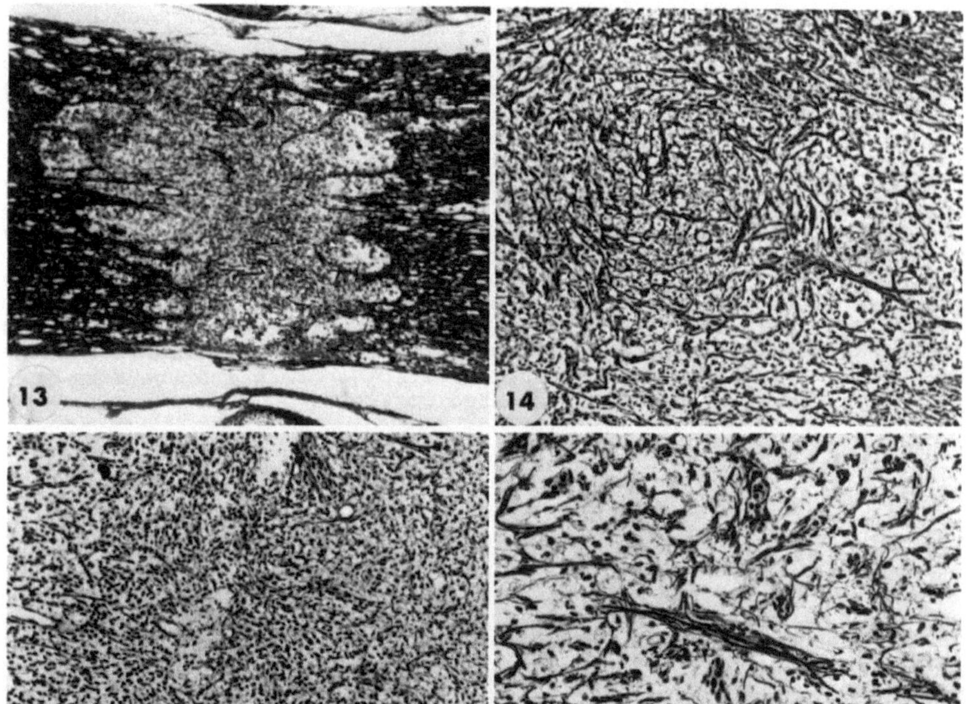
13
14
15
16
17
18

dently by each of the authors using a double-blind experimental protocol.

The results of the study are illustrated in Figs. 13-18. In every case the drug-treated animals showed remarkably greater invasion of the lesion by nerve fibers than did the vehicle-treated controls. In fact, there was no obvious difference between vehicle-treated control specimens (Figs. 13, 15, 17) and untreated animals (compare Figs. 7 and 11 with Figs. 13 and 17). The nerve fibers in the lesions of treated rats, on the other hand, grew in such profusion that they were no longer oriented longitudinally but were rather haphazardly disposed (Fig. 14). Fibers were frequently undulating and varicose and were often arranged in small bundles containing three to six fine axons (Figs. 16 and 18). When the slides were coded and randomized there was no difficulty in distinguishing between the extensive axonal invasion of the lesion in drug-treated animals and the more modest nerve ingrowth in vehicle-treated controls.

Concluding Statement

The foregoing observations indicate that the success of axonal regeneration after spinal cord injury is dependent on the histological composition of the lesion. Axons will grow into sites of spinal cord injury provided that region is populated by cells found normally in nerve tissue (e.g., blood vessels, astrocytes, and ependymal cells) but will refuse to enter regions populated solely by cells not normally found in nerve tissue (e.g., monocytes, lymphocytes, and fibroblasts). Past efforts at repair of spinal cord injury have emphasized the prevention of tissue necrosis and the elimination or minimization of collagenous and glial scarring. It now seems likely that these approaches must be supplemented by procedures designed to repopulate the injured tissue with appropriate cells.

As a result of these histological considerations it is now feasible to initiate a more systematic and rational approach to the treatment of spinal cord injury. For this purpose, we must first develop more reproducible, sensitive, and quantifiable models of spinal cord injury than are now available. Secondly, we must use these models to evaluate rigorously and individually the efficacy of treatments designed (i) to reduce necrosis, (ii) to stimulate growth and migration of vessels, ependymal cells, and astrocytes into the lesion, and (iii) to enhance axonal outgrowth. In this way, we will obtain sufficient information to enable us to design the complex therapeutic protocols that will undoubtedly be necessary to retard ischemia, prevent collagenous scarring, promote the oriented growth of vascular, ependymal, and astrocytic elements, and enhance axonal regeneration in human spinal cord injuries.

References

Balentine JD (1978) Pathology of experimental spinal cord trauma: the necrotic lesion as a function of vascular injury. Lab Invest 39:236-253

Benfey M, Aguayo AJ (1982) Extensive elongation of axons from rat brain into peripheral nerve grafts. Nature 296:150-152

Chaban GH, Barrett CP, Donati EJ, Guth L (to be published). Tissue necrosis after experimental optic nerve injury: effect of position of lesion and magnitude of crushing force.

David S, Aguayo AJ (1981) Axonal elongation into PNS "bridges" after CNS injury in adult rats. Science 214:931-933

De La Torre JC (1981) Spinal cord injury: review of basic and applied research. Spine 6:315-335

Faden AI, Jacobs TP, Smith MT, Holoday JW (1983) Comparison of thyrotropin-releasing hormone (TRH), naloxone, and dexamethasone treatments in experimental spinal injury. Neurology 33:673-678

Gilmore SA, Leiting JE (1980) Changes in the central canal area of immature rats following spinal cord injury. Brain Res 201:185-189

Guth L, Albuquerque EX, Deshpande SS, Barrett CP, Donati EJ, Warnick JE (1980) Ineffectiveness of enzyme therapy on regeneration in the transected spinal cord of the rat. J Neurosurg 52:73-86

Guth L, Barrett CP, Donati EJ, Deshpande SS, Albuquerque EX (1981) Histopathological reactions and axonal regeneration in the transected spinal cord of hibernating squirrels. J Comp Neurol 203:297-308

Hall ED, Braughler JM (1982) Thyrotropin-releasing hormone for spinal trauma. N Engl J Med 306:429-430

Kao CC (1974) Comparison of healing process in transected spinal cords grafted with autogenous brain tissue, sciatic nerve, and nodose ganglion. Exp Neurol 44:427-439

Krikorian JG, Guth L, Donati EJ (1981) Origin of the connective tissue scar in the transected rat spinal cord. Exp Neurol 72:698-707

Matthews MA, St.Onge MF, Faciane CL, Gelderd, JB (1978) Spinal cord transection: a quantitative analysis of elements of the connective tissue matrix formed within the site of lesion following administration of Piromen, cytoxan, or trypsin. Neuropathol Appl Neurobiol 5:161-180

Matthews ML, St.Onge MF, Faciane CL (1979) An electron microscopic analysis of abnormal ependymal cell proliferation and envelopment of sprouting axons following spinal cord transection in the rat. Acta Neuropathol 45:27-36

Michel ME, Reier PJ (1979) Axonal-ependymal associations during early regeneration of the transected spinal cord in Xenopus laevis tadpoles. J Neurocytol 8:529-548

Ramon y Cajal S (1928) Degneration and regeneration of the nervous system. Oxford University Press, London

Richardson PM, McGuiness UM, Aguayo AJ (1980) Axons from CNS neurons regenerate into PNS grafts. Nature 284:264-265

Richardson PM, McGuiness UM, Aguayo AJ (1982) Peripheral nerve autografts to the rat spinal cord: studies with axonal tracing methods. Brain Res 237:147-162

Selzer ME (1978) Mechanisms of functional recovery and regeneration after spinal cord transection in larval sea lamprey. J Physiol (Lond) 277:395-408

Singer M, Nordlander RH, Egar M (1979) Axonal guidance during embryogenesis and regeneration in the spinal cord of the newt: the blueprint hypothesis of neuronal pathway patterning. J Comp Neurol 185:1-22

Stensaas LJ (1983) Regeneration in the spinal cord of the newt Notophthalmus (Triturus) pyrrhogaster. In: Kao CC, Bunge RP Reier PJ Grillner S (eds) Fundamentals of spinal cord reconstruction. Raven, New York, pp 121-150

Windle WF (1956) Regeneration of axons in the vertebrate central nervous system. Physiol Rev 36:427-440

Wood MR, Cohen MJ (1981) Synaptic regeneration and glial reactions in the transected spinal cord of the lamprey. J Neurocytol 10:57-79

Gangliosides Enhance Mechanisms of Recovery from Neural Damage by a Dual Mechanism

A. Gorio[1], D. Janigro[1], F. Di Gregorio[1], G. Ferrari[1], G. Jonsson[2], F. Vyskocil[3], and R. Zanoni[1]

1 Department of Cytopharmacology, Fidia Research Laboratories, Via Ponte della Fabbrica, 3a, 35031 Abano Terme, Italy
2 Department of Histology, Karolinska Institutet, 10401 Stockholm, Sweden
3 Czechoslovak Academy of Sciences, Prague, Czechoslovakia

Introduction

Recovery from lesions to the nervous system is a process involving several steps which begins immediately after the lesion and may last several months (in humans), if it occurs at all. There are four major parts in such a process which are usually taken into medical consideration: (1) Immediate care after the lesion - often patients suffer worse damage from first aid at the site of the accident than from the accident itself. For instance, the spinal cord often remains intact even after a crush lesion which may disrupt vertebrate column bones; however, the first aid can irreversibly damage the spinal cord if it is not done properly. (2) Immediate treatment after hospitalization, the aim of which is to counteract the progressive neural tissue deterioration caused by the trauma initially. (3) Regeneration of the lesioned neural processes for reconstructing the lesioned and consequently degenerated pathways. (4) Rehabilitation of the patient. Pharmacological treatment to improve recovery has as its obvious targets points 2 and 3. While point 1 is merely related to first aid medical education, point 4 may or may not be related either to neuronal regeneration or to network reorganization and its pharmacological enhancement is still being debated. The pathophysiological consequences of neural trauma are sequentially very complex (see other chapters in this volume), but they seem to stem from a progressive tissue ischemia. It is clear that blood flow dramatically decreases at the site of the lesion and that the injured area suffers a high degree of hypoxia, neurons loose K^+, accumulate Na^+ and Ca^{++}, and then degenerate. In the proximity of the lesioned area, surviving axons become unexcitable, probably because the composition of the extracellular fluids is grossly altered (Young et al. 1982 a, b). If such an abnormal environment, characterized by poor levels of oxygen and ionic perturbations, persists in the surroundings of the lesioned area, regeneration and re-establishment of excitability may not be possible. Hypoxia may cause massive accumulation of extracellular K^+ (up to 80 mM), accompanied by a decrease in Na^+ and Cl^-. In these conditions neurons are depolarized. Early restoration of oxygen may return resting potential to normal values and restore excitability; this process may occur in several minutes (Sykova 1983). It is conceivable that restoration of proper ionic balances and the counteraction of the spreading of such deleterious events are targets for pharmacological agents aiming to improve recovery after trauma. As mentioned above, a normalized extracellular environment may well improve neuronal regeneration. Of relevance in this connection is the widely reported concept of abortive regeneration, first described by Ramon y Cajal (1928). He observed that lesioned CNS neurons regenerate their axons only for a short distance and for a brief time, and

Experimental Brain Research, Suppl. 13

that such processes are then retracted and regeneration does not occur (Ramon y Cajal 1928). It is conceivable that upon entering a highly ischemic and ionically disbalanced region, regrowing axons fail to continue growth and are successively retracted.

In the following pages we will describe some experimental evidence supporting the use of gangliosides as stimulatory agents of repair after damage to the nervous system. GM_1 is the elected ganglioside for CNS experimentation because its bioavailability in the CNS is twice that of the other gangliosides which constitute the mixture. The composition of the ganglioside mixture is as follows (%): GM_1 21, GD_{1a} 39.7, GD_{1b} 16, GT_{1b} 19.

Motor Neuron Sprouting

Ceccarelli and collaborators showed, some years ago, that daily treatment with gangliosides enhanced reinnervation of cat nictitating membrane after pre- or postganglionic denervation (Ceccarelli et al 1976). Following this finding, it was investigated whether gangliosides influenced reinnervation by accelerating the rate of growth or by affecting sprouting. We designed two types of study: one for monitoring the time-course of reinnervation and the second for examining the extent of reinnervation capacity of the nervous system in the presence or absence of ganglioside treatment. The first kind of experiment was performed by crushing the sciatic nerve and monitoring the dynamics of nerve re-entry into the EDL muscle. Motor axons begin reinnervation of the denervated muscle 2 weeks after denervation; this early stage of reinnervation is characterized by sprouting of the regrowing axons as soon as they re-enter the muscle and by the consequent increase in the extent of polyneuronal innervation, which gradually reaches its maximum 25 days after lesion of the nerve, i.e., 11-12 days after the beginning of muscle reinnervation (Gorio et al. 1980; Carmignoto et al. 1983; Gorio et al. 1983a). Daily treatment with gangliosides did not reduce the time required to re-enter the muscle by the regenerating nerves; in other words, it did not affect the rate of neurite elongation. However, sprouting and polyneuronal innervation were greatly affected since both the onset and the maximum extent of polyneuronal innervation were achieved much earlier compared to controls, as if the sprouting wave was anticipated to a large extent in the treated animals compared to controls. The process of synaptic repression was not affected, and the redundant nerve terminals were eliminated in the same fashion in treated animals and controls. Such a temporal evolution of the effect suggested that ganglioside action was probably mediated by the presence of a growth factor; indeed, if gangliosides had increased sprouting by themselves, we should have found some interference with the repression phase (Gorio et al. 1980). The second experimental model showed that ganglioside treatment enhanced the sprouting capacity of neurons (Gorio et al. 1983b).

These data obtained while monitoring the reinnervation of rat EDL and soleus muscles indicate that ganglioside treatment affects both the dynamics and te extent of neuronal sprouting of damaged and intact neurons.

Neuritogenesis and Regeneration In Vitro

The experiments mentioned above show quite clearly that gangliosides affect neuronal regeneration, perhaps by enhancing the efficacy of some growth-promoting factor released by the denervated tissue. An initial system for testing such an hypothesis is the pheochromotcytoma cell line (PC12), a tumor line derived from the adrenal medulla. These cells reduplicate in vitro; however, they stop dividing and differentiate, forming neurites, when exposed to nerve growth factor. In such a condition if gangliosides were capable of enhancing differentiation and neurite regeneration, we could then test whether the treatment affected NGF fate (Ferrari et al. 1983). Indeed, it was found that ganglioside mixture or single gangliosides enhanced neuritogenesis in the presence or absence of serum, but they failed to enhance neuritogenesis if NGF was omitted from the bathing medium, indicating that they did not act as a substitute for the specific growth factor.

In addition it was found that gangliosides enhanced neurite regeneration. PC12 were treated with NGF for about 10 days, allowing a full expression of the neuritogenesis program, and were then resuspended and replated. Such a mechanical manipulation chopped away their neurites. The replated cells gradually regrew their neurites in several hours, and gangliosides accelerated the process drastically. This effect was not due to alterations of NGF fate, since binding experiments, performed using iodinated NGF, did not indicate any changes in binding and internalization of NGF after ganglioside treatment. Taken all together, these experiments indicate that a specific growth factor is responsible for the processes, perhaps occurring at membrane level, corroborate the effect of such a growth factor. It seems that gangliosides act upon this mechanism.

Counteraction of Lesion-Induced Damage to the Nervous System

There are now several pieces of evidence obtained in our laboratories and elsewhere to indicate that gangliosides have strong counteracting effects on lesion-induced damage. Serotoninergic pathways of newborn rats can be selectively damaged by a single intraperitoneal injection of the neurotoxin 5,7-HT (Jonsson et al. to be published b). A few hours after birth, pups are injected with the neurotoxin at a dose of 50 mg/kg; its degenerative promoting action is very rapid, and nerve terminals most distal from the cell body degenerate. 5-HT levels in cortex and spinal cord decrease drastically in a matter of a few hours, while such a dose of toxin spares the terminals present in the pons-medulla, mesencephalon, and striatum, which are closer to the raphe nuclei (Figs. 1, 2). The primary effect of the toxin is followed by a slower retrograde degeneration, which is also termined secondary degeneration (Fig. 1). Such an effect is quite dangerous since it may cause neuronal cell death, which often occurs with higher doses of 5,7-HT. Four weeks after the lsion, the levels of serotonin in frontal and occipital cortex are still very low, and no regeneration of the lesioned axons has been observed in these distal regions (Fig. 1); however, the proximal axons, which were spared by the lesion, outgrow extensively so that the mesencephalon, pons-medulla, and striatum are hyperinnervated (Fig. 3). Such a process is termed "pruning". If a group of lesioned animals is treated daily for 4 days subsequently to the lesion, with GM_1 at a dose of 30 mg/kg, the process described above is dramatically different in several aspects. The primary effect of the neurotoxic lesion is normally

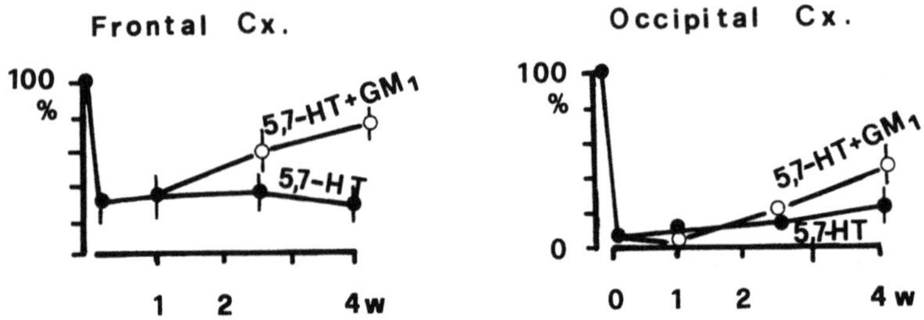

Fig. 1. Levels of 5-HT in the frontal and occipital cortex of 2-day, 1-week, 2.5-week, and 4-week-old rats after neonatal treatment with 5,7-HT and/or GM_1. Rats were killed 4 h after the last GM_1 injection. Each point represents the mean ± SEM (Jonsson et al. 1984a)

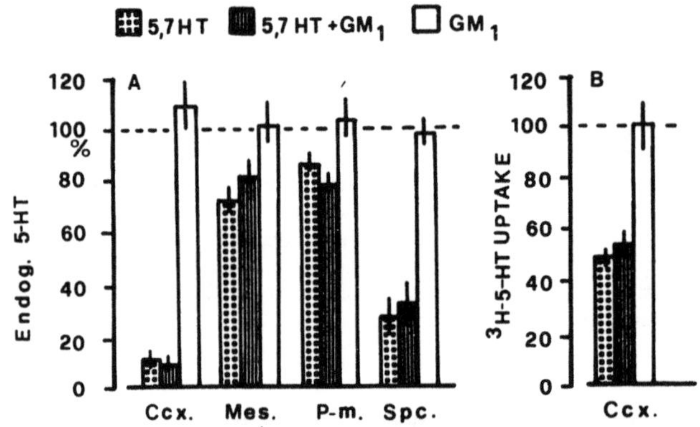

Fig. 2. 5-HT levels (A) and in vitro uptake of ^{3}H-5-HT (0.05 µM, 10 min) (B) in slices from cerebral cortex after 5,7-HT (50 mg/kg s.c., 20 h) and/or GM_1 (2 x 30 mg/kg s.c., 14-h interval) treatment of newborn rats. The first GM_1 dose was injected 2 h after 5,7-HT. each bar represents the mean ± SEM expressed as % of saline-injected control. A) 5-HT control values; CCx (cerebral cortex): 60 ± 5.2; Mes (mesencephalon): 136 ± 19; P-m (pons-medulla): 180 ± 13; Spc. (spinal cord): 192 ± 21 ng 5-HT/g.B) H-5-HT uptake in control: 10.3 ± 0.97 nCi/slice (Jonsson et al. 1984a)

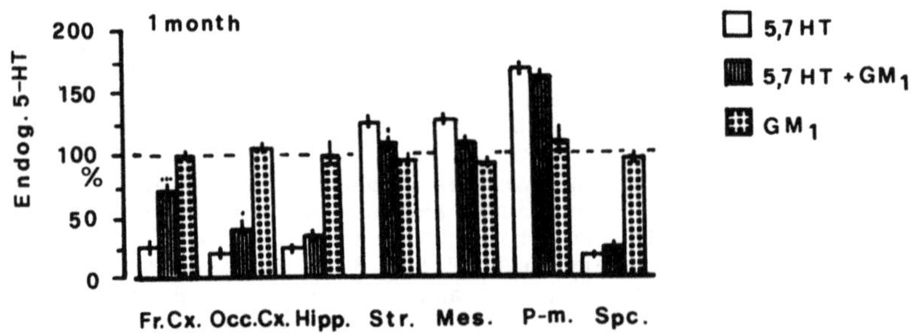

exerted; however, the lesioned axons regrow rapidly and 4 weeks afterward the level of 5-HT is almost back to normal in frontal cortex (Fig. 1). This effect is accompanied by a lesser outgrowth of the proximal axons, as if the stimulated regeneration of the distal ones has inhibited their growth (Fig. 3). The enhanced reinnervation of the cortex by 5-HT axons is correlated by an increased uptake of serotonin, and a higher number of 5-HT terminals determines a higher number of active sites for uptake of the transmitter by comparison with the control animals, whose regeneration of 5-HT axons is much delayed (Fig. 4). Quantitative immunocytochemistry, performed using a monoclonal antibody to 5-HT, confirms the biochemical data even at much higher levels of resolution, allowing us to analyze the presence of 5-HT axons in much smaller areas than would be possible biochemically (Jonsson et al. to be published a). Such experiments show that GM_1does not interfere wtih the primary mechanisms of the neurotoxins, but counteracts the retrograde degeneration of the lesioned axons and allows their regrowth.

These observations on the counteracting action of gangliosides are strongly reinforced by some compartmental studies showing that GM_1 reduces the degree of initial impairment after lesion. Unilateral lesions of rat entorhinal cortex caused a loss of the learned alternation behavior, which was recovered in about 2 weeks. If lesioned animals were treated wtih gangliosides the initial impairment (48 h after lesioning) was strongly reduced while the rate of recovery seemed identical (Karpiak 1983). The severity of postlesion behavioral loss was also tested after bilateral lesion of the entorhinal cortex. The tested behavior was open field activity, which increases greatly after such a lesion. In the animals treated with gangliosides, this increase in open field activity was very much reduced (Fass et al. to be published).

Both reported effects were observed acutely 24 or 48 h after lesions; therefore they have to be interpreted as indicating that ganglioside or, namely, GM_1 effets occur by counteracting the noxious secondary effects of the lesions instead of by enhancing sprouting, which may occur a few days later.

Ganglioside Dual Mode of Action

The incorporation of GM_1 into purified neuronal membranes "in vitro" causes several microenvironmental changes and the activity of several membrane enzymes is strongly enhanced (Rapport and Gorio 1981). In particular, 5×10^{-8} M GM_1 increases by about 40% the (Na^+, K^+)

◄ Fig. 3. 5-HT levels in the brains of 1-month-old rats after 5,7-HT and/or GM_1 treatment in the neonatal stage. Each bar represents the mean ± SEM. Control values: Fr.Cx. (frontal cortex): 159 ± 4.6; Occ.Cx. (occipital cortex): 87 ± 3.3; Hipp (hippocampus): 130 ± 14; Str (striatum): 137 ± 7.5; Mes (mesencephalon): 352 ± 16; P-m. (pons-medulla): 284 ± 9.4; Spc. (thoracic-lumbar spinal cord - thoracic-lumbar): 236 ± 15 ng 5-HT/g. Statistical differences between 5,7-HT + GM_1 groups: x $0.05 > p$ 0.01; xxx $p < 0.001$ (Student's t-test) (Jonsson et al. 1984a)

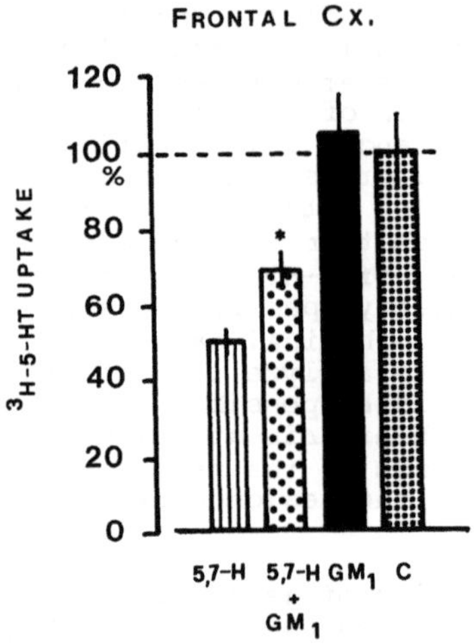

Fig. 4. In vitro uptake of ^{3}H-5-HT (0.05 µm, 10 min) in slices from the frontal cortex of 2-month-old rats after 5,7-HT and/or GM_1 treatment in the neonatal stage. 5,7-HT and GM_1 was administered as in Fig. 2. Each bar represents the mean ± SEM. Control value: 14.6 ± 1.56 nCi/slice. Statistical difference between 5,7-HT and 5,7 HT + GM_1 groups: x $0.05 > p > 0.01$ (Student's t-test) (Jonsson et al. 1984a)

ATPase Vmax; higher oncentrations of GM_1 have a lesser effect and may even become inhibitory (Leon et al. 1981). To check such in vitro biochemical effects, a series of expeiments was designed to examine the activity of gangliosides on membrane ionic processes and pump activities. The Na^+/K^+ pump is activated by intracellular sodium but it requires the presence of extracellular potassium ions to function; therefore if K^+ ions are omitted from the incubating media, the pump cannot function. In such conditions cells accumulate Na^+ and loose K^+. Mouse diaphragm neuromuscular preparations were incubated in Kreb's solution lacking K^+; the response to such treatment is a sudden hyperpolarization which is followed by depolarization to -61.3 mV after several hours of K^+ free incubation. When K^+ is reintroduced in the bathing medium there is activation of Na^+/K^+ pump, which causes membrane repolarization to -71.6 mV, and re-establishment of the proper ionic balances. The restoration of the normal resting potential is preceded by a transient hyperpolarization to -80.7 mV. However, if the nerve-muscle preparation was incubated for 2 h with 5×10^{-8} gangliosides, the transient hyperpolarization was -90.4 mV. Therefore the net charge flow which is observed in these circumstances is increased. Such an effect is explicable by a higher activation of the Na^+,K^+)ATPase induced by gangliosides. The effect is very different if gangliosides are added to the bathing medium 5-6 h prior to K^+ readmission or if the animals are treated in vivo at a dose of 1 or 10 mg/kg for 3 days, one injection/day. In fact, such neuromuscular preparations are very resistant to the lack of extracellular K^+, and no depolarization and consequent hyperpolarization are observed. This resistance to the ionic consequences of the lack of K^+ was correlated to an increased resistance to hypoxia in the same experimental circumstances (Vyskocil et al. to be published). Therefore these results suggest that the early effects of gangliosides are to activate membrane enzymes such as (Na^+,K^+)ATPase, while the late ones are protective against ionic disbalances, perhaps via activation of some intracellular mechanisms. An in vitro preparation which would allow us to examine in greater detail such aspects of ganglioside action is the hippocampal slice preparation. Slices 450 µm thick were cut transverse to the longitudinal axis of the hippocampus on a tissue chopper (Sorval) and transferred to the recording chamber (Janigro et al. to be published). The Kreb's solution was continuously bubbled with 95% O_2 and 5% CO_2. Intracellular recordings were made with thin wall microelectrodes of a DC-resistance ranging from 80 to 100 MΩ in CA3 pyramidal neurons. When such slices are exposed to an O_2 level reduced from 95% to 12%, there is a transient increase of excitation which is followed by a decreased membrane resistance and large neuronal depolarization with consequent tissue unexcitability. Upon a prompt readmission of the normal oxygen level there is a restoration of the resting potential, preceded by a transient hyperpolrization. The magnitude of this hyperpolrization is related to the amount of cations accumulated in the cell. The parameters examined were the latency between the hypoxic episode and the maximal depolariztion, magnitude of depolarization, and amplitude of hyperpolrization. We found that if the preparation were incubated with 5 x 10 M GM_1 for a minimum of 4-5 h, the latency increased from 3.15 to 7.3 min on average, the depolarization decreased from -48.7 to 31.8 mV, and consequently the hyperpolrization decreased from 25.5 to 15.5 mV. No effect of GM_1 on any neurophysiological parameters, such as input resistance, resting potential, action potentials, and synaptic potentials, was ever observed in normal conditions (Janigro et al. to be published).

The neuromuscular preparation and the hippocampal slice experiments suggest that gangliosides may act by a dual mechanism: an early and a late one. The early one occurs at the time of incorporation in neuronal membranes, triggers membrane enzyme activities such as (Na^+, K^+)ATPase, and is compatible with the stimulated neuronal sprouting and neuritogenesis; the late one, by contrast, may be evident only after several hours of ganglioside incorporation in neuronal membrane and may be due to intracellular events, which may be triggered by the initial incorporation of gangliosides into the neuronal membrane or by their internalization.

References

Carmignoto G, Finesso M, Siliprandi R, Gorio A (1983) Muscle reinnervation I. Restoration of transmitter release mechanisms. Neuroscience 8:393-401

Ceccarelli B, Aporti F, Finesso M (1976) Effects of brain gangliosides on functional recovery in experimental regeneration and reinnervation. In: Porcellati G, Ceccarelli B, Tettamanti G (eds) Advance in experimental medicine and biology, vol 71. Plenum, New York, p 275

Fass B, Ramirez JJ (to be published) Effects of ganglioside treatments on lesion-induced behavioral impairments and sprouting in the CNS. J Neurosci Res

Ferrari G, Fabris M, Gorio A (1983) Gangliosides enhance neurite outgrowth in PC12 cells. Dev Brain Res 8:215-221

Gorio A, Carmignoto G, Facci L, Finesso M (1980) Motor nerve sprouting induced by gangliosides treatment. Brain Res 197:236-241

Gorio A, Carmingnoto G, Finesso M, Polato P, Nunzi MG (1983a) Muscle reinnervation II. Formation and repression. Neuroscience 8:403-416

Gorio A, Marini P, Zanoni R (1983b) Muscle reinnervation III. Motoneuron sprouting capacity, enhancement by exogenous gangliosides. Neuroscience 8:417-429

Janigro D, Di Gregorio F, Vyskocil F, Gorio A (1984) Ganglioside dual mode of action: a working hypothesis. J Neurosci Res 121:499-509

Jonsson G, Gorio A, Hallmann H, Janigro D, Kojima H, Zanoni R (1984a) a) Effect of GM_1 ganglioside on neonatally neurotoxin induced degeneration of serotonin neurons in the rat brain. Dev Brain Res 16:171-180

Jonsson G, Gorio A, Hallmann H, Kojima H, Janigro D, Luthman J, Zanoni R (1984b) Effects of GM_1 ganglioside on developing and mature serotonin and noradrenaline neurons lesioned by selective neurotoxins. J Neurosci Res 12:459-476

Karpiak SE (1983) Ganglioside treatment improves recovery of alternation behavior after unilateral entorhinal cortex lesion. Exp Neurol 81:330-339

Leon A, Tettamanti G, Toffano G (1981) Changes in functional properties of neuron membranes by insertion of exogenous ganglioside. In: Rapport MM, Gorio A (eds) Gangliosides in neurological and neuromuscular function, development, and repair. Raven, New York, p 45

Ramon y Cajal S (1928) Degeneration and regeneration of the nervous system. Translator Oxford MRM. University Press, London

Rapport MM, Gorio A (eds) (1981) Gangliosides in neurological and neuromuscular function, development, and repair. Raven, New York

Sykova E (1983) Extracellular K^+ accumulation in the central nervous system. Prog Biophys Mol Biol 42:135-189

Vyskocil F, Di Gregorio F, Gorio A (to be published) The fascilitating effect of gangliosides on electrogenic (Na^+/K^+)pump and resistance of membrane potential to hypoxia in neuromuscular preparations. Pflügers Arch

Young W, Yen V, Blight A (1982) Extracellular calcium ionic activity in experimental spinal cord contusion. Brain Res 253:105-113

Young W, Koreh I, Yen V, Lindsay A (1982) Effect of sympathectomy on extracellular potassium ionic activity and blood flow in experimental spinal cord contusion. Brain Res 253:115-124

GM_1 Ganglioside Treatment Enhance Regrowth of Central and Peripheral Noradrenaline Neurons After Selective 6-Hydroxydopamine-Induced Lesions

G. Jonsson[1], A. Gorio[2], H. Hallman[1], D. Janigro[2], H. Kojima[1], and J. Luthman[1]

1 Department of Histology, Karolinska Institutet, 10401 Stockholm, Sweden

2 Department of Cytopharmacology, Fidia Research Laboratories, Via Ponte della Fabbrica 3a, 35031 Abano Terme, Italy

Introduction

There are several lines of evidence indicating that exogenous administration of gangliosides can promote regrowth processes and functional recovery after lesions of nervous tissue both in CNS and PNS (Ceccarelli et al. 1976; Gorio et al. 1980, 1983; Toffano et al. 1983; Wojcik et al. 1982). Gangliosides are normally present as integral components of the plasma membrane of most vertebrate cells, and are especially plentiful in that of neurons, where they are considered to be associated with a variety of membane-bound processes (see Ando 1983). It has, moreover, been demonstrated that exogenous gangliosides can be incorporated into cell membranes, although there is no direct evidence to indicate that they can serve as neuronotrophic factors. On the other hand, gangliosides have been implicated to act as receptor molecules for several bioactive compounds (see Ando 1983). Very little is so far known as to the mode of action of gangliosides in stimulating regrowth processes, although it is generally considered that they act by facilitating the effect of endogenous neuronotrophic factors (see Ferrari et al. 1983). The purpose of the present paper is to summarize recent studies from our laboratories analyzing the effect of systemic treatment with the monosialoganglioside GM_1 on sprouting-regeneration processes of selectively lesioned noradrenaline (NA) neurons in CNS and PNS. The catecholamine neurotoxin 6-hydroxydopamine (6-OH-DA) was used to induce the NA lesions (see Jonsson 1983), and neuro- and histochemical techniques were employed to evaluate the effect GM_1 had on such selective lesions.

Materials and Methods

Male albino rats (Sprague-Dawley, 150g body wt.) and mice (N.M.R.I., 25g body wt.) were used. Chemical sympathectomy was induced in mice by administering 6-OH-DA (25 mg/kg i.v.). In acute experiments a a single injection of GM_1 (20 mg/kg i.p.) was administered 1 h before 6-OH-DA, while in chronic experiments the animals received additional daily injections of GM_1 for 1 week. In rats a local NA denervation was produced by incortical infusion of 6-OH-DA according to the microperfusion method of Kasamatsu et al. (1979, 1981). A cannula (20-gauge hypodermic needle) was implanted under stereotaxic conditions directly into the cortical tissue (1-1.5 mm deep into the gray matter of parietal cortex) and connected to an osmotic minipump (Alzet 2001; delivery rate 1 μl/h filled with 2 mM 6-OH-DA in 0.9% NaCl). The infusion of 6-OH-DA was terminated after 3 or 7 days by removal of the minipump-cannula assembly. In these experiments the

Experimental Brain Research, Suppl. 13

effect of various preand posttreament protocols of GM_1 (30 mg/kg i.p.) was investigated.

Chemical NA Assay

Endogenous NA in brain tissue pieces was determined by liquid chromatography with electrochemical detection (LCEC) according to Keller et al. (1976). The values were expressed as ng NA/g wet weight of the tissue (corrected for recovery).

^{3}H-NA Uptake In Vitro

Slices (0.5 mm thick) from cerebral cortex and isolated irides were incubated in vitro in Krebs-Ringer bicarbonate buffer (pH 7.4) containing 0.05 μM l-^{3}H-NA for 10 or 15 min at +37°C. Radioactivity (>95% unchanged ^{3}H-NA) taken up and retained by the tissue was determined by liquid scintillation spectrometry (Lidbrink and Jonsson 1974; Olson et al. 1968). ^{3}H-NA uptake was expressed as nCi/slice or iris after correction for "extraneuronal uptake", thus representing ^{3}H-NA uptake in NA nerve terminals.

NA Fluorescence Histochemistry

Mouse irides were prepared as whole mounts and air-dried (Malmfors 1965), while rat brains were freeze dried. After drying, the tissue was exposed to formaldehyde gas (+80°C for 1 h). The brains were serially sectioned (10 μm) and analyzed by fluorescence microscopy (Corrodi and Jonsson 1967; Fuxe et al. 1970). The irides were analyzed by an interactive computerized image-analysis system (IBAS, Kontron/Zeiss) for the determination of NA nerve density. An image-intensified black-and-white TV camera was attached to the fluorescence microscope used to demonstrate NA nerve terminals. After various contrast enhancement steps of digitized images, a gray level was chosen to select for fluorescent NA nerves excluding virtually all background. The binary picture was generated and the computer calculated the area covered by fluorescent nerves in % of the field of view (measuring field: 255 x 255 μm). Four randomly sampled areas were measured and the values averaged to represent one observation. The values obtained can be considered as a measure of the relative NA nerve density per unit area.

Results

Effect of GM_1 on 6-OH-DA-Induced NA Lesions in Mouse Iris

The sympathetic adrenergic nerves innervating mouse iris form a dense plexus of evenly distributed varicose fibers. The area covered by these NA nerve terminals contributes about 30% of the total area as determined by computerized image analysis of irides processed for fluorescence histochemistry according to Falck-Hillarp. Administration of 6-OH-DA (25 mg/kg i.v.) causes rapid (within a few hours), selective, and almost complete degeneration of the NA nerve terminals in the iris, as reflected by a very pronounced (more than 90%) reduction of NA nerve density and in vitro uptake of ^{3}H-NA 16 h af-

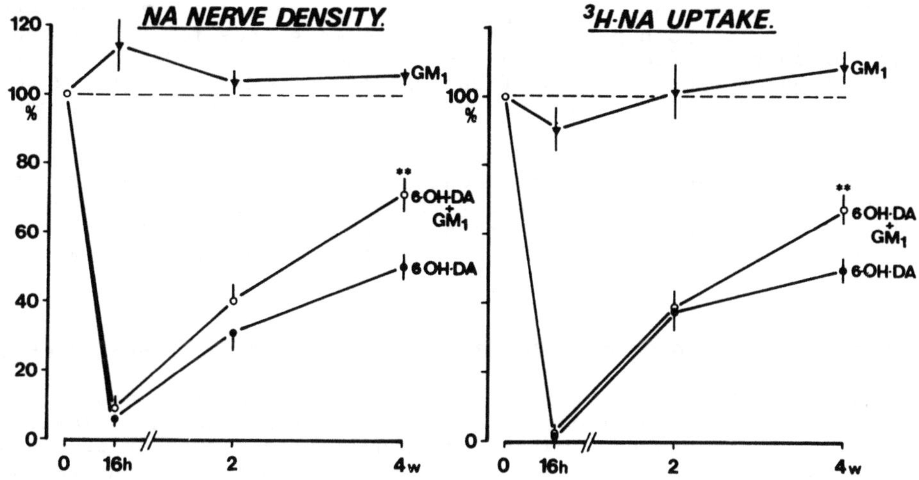

Fig. 1. Effect of 6-OH-DA (25 mg/kg i.v.) and/or GM_1 (20 mg/kg i.p.) on the NA nerve density and ^{3}H-NA uptake in vitro in mouse irides at various times after the 6-OH-DA administration (16 h, 2 weeks, and 4 weeks). The first GM_1 injection was given 1 h before 6-OH-DA. Animals killed 16 h after 6 OH-DA received only one GM_1 injection, whereas mice killed 2 or 4 weeks after 6-OH-DA received daily injections of GM_1 for 1 week. NA nerve density was determined in irides processed for fluorescence histochemical demonstration of NA employing computer-assisted image analysis. Each point represents the mean $\pm$ SEM (n = 4-6), expressed as % of control. ** = $0.001 > p > 0.01$ (Student's t-test; unpublished data)

ter 6-OH-DA (Fig. 1). In the acute phase after 6-OH-DA strongly fluorescent NA accumulations in non-terminal NA axons can be found which certainly reflect the degenerative effects of 6-OH-DA on distally located nerve terminal arbours (see Jonsson 1983). GM_1 treatment had no significant effect on the 6-OH-DA induced reduction of NA nerve density and ^{3}H-NA uptake when analyzed acutely (16 h). No notable difference in the number of fluorescent NA accumulations after 6-OH-DA was found after combined GM_1 + 6-OH-DA treatment. These findings thus indicate that GM_1 treatment does not interfere with the primary, target-directed neurodegenerative actions of 6-OH-DA on NA nerve terminals.

The mode of 6-OH-DA administration used produces only an NA denervation at the nerve terminal level in effector organs, while the ganglionic NA cell bodies remain intact (Jonsson and Sachs 1970). With time there is thus a regeneration of NA nerves in the denervated organs and it takes about 2-3 months to restore the adrenergic nerve plexus (Jonsson and Sachs 1970, 1972). For the present study it was observed that there was about a 50% restoration of NA nerve density and ^{3}H-NA uptake in the iris 4 weeks after 6-OH-DA (Fig. 1). The regenerating NA nerve terminals had a similar fluorescence morphology and ^{3}H-NA uptake kinetics compared to normal, mature terminals, although it was noted that they displayed a lower fluorescence intensity, indicating reduced NA levels. This is probably related to a lag in the restoration of NA storage granules (cf. Jonsson and Sachs 1972). Analysis in the chronic stage demonstrated that GM_1 treatment had a significant enhancing effect on the recovery of both

NA nerve density and ^{3}H-NA uptake 4 weeks after 6-OH-DA administration (Fig. 1). Both parameters analyzed were thus found to be about 70% of control after GM_1 + 6-OH-DA as compared to 50% of control after 6-OH-DA alone. No significant effect of GM_1 was notet 2 weeks after 6-OH-DA. There was no obvious difference in the fluorescence morpholgy of NA nerve terminals regenerated after 6-OH-DA alone as compared to those regenerated after GM_1 + 6-OH-DA. Moreover, at no timepoint studied was GM_1 treatment alone found to have any significant effect either on NA density or on ^{3}H-NA uptake. These results clearly demonstrate that GM_1 enhances regrowth of sympathetic adrenergic nerves in mouse iris after 6-OH-DA induced sympathectomy.

Effect of GM_1 on 6-OH-DA Induced NA Lesions in Rat Cerebral Cortex

Infusion of 2 mM 6-OH-DA for 3 or 7 days unilaterally into the gray matter of the cerebral cortex using an osmotic minipump (delivery rate: 1 μl/h) caused a marked and longlasting reduction of endogenous NA levels and density of NA nerve terminals demonstrated by fluorescence histochemistry in a restricted area around the infusion point (Figs. 2, 3). The pattern of the 6-OH-DA induced alteration of endogenous NA was almost identical with that seen after measuring the spatial distribution of ^{3}H-NA uptake. It was furthermore found that the spatial pattern of NA depletion was very similar after 6-OH-DA infusion for 3 or 7 days. The size of the general tissue damage in cerebral cortex due to cannulation was relatively limited, forming a cylinder with a radius of about 0.3-0.4 mm. The necrotic area was sharply demarcated and the cortical tissue outside it had an apparently unchanged appearance. There was no obvious difference sn size of the nonspecific damage after infusion of 2 mM 6-OH-DA or the solvent. The volume of the nonspecific tissue damage was calculated never to exceed 10% of the total volume of the slice sample containing it. Taken together, the neuro- and the histochemical data indicate that the 6-OH-DA infusion appears to produce a marked and symmetrical NA denervation in a semicylindrical volume of the cortex (radius of about 2 mm) in front of the point of infusion, while the denervation is more extended in a caudal (posterior) direction (see Fig. 3). The reason for this asymmetrical NA denervation around the infusion point is probably related to the particular organization of the cortical NA innervation which originates from the nucleus locus coeruleus in the brainstem (Ungerstedt 1971). The cortical NA axons have thus been shown to have a predominantly tangential orientation within the gray matter where they are oriented in an anteroposterior direction and give off collaterals along their whole course (Morrison et al. 1981). It is therefore most likely that the presently used local NA denervation technique causes degeneration of NA fibers in two ways: (i) a direct action of 6-OH-DA on NA terminals with an extension symmetrically around the infusion point related to 6-OH-DA diffusion; (ii) an anterograde degeneration of NA fibers due to damage of NA axons passing through the infused area. This latter effect will be most pronounced in the areas caudal to the 6-OH-DA infusion point and appear later. The view of a dual action of 6-OH-DA following infusion is supported by the observations that the NA depletion pattern was rather symmetrical around the infusion point acutely (Fig. 2), while in the chronic stage the NA depletion was asymmetrical, with considerably more marked NA depletion caudally than rostrally (Fig. 3). It should also be pointed out that the longlasting depletion of NA and reduction of histochemically demonstrable NA nerve terminals as well as diminution of ^{3}H-NA uptake is due to a 6-OH-DA induced degeneration of NA terminals and not merely to NA depletion and transient nondegenerative damage of the NA fibers (see Jonsson 1983; Nakai et al. 1984).

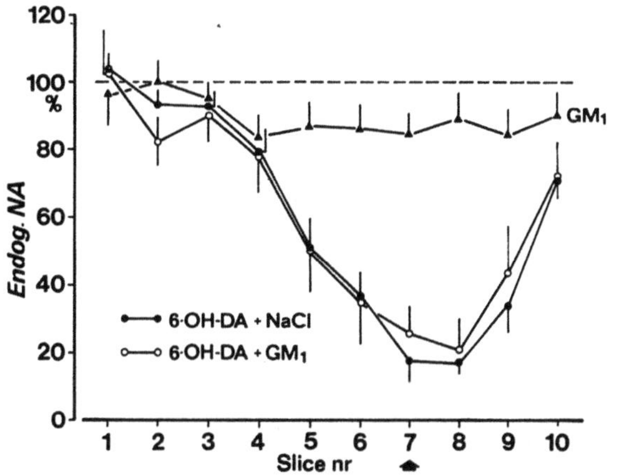

Fig. 2. Effect of GM_1 (30 mg/kg,i.p.) acutely on the spatial distribution of NA in cerebral cortex after unilateral infusion of 2 mM 6-OH-DA (1 µl/h) for 3 days. GM_1 or saline was injected daily for 6 days and this treatment was started 3 days before initiating the 6-OH-DA infusion. The animals were killed 1 day after terminating the 6-OH-DA infusion. An approximately 1 mm thick cortical slab was prepared by a tangential section and divided at regular intervals (1 mm) into ten slices (numbered in anteroposterior direction 1 to 10) which were subjected to NA assay. Slice No. 7 (indicated by arrow) of the right side contained the site of 6-OH-DA infusion. Each point represents the mean ± SEM (n = 4-5), expressed as % of noninfused control side. GM_1 values represent NA levels in the noninfused left side of GM_1-treated rats expressed as % of values obtained from the same side of saline-injected controls (data from Kojima et al. 1984)

With time there was a slow recovery of NA levels and fluorescent NA nerve terminals; this was initially most pronounced rostral to the point of 6-OH-DA infusion, but became evident throughout the whole NA-depleted region in later stages. This NA recovery is probably related to spontaneous regrowth (regeneration of damaged fibers and/or collateral sprouting) of NA nerve terminals into the denervated areas. Similar to the situation in PNS, GM_1 treatment had no effect on the spatial NA distribution when analyzed directly after termination of the 6-OH-DA infusion for 3 days. These data indicate that GM_1 treatment does not alter to any significant degree the primary neurodegenerative actions of 6-OH-DA on NA nerve terminals.

Analysis of the spatial NA distribution 2 (Fig. 3) or 6 weeks after ending the 6-OH-DA infusion revealed that GM_1 treatment caused a partial but significant reversal of the 6-OH-DA induced NA depletion. This reversal was most ponounced rostral to the point of 6-OH-DA infusion after 2 weeks, while it was evident throughout the whole NA-depleted area when analyzed 6 weeks after the 6-OH-DA infusion. It was found, however, that the GM_1-induced enhancement of NA recovery was quantitatively approximately similar 2 and 6 weeks after the 6-OH-DA infusion. A similar partial reversal was also found after GM_1 treatment in animals which had been surgically sympathectomized before starting the GM_1 and 6-OH-DA treatments. These latter results exclude the possibility that the partial NA reversal induced by GM_1 treatment is related to a stimulated ingrowth

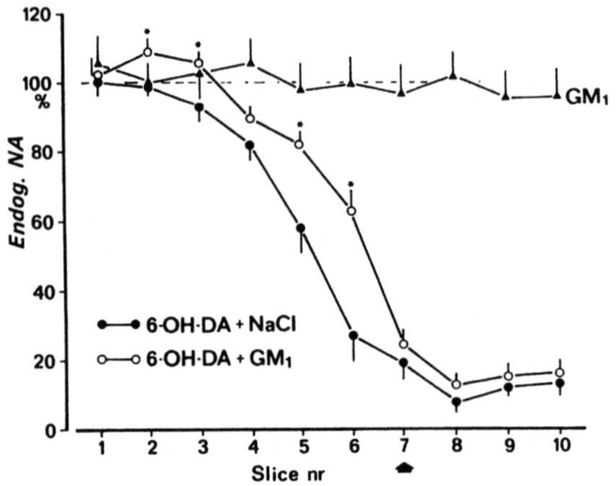

Fig. 3. Effect of GM_1 (30 mg/kg i.p.) on the spatial distribution of NA in cerebral cortex in anteroposterior direction after unilateral infusion of 2 mM 6-OH-DA for 7 days. Daily injections of GM_1 were started 3 days before initiating the 6-OH-DA infusion and continued during the whole 6-OH-DA infusion period. The rats were killed 2 weeks after termination of the 6-OH-DA administration. Each point represents the mean $\pm$ SEM (n = 6-7), expressed as % of noninfused control side. GM_1 values represent NA levels in the noninfused side after GM_1 treatment, expressed as % of values obtained from the same side of saline-treated controls. * = $0.05 > p > 0.01$ (Student's t-test; data from Kojima et al. 1984)

of sympathetic NA fibers which has been shown to occur after certain brain lesions (Stenevi and Björklund 1978; Crutcher et al. 1979). In agreement with the neurochemical data, the fluorescence histochemical analysis showed that GM_1 treatment had a clear counteracting effect on the 6-OH-DA induced reduction of NA nerve density as analyzed in the chronic stage. This effect was most clear-cut rostral to the point of 6-OH-DA infusion, although there appeared to be no difference in the fluorescence morphology between NA nerve terminals demonstrated after 6-OH-DA alone and after GM_1 + 6-Oh-DA. In the partially denervated areas strongly fluorescent, giant varicosities, often appearing in continuity with normal fine NA terminals, could be observed. These "growth-cone"-like structures have previously been observed during conditions of regrowth of NA nerve terminals (see Björklund and Lindvall 1979; Nakai et al. 1984) and in the present experiments were more frequently observed after GM_1 + 6-OH-DA treatment than after 6-OH-DA alone.

It was consistently observed that it was necessary to pretreat the animals with GM_1 before initiating the 6-OH-DA infusion in order to obtain the reversal effect of GM_1 , whereas different schedules of posttreatment were never observed to alter significantly the recovery of cortical NA levels after 6-OH-DA infusion. In none of the experiments was GM_1 treatment found to alter the steady-state levels of NA. In conclusion, the present results have demonstrated that exogenous GM_1 treatment can produce a partial reversal of local NA denervation in rat cerebral cortex induced by 6-OH-DA as observed both neuro- and histochemically, and that this effect of GM_1 is probably related to an enhanced regrowth of NA nerve terminals.

Discussion

It has been shown in a large number of studies that 6-OH-DA properly used is a reasonably specific catecholamine neurotoxin that can be used to selectively destroy NA neurons (see Jonsson 1983). The specificity of 6-OH-DA is related to its selective accumulation in NA neurons, where it undergoes complex autooxidation reactions with formation of potent cytotoxic products that initiate the degeneration. The irreversible 6-OH-DA induced damage is produced within a couple of hours and the neurotoxin appears to act largely in an "all-or-nothing" fashion. There is thus no evidence available to indicate that 6-OH-DA produces marked NA depletion only and neuronal damage of a transient nature that is not associated with true structural degeneration.

The results presented show that exogenous GM_1 can significantly enhance regrowth of NA nerve terminals in both CNS and PNS after a 6-OH-DA induced lesion. The most straightforward explanation for this is that GM_1 has a regrowth stimulatory effect, and this would be in line with previously reported results (Ceccarelli et al. 1976; Gorio et al. 1980, 1983; Toffano et al. 1983; Wojcik et al. 1982; Agnati et al. 1983). However, considering the lesion model used in the present studies it can be argued that the regrowth stimulatory effect is only apparent and in reality is due to (a) interference of GM_1 with the neurotoxic potency of 6-OH-DA, (b) GM_1 increasing NA steady-state levels, leading to increased detectability of NA terminals, and/or (c) to a preventing effect of GM_1 against possible retrograde degeneration processes following the 6-OH-DA induced nerve terminal damage. Although none of these possibilities can be ruled out completely, the data available indicate that alternatives (a) and (b) are rather unlikely. There were no data to indicate that GM_1 alters the primary neurodegenerative actions of 6-OH-DA and, moreover, in none of the experiments were there any indications that GM_1 affects the steady-state levels of NA. This is also in agreement with recently obtained results in our laboratory were the effect of GM_1 was studied in developing rats (Jonsson et al. 1984). Concerning alternative (c), it is conceivable that the primary neurotoxin-induced nerve terminal damage is accompanied by more or less pronounced retrograde degeneration. If GM_1 reduces the extent of this process, such an action would result in a regrowth enhancement effect as analyzed in the chronic stage. Support for such an action of GM_1 is provided by the present findings that the regrowth enhancement effect was found to be approximately similar quantitatively in the cortical lesion model when analyzing the 6-OH-DA induced damage 2 and 6 weeks after induction of the damage.

The results available so far thus indicate that the presently observed counteracting effects of GM_1 are due either to a regrowth stimulation and/or to a protection against retrograde degeneration. The present results do not clearly differentiate between these two possible modes of action of GM_1 and consequently it is feasible that the effect of GM_1 is due to a combination of these two modes of action.

Acknowledgments. The present studies have been supported by grants from the Swedish MRC (04X-2295), Karolinska Institutet, Expressen's Prenatal Research Foundation, Bergwall's and Jeansson's Foundations. The skillful technical assistance of Ms.E.Lindqvist, Ms. B. Kaeller, and Ms. B. Drevingar is gratefully acknowledged. The authors are also very grateful to Ms. Ida Engqvist for expert secretarial help.

References

Agnati LF, Fuxe K, Calza L, Benefenati F, Cavicchioli L, Toffano G, Goldstein M (1983) Gangliosides increase the survival of lesioned nigral dopamine neurons and favour the recovery of dopaminergic synaptic function in striatum of rats by collateral sprouting. Acta Physiol Scand 119:347-363

Ando S (1983) Gangliosides in the nervous system. Neurochem Int 5: 507-537

Björklund A, Lindvall O (1979) Regeneration of normal terminal innervation patterns by central noradrenergic neurons after 5,7-dihydroxytryptamine-induced axotomy in the adult rat. Brain Res 171:271-293

Ceccarelli B, Aporti F, Finesso M (1976) Effects of brain gangliosides on functional recovery in experimental regeneration and reinnervation. In: Porcellati G, Ceccarelli B, Tettamanti G (eds) Advances in experimental medicine and biology, vol 71. Plenum, New York, pp 275-293

Corrodi K, Jonsson G (1967) The formaldehyde fluorescence method for the histochemical demonstration of biogenic monoamines. A review on the methodology. J Histochem Cytochem 15:65-78

Crutcher KA, Brothers L, Davis JN (1979) Sprouting of sympathetic nerves in the absence of afferent input. Exp Neurol 66:778-783

Ferrari G, Fabris M, Gorio A (1983) Gangliosides enhance neurite outgrowth in PC12 cells. Dev Brain Res 8:215-221

Fuxe K, Hoekfelt T, Jonsson G, Ungerstedt U (1970) Fluorescence microscopy in neuroanatomy. In: Nauta WJH, Ebbesson SOE (eds) Contemporary research in neuroanatomy. Springer, Berlin Heidelberg New York, pp 275-314

Gorio A, Carmignoto G, Facci L, Finesso M (1980) Motor nerve sprouting induced by ganglioside treatment. Possible implications for gangliosides on neuronal growth. Brain Res 197:236-241

Gorio A, Marini P, Zanoni R (1983) Muscle reinnervation. III. Motorneuron sprouting capacity, enhancement by exogenous gangliosides. Neuroscience 8:417-429

Jonsson G (1983) Chemical lesioning techniques: monoamine neurotoxins. In: Björklund A, Hökfelt T (eds) Handbook of chemical neuroanatomy, vol 1. Elsevier, Amsterdam, pp 463-507

Jonsson G, Sachs Ch (1970) Effects of 6-hydroxydopamine on the uptake and storage of noradrenaline in sympathetic adrenergic neurons. Eur J Pharmacol 9:141-155

Jonsson G, Sach Ch (1972) Neurochemical properties of adrenergic nerves regenerated after 6-hydroxydopamine. J Neurochem 19: 2577-2585

Jonsson G, Gorio A, Hallman H, Janigro D, Kojima H, Zanoni R (1984) Effect of GM_1 ganglioside on neonatally neurotoxin induced degeneration of serotonin neurons in the rat brain. Dev Brain Res 16:171-180

Kasamatsu T, Pettigrew JD, Ary M (1979) Restoration of visual cortical plasticity by local microperfusion of norepinephrine. J Comp Neurol 185:163-182

Kasamatsu T, Itakura T, Jonsson G (1981) Intracortical spread of exogenous catecholamines: effective concentration for modifying cortical plasticity. J Pharmacol 217:841-850

Keller R, Oke A, Mefford J, Adams RN (1976) Liquid chromatographic analysis of catecholamines - routine assay for regional brain mapping. Life Sci 19:995-1004

Kojima H, Gorio A, Janigro D, Jonsson G (1984) GM_1 ganglioside enhances regrowth of noradrenaline nerve terminals in rat cerebral cortex lesioned by the neurotoxin 6-hydroxydopamine. Neuroscience 13:1011-1022

Lidbrink P, Jonsson G (1974) Noradrenaline nerve terminals in the cerebral cortex: effects on noradrenaline uptake and storage following axonal lesion with 6-hydroxydopamine. J Neurochem 22: 617-626

Malmfors T (1965) Studies on adrenergic nerves. Acta Physiol Scand (Suppl) 64:248

Morrison JH, Molliver ME, Grzanna R, Coyle JT (1981) The intracortical trajectory of the coeruleo-cortical projection in the rat: a tangentially organized cortical afferent. Neuroscience 6: 139-158

Nakai K, Jonsson G, Kasamatsu T (1984) Reinnervation of the cat visual cortex by central norepinephrine fibers following localized lesions with 6-hydroxydopamine. J Neurosci (to be published)

Olson L, Hamberger B, Jonsson G, Malmfors T (1968) Combined fluorescence histochemistry and ^3H-noradrenaline measurements of adrenergic nerves. Histochemie 15:38-45

Stenevi U, Björklund A (1978) Growth of vascular sympathetic axons into the hippocampus after lesion of septohippocampal pathway: a pitfall in brain lesion studies. Neurosci Lett 7:219-224

Toffano G, Savoini G, Moroni F, Lombardi G, Calza L, Agnati LF (1983) GM_1 ganglioside stimulates the regeneration of dopaminergic neurons in the central nervous system. Brain Res 261:163-166

Ungerstedt U (1971) Stereotaxic mapping of the monoamine pathways in the rat brain. Acta Physiol Scand (Suppl) 367:1-48

Wojcik M, Ulas J, Oderfelt-Nowak B (1982) The stimulating effect of ganglioside injections on the recovery of choline acetyltransferase activities in the hippocampus of the rat after septal lesions. Neuroscience 7:495-500

Polyamines Enhance Survival and Regeneration of Sympathetic Neurons

G. M. Gilad, M. Dornay, L. E. Varon, and V. H. Gilad

Center for Neurosciences and Behavioral Research, The Weizmann Institute of Science, Rehovot 76100, Israel

Introduction

Neurons of the mammalian nervous system are only generated during a limited period of embryonic development and cannot be replaced once this generative period has ended. The generative period is shortly followed by a period of naturally occurring neuronal cell death which is an important factor in the regulation of final neuronal population size (Hamburger and Levi-Montalcini 1949; Hamburger and Oppenheim 1982). Later in life, if axonal injury occurs, recovery is limited by the capacity of the parent neuron to survive and to regenerate a new axon (Cajal 1928; Grafstein and McQuarrie 1978; Leiberman 1971) and by the degree to which undamaged neurons can form new synaptic contacts or use their existing ones more effectively.

Following axonal injury the survival of the parent nerve cell body is a precondition for a successful axonal regeneration. It was recognized nearly a century ago that axonal injury elicits a series of morphological changes in the parent cell body (Nissl 1982) termed the axon reaction (Leiberman 1971). These changes are the reflection of alterations in underlying biochemical mechanisms. The current view is that a selective reordering of priorities in protein biosynthesis takes place within the injured neuron, favoring the synthesis of some proteins while others are diminished (Grafstein and McQuarrie 1978; Leiberman 1971).

We have proposed (Gilad and Gilad 1983b) that this shift in protein biosynthesis occurs very early after the injury, and represents an early commitment of the injured neuron toward recovery and regeneration or, alternatively, degeneration and death. We have recently reported that even within minutes after axonal injury there is a transient increase in the enzymatic activity of polyamine (PA)-dependent and calcium-activated transglutaminase (Gilad and Gilad 1984), which can catalyze the crosslinking of cellular proteins (Folk 1980). The facts that PAS are involved in the regulation of RNA and protein synthesis (Russell 1979), and that changes in RNA and protein metabolism are the most conspicuous event of the axon reaction (Brattgard et al. 1957; Watson 1968), have suggested that changes in PA metabolism may be important determinants in this reaction (Gilad and Gilad 1983a). Indeed, it has been recently demonstrated that within hours after axonal injury there is an increased production of PA in the parent neuron (Gilad and Gilad 1983b,c; Kreutzberg et al. 1984) and that in sympathetic neurons this polyamine response is critical for neuron survival (Gilad and Gilad 1983c). In addition, the activity of ornithine decarboxylase, the enzyme catalyzing the first

Experimental Brain Research, Suppl. 13

step in PA biosynthesis (Russell 1979), is elevated in sympathetic ganglia during the perinatal period (Gilad and Gilad 1983b), when neuronal cell division still takes place and before naturally occurring neuronal death starts (Hendry 1977). In the present study, therefore, we sought to determine the effect of treatment with exogenous PA on neuronal survival during development and on survival and regeneration after axonal injury.

Materials and Methods

Principle of the study

The rationale for treatment with PA lies in the demonstration that when increased demand for PA arises, not only does PA biosynthesis increase, but also the capacity to take up extracellular PA is greatly enhanced (Janne et al. 1981).

After preganglionic nerve crush, regeneration was evaluated measuring choline acetyltransferase (CAT) activity as a marker for cholinergic terminals in the superior cervical ganglion (SCG). After postganglionic nerve crush, measurements of [^{3}H] norepinephrine (NE) uptake served to evaluate regeneration of noradrenergic terminals in the pineal. Recovery from eye ptosis was evaluated after both pre- and postganglionic nerve crush. During development neuronal cell counts and measurements of tyrosine hydroxylase (TH) activity served to evaluate the effect of PA on sympathetic neurons in the SCG.

In regeneration studies PA treatment was started at the time of operation and lasted for 7 days, and measurements were taken 30 and 60 days later. PA treatment in development was started on the 2nd day after birth and lasted until the 9th day, and measurements were taken 35 days later.

Methods

Experiments were performed on 3-month-old Sprague-Dawley rats of both sexes while under fluothane anesthesia (2.5% in 100% O_2). The preganglionic nerve was crushed 4 mm caudal to the SCG, and the two major postganglionic nerves at a distance of 5 mm from the ganglion. 10 mg/kg of spermine, spermidine, or putrescine, or a combination of the three polyamines (10 mg/kg each), was dissolved in 250 µl saline [0.9% (wt/vl) NaCl] and injected (intraperitoneally) daily.

At different postoperative times rats were killed by decapitation and ganglia and pineal glands removed. The enzyme activities of CAT and TH were assayed as previously described (Gilad and Kopin 1979). The uptake of NE was measured according to Heikkila et al. (1975). Protein concentrations were determined according to the method of Lowry et al. (1951).

For histology, ganglia were removed and fixed for 7 days in a saline solution containing 4% (wt/vl) paraformaldehyde. They were then embedded in paraffin, cut serially into 6 µm thick sections, and stained with cresyl violet. Neuronal cell counts were performed according to the "total count method" (Konigsmark 1970).

Recovery from eye ptosis was evaluated and rated by three independent observers unfamiliar with the assignment of animal groups.

Results

Effect of Polyamine Treatment on Recovery After Preganglionic Nerve Crush

Table 1 summarizes the effects of PA treatment on the changes in CAT activity and protein content of the SCG 30 and 60 days after preganglionic nerve crush. After 30 days CAT activity was 47.6% higher in combined PA-treated than in saline-treated rats. After 60 days no difference was detected between PA-treated and saline--treated groups. No differences in protein content were detected.

Thirty days after preganglionic nerve crush, recovery from eye ptosis was enhanced in PA-treated rats, as evidenced by a much wider opening of the eyelid in these rats compared with the saline-treated ones. At 60 days postoperatively this difference was diminished and both groups showed comparable eye opening which approached normality.

Effect of Polyamine Treatment on Recovery After Postganglionic Nerve Crush

At 40 days after postganglionic nerve crush NE uptake in the pineal was still very low, but significantly higher in PA-treated rats (Table 2). Protein content of the pineal was also increased by PA treatment and at 40 days it was no different from control values (Table 2). The effect of PA on protein content of the pineal indicates a direct effect on the denervated target organ.

Table 1. Effect of polyamine (PA) treatment on choline acetyltransferase (CAT) activity and protein content of the superior cervical ganglion 30 and 60 days after preganglionic nerve crush (results mean ±SEM values of six rats, are expressed as a percentage of values in unoperated saline-injected controls)

Treatment	CAT activity		Protein content	
	30 days	60 days	30 days	60 days
Control	100±5.6	100±4.0	100±5.0	100±4.0
Crush + saline	21±1.6**	55±4.0**	71±8.0	106±9.0
Crush + PA	31±3.0*	46±5.0	82±7.0	96±10.0

A polyamine combination (putrescine, spermidine, and spermine 10 mg/kg each) or saline [0.9% (wt/vol) NaCl] was injected once a day for 7 days, starting at the time of operation

Control values: CAT activity=105±6 nmol/ganglion/h; protein content=0.20±0.01 mg/ganglion

* Significantly ($p<0.5$) different fromsaline treatment

Significantly (+=$p<0.05$;++=$p<0.001$) different from control

Table 2. Effect of polyamine treatment on [^{3}H] norepinephrine (NE) uptake and protein content in the pineal 40 days after postganglionic nerve crush (results, mean ±SEM values of five or six rats, are expressed as a percentage of values in unoperated saline-injected controls)

Treatment[a]	NE uptake[b]	Protein content[b]
Control	100±7	100±6
Crush + saline	10±1	74±5
Crush + PA	14±1*	99±7*

Polyamines (PA) (putrescine, spermidine, and spermine 10 mg/kg each) or saline [0.9% (wt/vol)] NaCl were injected once a day for 7 days, starting at the time of operation

Control values: NE uptake=300±16 fmol/pineal/4 min; protein content=0.21±0.01 mg/pineal

* Significantly ($p<0.05$) different from saline treatment

\+ Significantly (+=$p<0.01$; ++=p 0.001) different from control

Recovery from eye ptosis was enhanced after PA treatment, indicating that the rate of growth of regenerating postganglionic axons innervating the iris was also enhanced.

Effect of Polyamine Treatment During Development

Treatment during development with spermine, spermidine, or putrescine, or with a combination of PAS, led to a 35%-40% increase both in the number of nerve cells and in TH activity in the adult SCG (Table 3). No change in CAT activity was detected. A significant increase of 15% in protein content was observed after putrescine treatment.

Discussion

Increased demand for PA appears to be an essential part of the axon reaction in mammalian peripheral neurons (Gilad and Gilad 1983c; Kreutberg et al. 1984). Such increased demand is evident by virtue of increased biosynthesis of PA (Gilad and Gilad 1983b; Kreutzberg et al. 1984), and probably, like in other cells, an increased capacity to take up extracellular PA (Jänne et al. 1981). The results of the present study demonstrate that treatment with exogenous PA can enhance regeneration and functional recovery following axonal injuries of rat sympathetic neurons. Enhanced functional recovery was reported in injured rat sciatic nerve after treatment with spermine (Sebille and Bondoux-Jahan 1980). The effect of PA can be explained as the result of a direct action on the injured nerve cell itself, whereby the increased demand can be met by increased supply of PA now available via the uptake system. Alternatively, the PA effect

Table 3. Effect of postnatal polyamine treatment on tyrosine hydroxylase (TH) activity, choline acetyltransferase (CAT) activity, protein content, and nerve cell number in the superior cervical ganglion at 35 days after birth (results, mean ±SEM values of five to ten rats, are expressed as a percentage of values in saline-injected controls)

Treatment[a]	Nerve cell number[b]	TH activity[b]	CAT activity[b]	Protein content[b]
Control	100±7	100±5	100±5	100±3
Putrescine	140±8*	151±10**	98±4	115±4*
Spermidine	135±9*	154±12**	105±5	110±5
Spermine	140±10*	165±14**	110±7	104±3
PA	135±10*	—	—	—

Each polyamine (10 mg/kg), a polyamine combination (PA) (putrescine, spermidine, and spermine 10 mg/kg each), or saline [0.9% (wt/vol) NaCl] was injected daily from 2 days to 9 days after birth

Control values: nerve cell number=22100±1000; TH activity=1.6±0.1 nmol/ganglion/h; CAT activity=112±3 nmol/ganglion/h; protein content=0.19±0.01 mg/ganglion

* $p<0.5$; ** $p<0.01$

could be indirect through an interaction with other cells, which in turn can exert a beneficial effect on the injured neuron. Such an indirect effect may occur by interaction with the target tissue of innervation. Increased PA production has been observed in denervated muscle after axotomy (Hopkins and Manchester 1981; Kaminska et al. 1982). This finding, together with the demonstration that the content of nerve growth factor (NGF) increases in the denervated iris (Ebendal et al. 1980), suggests that PA can act indirectly by supporting metabolic activity in the target tissue directed at producing an increased amount of NGF. In turn, more NGF will become available to the injured neuron, where it can act to increase survival, regeneration, and recovery (Levi-Montalcini and Angeletti 1968), possibly by increasing intraneuronal PA production (Hendry and Bonyhady 1980). If the beneficial effect of PA treatment indeed occurs via enhanced production of NGF, and if such a beneficial effect holds true for other types of neurons, then PA treatment might circumvent the need for supplying specific growth factors if the need arises. These questions are currently being investigated. During development PAs could also exert their effect either directly on the sympathetic neuron or indirectly, as previously explained. Wherever the site of action is, the PA effect occurs during a limited period which lasts only a few days both during postnatal development and after axonal injury.

Acknowledgments. This research was supported by grants from the Muscular Dystrophy Association, the American Paralysis Association, and the Herman Goldman Foundation. Gad M. Gilad is an incumbent of the Paul and Gabriella Rosenbaum Foundation, Chicago, Illinois. We thank Ms. Jackie Biran for typing the manuscript.

References

Brattgard SO, Hyden H, Sjöstrand J (1957) The chemical changes in regenerating neurons. J Neurochem 1:316-325

Cajal SRY (1928) Degeneration and regeneration of the nervous system. University Press, London

Ebendal T, Olson L, Seiger A, Hedlund KO (1980) Nerve growth factors in the rat iris. Nature 286:25-28

Edds MV (1953) Collateral nerve regeneration. Q Rev Biol 28:260-276

Folk JE (1980) Transglutaminases. Annu Rev Biochem 49:517-531

Gilad GM, Gilad VH (1983a) Sequential biochemical alterations in sympathetic neurons following axonal injury. In: Haber B, Perez-Polo JR, Hashim GA, Stella AMG (eds) Nervous system regeneration. Liss, New York, p 437

Gilad GM, Gilad VH (1983b) Early rapid and transient increase in ornithine decarboxylase activity within sympathetic neurons after axonal injury. Exp Neurol 81:158-166

Gilad GM, Gilad VH (1983c) Polyamine biosynthesis is required for survival of sympathetic neurons after axonal injury. Brain Res 273:191-194

Gilad GM, Gilad VH (1984) Calcium-dependent transglutaminase of rat sympathetic ganglion in development and after nerve injury. J Neurochem (to be published)

Gilad GM, Kopin IJ (1979) Neurochemical aspect of neuronal ontogenesis in the developing rat cerebellum: changes in neurotransmitter and polyamine synthesizing enzymes. J Neurochem 33:1195-1204

Grafstein B, McQuarrie IG (1978) Role of the nerve cell body in axonal regeneration. In: Cotman CW (ed) Neuronal plasticity. Raven, New York, p155

Hamburger V, Levi-Montalcini R (1949) Proliferation, differentiation and degeneration in the spinal ganglia of the chick embryo under normal and experimental conditions. J Exp Zool 111:457-502

Hamburger V, Oppenheim RW (1982) Naturally occurring neuronal death in vertebrates. Neurosci Comment 1:39-55

Heikkila RE, Orlowsky H, Mytilineou C, Cohen C (1975) Amphetamine: evaluation of d- and l-isomers as releasing agents and uptake inhibitors for ^{3}H-dopamine and ^{3}H-norepinephrine in slices of rat neostriatum and cerebral cortex. J Pharmacol Exp Ther 194:47-56

Hendry IA (1977) Cell division in the developing sympathetic nervous system. J Neurocytol 6:299-309

Hendry IA, Bonyhady R (1980) Retrogradely transported nerve growth factor increases ornithine decarboxylase activity in rat superior cervical ganglion. Brain Res 200:39-45

Hopkins D, Manchester KL (1981) The influence of nerve section on the metabolism of polyamines in rat diaphragm muscle. Biochem J 196:603-610

Jänne J, Alhonen-Honigsto L, Seppänen P, Höllta E (1981) Cellular compensatory mechanisms during chemically induced polyamine deprivation. In: Calderera CM, Zappia V, Bachrach U (eds) Advances in polyamine research, vol 3, Raven, New York, p85

Kaminska AM, Stern LZ, Russell DH (1982) Polyamine metabolism in muscle: Differential response to tenotomy and denervation in the soleus and gastrocnemius muscle of adult rats. Exp Neurol 78:331-339

Konigsmark BW (1970) Methods for the counting of neurons. In: Nauta WJH, Ebbesson SDE (eds) Contemporary research methods in neuroanatomy. Springer, Berlin Heidelberg New York, p315

Kreutzberg GW, Tetzlaff W, Gräber M (1984) Reaction of neurons and their microenvironment to axotomy. In: Gilad GM, Gorio A, Kreutzberg GW (eds) Recovery from neural trauma. Springer, Berlin Heidelberg New York (to be published)

Leiberman AR (1971) The axon reaction: a review of the principal features of perikaryal responses to axon injury. Int Rev Neurobiol 14:49-124

Levi-Montalcini R, Angeletti PU (1968) Nerve growth factor. Physiol Rev 48:534-569

Lowry OH, Rosebrough NJ, Farr AL, Randall RJ (1951) Protein measurement with the Folin phenol reagent. J Biol Chem 193:265-275

Nissl F (1892) Ueber die Veraenderungen der Ganglienzellen am Facialiskern des Kaninchens nach Ausreissung der Nerven. Allg Z Psychiatr 48:197-198

Russell DH (1979) (ed) Polyamines in normal and neoplastic growth. Raven, New York

Sebille A, Bondoux-Jahan M (1980) Motor function recovery after axotomy: enhancement by cyclophosphamid and spermine in rat. Exp Neurol 70:507-555

Watson WE (1968) Observations on the nucleolar and total cell body nucleic acid of injured nerve cells. J Physiol (Lond) 196:655-676

Reconstruction Models to Enhance Recovery

Nerve Regeneration Chamber: Physical and Molecular Influences

S. Varon and L. R. Williams

Department of Biology, School of Medicine, University of California, San Diego, La Jolla, CA, USA

Introduction

Axons from the adult mammalian central nervous system (CNS) retain the ability to regenerate for considerable distances if they are presented with peripheral nerve as a terrain in which to regrow (Benfey and Aguayo 1982). This competence of nerve tissue to allow or promote the regeneration of both peripheral and central axons may result from the properties and/or physical organization of cells unique to nerve, such as the Schwann cells, and it is likely to be expressed via molecular agents present in the fluids, or bound to cell surfaces and the extracellular matrix of the nerve tissue itself (cf. Varon et al. 1983, 1984). Nerve components that are critical for axonal regeneration can be investigated most effectively by combining the use of in vitro and in vivo model systems.

In collaboration with Dr. Lundborg and coworkers in Sweden, we have developed an in vivo chamber model which is particularly suited for experimental analyses of peripheral nerve regeneration in the adult rat (Lundborg et al. 1982a, b; Varon and Lundborg 1983). A sciatic nerve is transected and its proximal and distal stumps are sutured into the opposite ends of a cylindrical silicone tube, the internal diameter of which (1.2 mm) matches the diameter of the nerve inserts. Each stump occupies only 2 of the 14 mm length of the silicone tube, leaving a 10 mm long empty space between them (the "chamber"). Within one month, this gap is bridged by a new tissue structure, the cross-section of which reveals the typical organization of a regenerating nerve. The regenerating axons, after crossing the chamber gap, will continue to grow into and down the distal nerve segment to reach and functionally connect with their end organs (Lundborg et al. 1982a,b). Nerve regeneration inside the chamber, thus far, has failed to occur when (i) the interstump gap is increased beyond 10 mm and (ii) the distal end of the chamber receives no insert, or inserts other than nerve tissue (Lundborg et al. 1982b; Williams et al. 1984).

This chamber model has been analyzed in some detail with regard to the spatial and temporal sequences in which the regeneration of a new nerve proceeds within the chamber itself (Williams et al. 1983). Fluid fills the chamber over the first day. Then, a coaxial acellular fibrin matrix develops over the first week to form a continuous bridge, still surrounded by fluid, across the interstump gap. Cells and new vessels advance from both stumps into the fibrin matrix over the second week, to generate a fluid-surrounded

Experimental Brain Research, Suppl. 13

cellular structure. Only then will the axons grow out of the proximal stump, extend through the cellular bridge over the third week, and reach the distal stump by the end of it. Myelination follows the axonal advance with a modest delay.

The chamber fluid has been found to contain trophic factors for the very neurons (sensory, motor, sympathetic) which contribute axons to the sciatic nerve (Longo et al. 1983). Trophic factors reach their highest levels in the early postoperative times, suggesting that their primary role may concern maintenance and repair of the damaged neurons. The late onset of axonal regrowth, on the other hand, correlates precisely with the appearance of neurite-promoting factors in the chamber fluid which are likely to reflect Schwann cell immigration (Longo et al. 1984). Temporal changes occur also within the chamber structure (Longo et al. 1984). Fibronectin (an extracellular matrix glycoprotein known to promote cell migration) is present in the chamber structure from the time the latter is first formed. In contrast, laminin - another extracellular matrix glycoprotein recently shown to have potent neurite promoting activity in vitro (Manthorpe et al. 1983; Wever et al. 1983) - only appears after the cells immigrate into the structure.

The above sequences of events suggest that each stage of the intrachamber regeneration process preconditions the next one. We speculate that: (1) the chamber fluid provides ingredients and regulation for the fibrin-fibronectin matrix formation, (2) the matrix favors the immigration of cells and vessels, and (3) the arriving cellular elements, in turn, "condition" the humoral and extracellular matrix environments to favor and promote the axonal regneration. It would follow that alterations of the timing, volume, and/or composition of the chamber fluid may have a profound impact on the final regeneration outcome. We review here some recent manipulations of the standard model (Williams and Varon 1985), which support such speculations.

Effects of Volume Increases and Fluid Prefilling on Chamber Performances

Standard chambers (11 µl capacity) are constructed from 14 mm long silicone tubing of 1.2 mm internal diameter (ID). Tubing with 1.8 mm ID can be stepped-down at each end with 2-mm pieces of the 1.2 mm ID material, in order to secure the same fit for the sciatic nerve stumps to be inserted later. The result is a chamber which has the same 10 mm distance between the two stumps as the standard chamber but an increased capacity of 25 µl. A similar stepped-down construction, starting from tubing with a 3.1 mm ID, can provide 10 mm long interstump spaces with 75 µl capacity. Chambers of all three sizes can still be accommodated into the posterior compartment of the hindlimb. Such 11, 25, and 75 µl chambers were implanted into adult rats either empty (E11, E25, and E75 chambers) or prefilled (PF11, PF25, and PF75 chambers) with bacteriostatic saline (0.9% NaCl + 0.9% butyl alcohol). Chambers were collected at 1 and 3 weeks postimplantation, fixed in aldehydes, and embedded in Epon-Araldite. One mm cross-sections were cut at 1, 3, 5, 7, and 9 mm from the proximal stump (sections S1 to S9), stained with toluidine blue for light microscopy, used to measure cross-sectional diamters, and scored for the presence or absence of: circumferential (perineurial-like) cells, vasculature, Schwann cells, and nonmyelinated and myelinated axons (Williams et al. 1983).

Standard E11 chambers fill with fluid by 12 h (Longo et al. 1983). E25 and E75 chambers do so by 30 and 50 h, respectively, indicating greater average filling rates for the largest chamber. Earlier inspection, e.g. at 12 h (Fig. 1), revealed that (i) fluid accumulates separately at each chamber end, (ii) proximal fluid is significantly more abundant than distal fluid, and (iii) mixing of the two fluid columns occurs only when they meet, at about the S7 level. Segregation of the proximal and distal fluid columns may reflect resistance by the interposed air cushion, surface tension on the chamber walls, or, possibly, some gelation of fluid contents. Circumstances are obviously different in the three PF chambers, which start already filled with fluid (saline). Here, contributions to the chamber from each end might still occur at different rates, but the two contributions presumably mix readily by diffusion across the pre-existing saline bridge. Information is yet to be collected about the extent, if any, to which additional bulk fluid and/or solutes may enter a chamber once filled (or prefilled) with fluid.

Fluid and solutes, behaviors in the early chambers may affect the subsequent construction of chamber matrices in at least two ways, i.e. by altering (i) the amount of relevant materials available for coalescence at different distances into the chamber (and at different times, as well), and (ii) the concentrations at which these materials will participate in the coalescence process. One could expect that in E chambers, for example, such materials will occur in greater amounts (but not necessarily at greater concentrations) at the proximal than the distal end of the chamber. Conversely, in PF chambers, relevant materials should be uniformly distributed along the chamber axis but increasingly diluted, at least initially, as the size of the chamber (hence, the prefilling fluid volume) is increased. The 1-week findings were consistent with both such presumptions. All six chamber sets had continuous, coaxial structures briding their interstump gap. All three E cham-

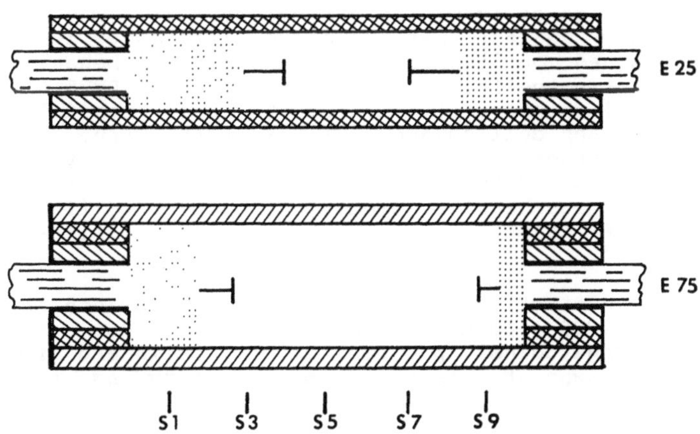

Fig. 1. Fluid accumulation into enlarged regeneration chambers (E25, E75), 12 h after implantation (n=4 ± SD). Bottom marks (S1 to S9) indicate the distance (mm) from the proximal nerve insert. Note (i) stepped-down construction of the chambers' ends, to accommodate the same nerve stumps, (ii) the segregation of proximal and distal early fluid columns, and (iii) the greater volume of the proximal fluids

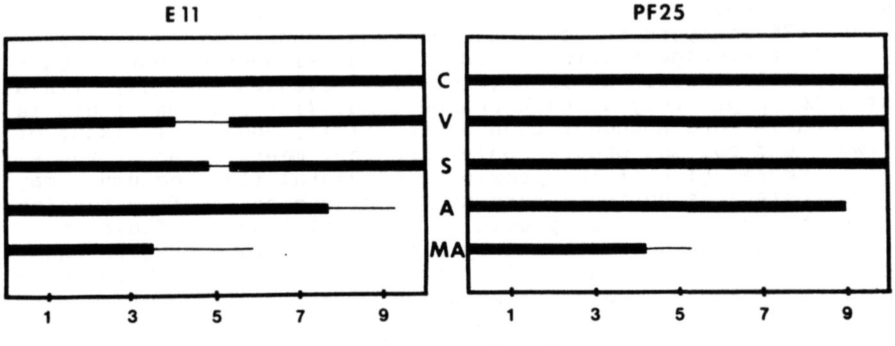

Fig. 2. Regeneration progress across E11 and PF25 chambers, 3 weeks postimplantation. Light microscopy cross-sections (at 1 to 9 mm from proximal stump) were scored for presence of: circumferential cells (C), vasculature (V), Schwann cells (S), nonmyelinated axons (A), and myelinated axons (MA). Thin bars = standard deviations (n=6)

bers exhibited a pronounced, proximodistal tapering of the structure diameters, while structures in PF chambers had more uniform diameters over the entire chamber length. Cross-sections of PF chamber matrices showed fibrin fibers which were more sparsely distributed and of thinner caliber than those found in E chamber structures. Moreover, the overall diameter of 1-week matrices increased with chamber diameter in the PF chambers but not the E chambers (except at their most proximal section) - a possible reflection of time differences in the coalescence process in the two chamber types.

The results of examining the chambers at 3 weeks postimplantation strongly suggest that the above differences in 1-week chamber matrices had considerable effects on the behavior of subsequently immigrating cells. With the exception of PF25 chambers, all experimental chambers revealed considerable retardation in vasculature and Schwann cell advance out of both nerve stumps, and thus in axonal regeneration progress. There were differences between the adverse consequences of restricting early fluid distribtuion in larger E chambers or immediately dispersing fluid contributions in larger PF chambers, but their detailed implications remain to be analyzed.

Most instructive, however, was the comparison between the experimental PF25 chamber and the standard E11 one. Figure 2 shows that the longitudinal progress made by cells and axons in the PF25 chambers was at least as satisfactory as that in the standard system.

Figure 3 illustrates cross-sections at midchamber (S5) of the two 3-week structures, displayed at the same magnification. No gross differences are apparent in the histological composition, but diameters of the PF25 structures were about twofold greater than the standard ones. A morphometric analysis of the two sections is shown in Table 1. The overall greater diameter of PF25 cross sections was even more evident with endoneurial cross-sectional

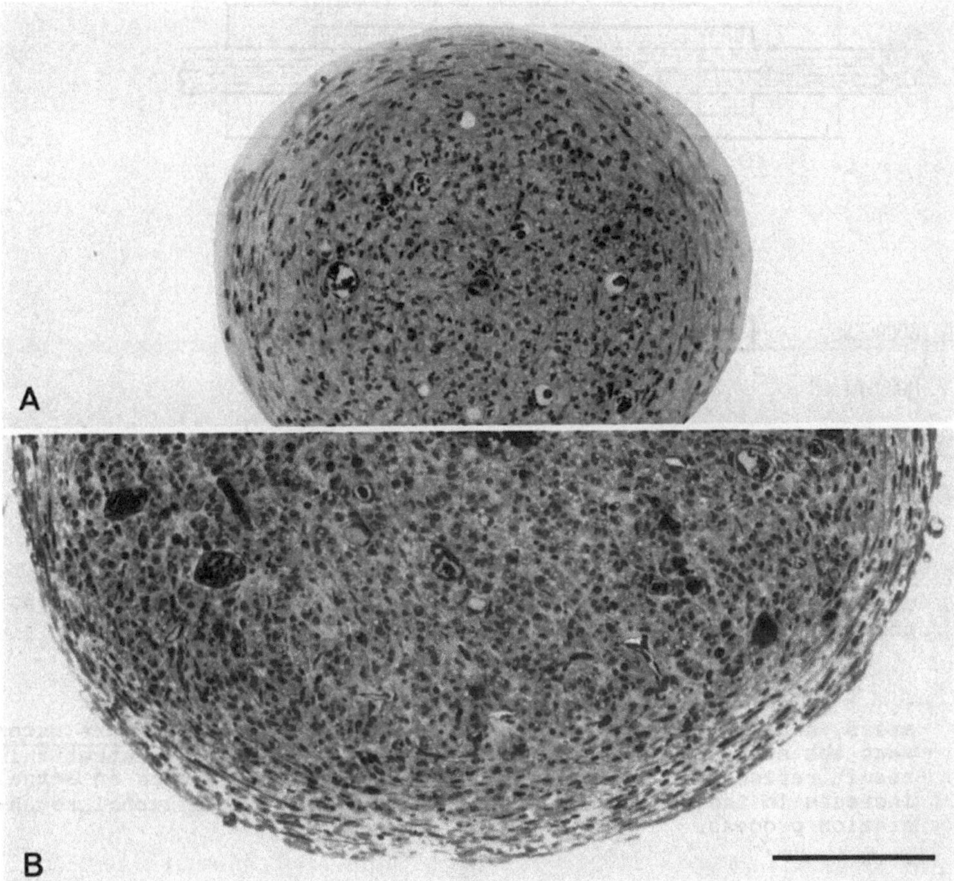

Fig. 3. Light micrographs of S5 cross-sections from E11 (A) and PF25 (B) chambers, at 3 weeks. The PF25 structures have twofold larger diameters and nearly threefold more axons. Scale bar = 100 µm

Table 1. Improvements by PF25 chambers over E11 ones in endoneurial and axonal contents of their regenerating nerve structures (axons counted were > 1 µm diameter)

S5 sections	E11 chambers	PF25 chambers
External diameters (mm)	0.26	0.55
Endoneurial cross-section (mm^2)	0.05	0.16
Axon number per section	1400	3700

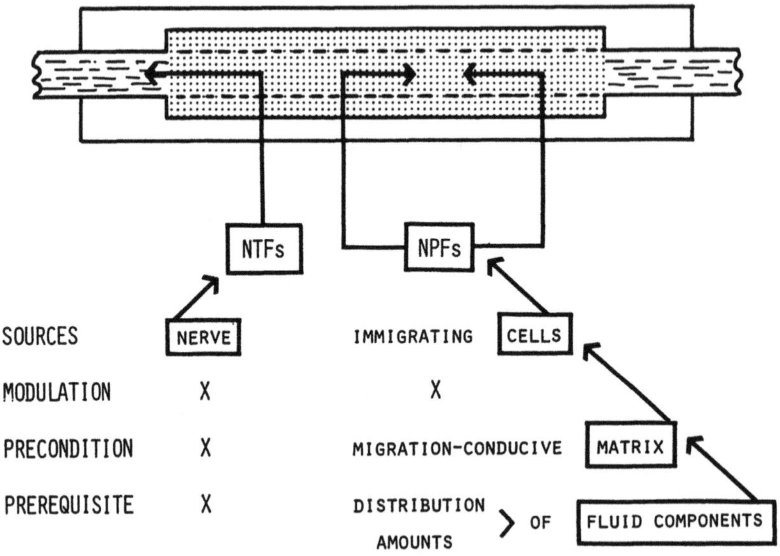

Fig. 4. Cellular and molecular influences currently recognized to control nerve regeneration in the chamber model. NTFs = neuronotrophic factors. NPFs = neurite promoting factors

areas, and PF25 structures contained nearly three times more axons than the standard ones. We do not know, at present, whether this result reflects an increase in intrachamber branching or an actual increase in the number of neurons recruited into the axonal regeneration process.

Projections and Speculations

The findings just reviewed show that experimental manipulation of the chambers can result in improving the regeneration process in these model systems. That further improvement in axonal numbers remains possible is suggested by the fact that, even in the PF25 chamber, the endoneurial cross-sectional area is still smaller in the central portions of the regenerating nerve than at its proximal and distal ends or in the corresponding nerve stumps (Williams and Varon 1985). Further experiments should (i) determine whether the regeneration process in the chambers found unsuccesssful thus far is irreversibly blocked or merely delayed, and (ii) define chamber diameters and time factors providing for an optimal number of regenerating axons.

More importantly, the new data gave us further insights on and potential access to additional, molecular controls of nerve regeneration, as shown diagrammatically in Fig. 4. These data strengthen the view that axonal regrowth (beside presumably requiring appropriate neuronotrophic support) is subordinate to influences (e.g., neurite promoting factors) supplied by immigrating cells, and that cell immigration in turn is subordinate to the "quality" of the matrix bridging the interstump gap. The retarded cell immigration observed in large E chambers and the largest PF chamber (Williams and Varon 1985) can be attributed to a restricted or an overdiluted contribution to the matrix by fluid components needed for either or both fibrin formation and migration-promoting competence (e.g., fibronectin). These conjectures are amenable to further experimentation, where use of the chamber prefilling modality would do away with a restricted distribution while composition changes in the saline used for prefilling may reveal the roles and possibly the nature of the relevant fluid components. Such knowledge will eventually make it possible to "prefabricate" surrogate matrices for nerve regeneration in the chamber model, and ultimately test their suitability for clinical interventions.

References

Benfey M, Aguayo AJ (1982) Extensive elongation of axons from rat brain into peripheral nerve grafts. Nature 296:150-152

Longo FM, Skaper SD, Manthorpe M, Williams LR, Lundborg G, Varon S (1983) Temporal changes in neuronotrophic activities accumulating in vivo within nerve regeneration chambers. Exp Neurol 81:756-769

Longo FM, Hayman EG, Davis GE, Ruoslahti E, Engvall E, Manthorpe M, Varon S (1984) Neurite promoting factors and extracellular matrix components accumulating in vivo within nerve regeneration chambers. Brain Res (to be published)

Lundborg G, Gelberman RH, Longo FM, Powell HC, Varon S (1982a) In vivo regeneration of cut nerves encased in silicone tubes: Growth across a six millimeter gap. J Neuropathol Exp Neurol 41(4):412-422

Lundborg G, Dahlin LB, Danielsen N, Gelberman RH, Longo FM, Powell HC, Varon S (1982b) Nerve regeneration in silicone chambers: influence of gap length and presence of distal stump. Exp Neurol 76:361-375

Manthorpe M, Engvall E, Ruoslahti E, Longo FM, Davis GE, Varon S (1983) Laminin promotes neuritic regeneration in vitro from cultured peripheral and central neurons. J Cell Biol 97:1882-1890

Varon S, Lundborg G (1983) In vivo models for peripheral nerve regeneration and the presence of neuronotrophic factors. In: Haber B, Perez-Polo JR, Hashim GA, Giuffrida Stella AM (eds) Nervous system regeneration. Liss, New York, pp 221-240

Varon S, Manthorpe M, Longo FM, Williams LR (1983) Growth factors in regeneration of neural tissues. In: Seil FJ (ed) Nerve, organ, and tissue regeneration: research perspectives. Academic, New York, pp 127-155

Varon S, Manthorpe M, Williams LR (1984) Neuronotrophic and neurite promoting factors and their clinical potentials. Dev Neurosci (to be published)

Wever U, Albrechtsen R, Manthorpe M, Varon S, Engvall E, Ruoslahti E (1983) Human laminin isolated in a nearly intact, biologically active form from placenta by limited proteolysis. J Biol Chem 258: 12654-12660

Williams LR, Varon S (1985) Modification of fibrin matrix formation in situ enhances nerve regeneration in silicone chambers. (to be published)

Williams LR, Longo FM, Powell HC, Lundborg G, Varon S (1983) Spatial-temporal aspects of peripheral nerve regeneration within a silicone chamber: parameters for a bioassay. J Comp Neurol 218: 460-470

Williams LR, Powell HC, Lundborg G, Varon S (1984) Competence of nerve tissue as distal insert promoting nerve regeneration in a silicone chamber. Brain Res 293:201-211

Experimental Problems in Spinal Cord Neural Reconstruction

J. C. de la Torre, M. T. Richard, and L. P. Ivan

Division of Neurosurgery and Experimental Surgery, University of Ottawa Health Sciences, Ottawa, Ontario K1H 8M5, Canada

Introduction

One of the most critical factors in studying biological problems is the development of the experimental model. Generally speaking, if the model does not complement the reality, the results obtained with it can be quite irrelevant despite the validity of the findings. This is especially true in spinal cord reconstruction studies following trauma. Regardless of the experimental approach used to study spinal cord regeneration, the investigation will almost always originate with the premise that neural conduction across or bypassing the lesion site needs to be established before complete functional return of the paralyzed structures can occur. How this may be achieved, and the pitfalls involved in structuring this plan, is the subject of this paper. For the purpose of such an analysis, two assumptions must be made:

1) the problem of functional spinal regeneration has a solution,
2) the solution is within the reach of our present technology.

The Experimental Problem

Figure 1 summarizes various experimental approaches that have attempted to re-establish a neural conducting bridge across or around a transected lesion site. For specific details of these models, the reader is referred to a recent review on this subject (de la Torre 1984). The partial success of some of these experimental approaches in aiding or supporting axonal growth across the lesion site suggests that re-establishment of neuritic connections is possible after they have been severed (Cotman and Hoff 1980).

There are, however, a number of complex elements that need to be experimentally tested when spinal cord reconstruction is contemplated in mammalian cordectomy models. Some of these elements are:

a) spinal tissue morphology during surgical repair and removal of scar and necrotic tissue from lesion site,
b) proper graft implant to bridge transected cord stumps with adequate vascular supply to bridge matrix,
c) biochemical stabilization of the damaged tissue,
d) pharmacological or physical manipulation to enhance axonal growth,
e) functional synaptic reconnection distal to the lesion.

Experimental Brain Research, Suppl. 13

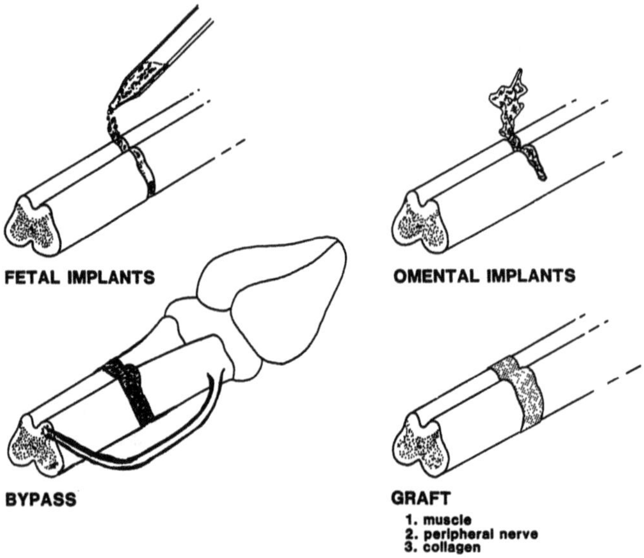

Fig. 1. Techniques for anastomosing or improving neural growth following a lesion or transection (or both) in mammalian spinal cord. Fetal implants of dissociated or embryonic cells explants have shown growth of catecholaminergic axons across the transection site. Omental apposition is reported to increase vascular flow and remove edematous fluid from the injured cord. Axonal bypass of the lesion site may be possible by cutting an intercostal nerve a few centimeters proximal to its dorsal root ganglion and placing the cut end below the lesion or by inserting a piece of sciatic nerve between two points in the cord tissue. This "viaduct" approach appears to provide a conduit for growing axons although their regeneration into the spinal tissue seems limited. Various graft implants have been recommended to bridge the site of transection. These grafts are usually composed of autologous tissue from muscle, peripheral nerve, iris, or noncellular collagen. Their potential for providing a scaffold to growing neurites and blood vessels has not been fully assessed

Assuming that the above elements are to some degree important in reconstructing the damaged neural circuitry - though not necessarily the only ones - we set out to study each problem with respect to its relative participation in spinal cord repair. Our objectives were to evaluate each factor individually in a series of subset experiments. Problems a-e therefore form the basis of a plan to examine and select optimal conditions that will lead to a final experimental study.

Findings from each experimental subset will be collected and eventually integrated into this final experiment. That study would determine whether significant sensory-motor differences are found between the experimental and control groups. The reason for not incorporating these problems into an initial study is because each element needs to be tested against itself. For example, finding the most ideal graft implant to bridge the transected cord stumps involves analyzing various graft materials to see which, if any, is superior. This approach tends to optimize the selection of one given element over another based on its relative value with respect to the animal model being tested.

The Experimental Model

An experimental biomedical model should ideally have two qualities: clinical relevancy and experimental replication. Sometimes, these two qualities appear irreconcilable and form the fulcrum of the "uncertainty principle", that is, the closer one condition is attained, the farther the other one becomes. Since all experimental biological models attempt to reflect some reality, it is often difficult to design a study which closely simulates the clinical picture while still being able to produce consistent results. We have noted in reviewing the literature that most studies dealing with spinal cord reconstruction begin by transecting the cord with a blade. This is followed by therapy or examination of the lesion. This experimental approach has several flaws which impose a number of problems when interpreting the results. First, most clinical injuries occur as a result of compression-acceleration or compression-dislocation of the spinal cord and not by clean transection of the tissue. Second, therapy in a clinical subject is rarely applied within 2 h following injury. Third, if the long range objectives sought from the experimental model are to reconstruct the damaged neural circuitry, then studying neural regeneration very shortly after transection of the cord is self-defeating because critical pathological problems, such as fibroblastic proliferation, gliosis, neovascularization, and formation of the scar tissue, do not occur until several days after trauma. For these reasons, we thought it would be more appropriate to develop a model that would simulate the clinical picture to a closer degree without sacrificing consistency of the experimental observations. In addition, we speculated that allowing stabilization of the initial injury for several weeks was more representative of the tissue pathology seen in clinical cases at the time of therapy and rehabilitation. To this end, we subjected all rats to a severe T_{10} acceleration-compression injury using the weight-drop method. This injury is known to induce irreversible paralysis and is preceded by hemorrhage, edema, and compression of the tissue in much the same way as a chronic spinal cord injury would develop (de la Torre and Boggan 1980). Fourteen days after this contusion trauma, the second phase of the injury was completed as the spinal cord was again exposed and totally transected at T_{10} in both the experimental and control animals.

The Experimental Design

Using the experimental model injury described above, we proceeded to analyze the first element in our list of problems. This problem involved comparing various surgical techniques to determine which approach produced the least tissue trauma during surgical repair of the spinal cord.

Specifically, we wanted to minimize the cellular changes and tissue edema that accompany the physical removal of scar and necrotic tissue from a spinal lesion site in preparation for a bridge graft insertion (Fig. 1).

Three surgical approaches were considered following exposure of the contused cord: (a) trimming of tissue from the proximal and distal cord stumps using a scalpel blade and conventional surgical procedure; (b) ice-cold irrigation of the proximal-distal stumps prior to trimming and lowering tissue temperature just slightly above freezing; trimming of proximal-distal stump tissue using a thin Gillete blade in a seesaw motion; (c) ice-cold saline irrigation of the proximal-distal stumps, trimming of the proximal-distal tissue using a Codman ML-10 CO_2 microsurgical laser at 6 W in a continuous mode.

The gap left by the removal of the necrotic tissue from the proximal-distal stumps was then filled in all animals with a cell-free, bovine-derived, collagen material as previously described by us (de la Torre et al. 1984a). Ninety days after the implantation of the collagen bridge, rats were killed and the cord region interphasing with the collagen bridge was processed for light and electron microscopy and for histofluorescence of catecholamines using the SPG method (de la Torre 1980). Light microscopic sections were taken using Nissl and Bodian silver stains while sequential sections were examined for catecholamine-containing varicosities (CCV), which are found within preterminal or terminal catecholaminergic axons. Sections were examined caudal to the tissue: graft junction and graded on a blind basis using four criteria (see Table 1):

1) vascular density (V),
2) number of neurites/high power field (N),
3) morphological intactness of the neural tissue (M),
4) number of CCV/high power field (C).

Since it is more difficult to quantitate or semiquantitate histologic changes using the electron microscope, ultrastructural evaluation of the tissue was not included in the analysis, but served only as a subjective reinforcement of the light microscopic results (de la Torre et al. 1984a).

Evaluation of Preliminary Findings

Table 1 shows the grading scale that was used to score mean morphological changes seen in sections taken at the collagen: distal tissue junction after trimming the proximal and distal cord stumps using either a scalpel blade, a thin blade plus ice-cold saline irrigation, or a laser plus ice-cold saline irrigation. Four subjective grades were used to evaluate the morphological changes (see Table 1, M, C, V, N). We assumed that seeing significant differences in the CCV and neuritic density near the collagen: distal tissue junction would imply less surgical damage was imparted on the proximal and distal cord stumps at the time they were trimmed. The reason is that regenerating CCV and other descending axonal fibers have to traverse the bridge implant in a cephalocaudal direction before they can be seen distal to the lesion. Figure 2 shows that less damage to the cord occurred when the tissue was trimmed with a thin blade and ice-cold saline irrigation as compared with using a scalpel blade and room-temperature saline irrigation. In addition, the extrusion of spinal cord tissue at the cut stump-ends produced by the annular compression of the dura following conventional blade transection appears reduced by the ice-cold saline irrigation. This may be due to the stump-ends hardening as the tissue is cooled down, thus providing more resistance and less compression damage by the blade. There is also less bleeding when ice-cold saline irrigation is continued following transection, possibly from vasoconstriction of the

TABLE 1. Morphological Evaluation of Trimmed Spinal Cord Tissue Using Three Surgical Approaches

Technique

a) scalpel blade and warm saline irrigation
b) thin blade and ice-cold saline irrigation
c) CO2 laser and ice-cold saline irrigation
d) laminectomy control, no injury

Scoring Code

M1-	>80 µm microcysts, extensive demyelination and/or necrosis		
M2-	50-80 µm microcysts, moderate demyelination and/or necrosis		
M3-	<50 µm microcysts, mild demyelination and/or necrosis		
C1-	0/HPF (high power field)		
C2-	1-10/HPF		
C3-	>10/HPF		
V1-	none-few	N1-	0-2/HPF
V2-	few-moderate	N2-	3-5/HPF
V3-	many	N3-	>5/HPF

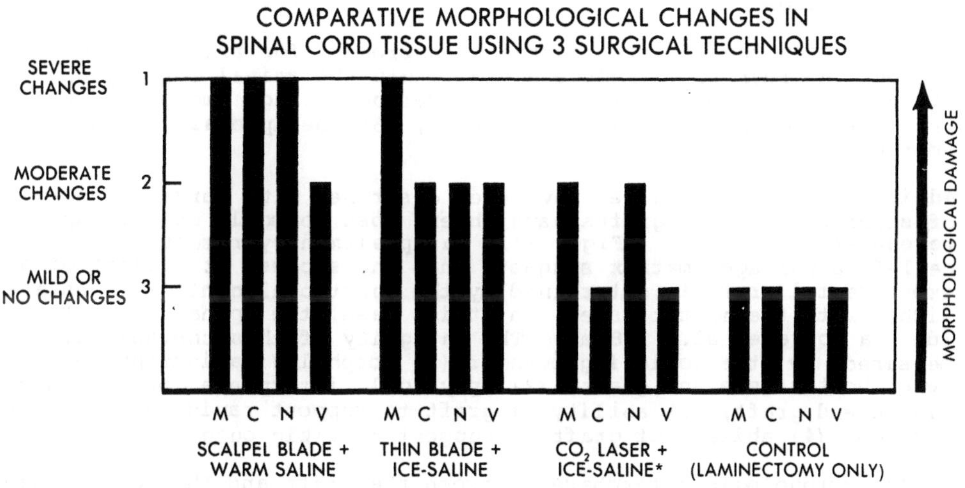

Fig. 2. Histogram shows histologic changes comparing three surgical approaches for trimming proximal and distal cord stumps prior to graft implantation. Cellular damage as a result of such tissue trimming was most pronounced using a scalpel blade + warm saline irrigation. Moderate cellular damage was seen using a thin blade + ice-cold saline irrigation. Least tissue changes were seen using a CO_2 laser + ice-cold saline irrigation. Analysis of variance indicated a significant difference (*$p<.05$) between scalpel blade and laser surgical approaches. A laminectomy only group receiving no injury or tissue trimming was used for sham control purposes

cut vessels. Ice-cold saline irrigation may also lower the metabolic activity of the tissue which could result in less oxygen demand and consequently less tissue injury from ischemia (Albin et al. 1966; Thienprasit et al. 1975; Meacham and McPherson 1973).

Moreover, Deschner and Gelderd (1984) recently reported that ice-cold saline perfusion of contused spinal cord reduced the amount of tissue cavitation and hemorrhage when compared to room-temperature saline perfusion of the same region.

Significantly less damage to the tissue occurred when a CO_2 laser and ice-cold saline irrigation was used to trim the cord stumps as compared to the scalpel blade method. Trimming the cord stumps with a CO_2 laser produced no discernible herniation of the spinal tissue when the cut surfaces were examined with the surgical microscope at 40x. This effect was seen regardless of the saline irrigation temperature. Since less bleeding was noted after cold rather than room temperature irrigation was used, we opted to keep the ice-cold irrigation maneuver in our surgical protocol.

We concluded from these findings that laser and ice-cold saline irrigation resulted in less cellular damage when compared to the other two methods (Fig. 2). Consequently, we have incorporated this technique on all our spinal cord regeneration studies when trimming of the proximal and distal stumps is required.

Other Problems

Having chosen the least traumatic approach for trimming the cord stumps in order to introduce a graft into the spinal cord gap, we are now in the process of examining various bridge implants. These are: (1) collagen matrix, (2) muscle, (3) peripheral nerve, (4) embryonic cells.

Additional tissue bridges have been described with varying degrees of success but these grafts have been most commonly used in spinal regeneration studies (Fig. 1). Our preliminary results using a cell-free collagen matrix suggest that the success or failure of an experimental graft is determined by the junctional contact it establishes with the host tissue, in this case, the spinal cord stumps (de la Torre et al. 1984a). The integrity of this contact may be measured by the following events: (1) morphologica interphase between host tissue and graft, (2) physical interphase between host tissue and graft, (3) ability of graft to support rapid neovascularization, (4) ability of graft to support neuritic outgrowth.

1) The morphologic interphase between the graft and the host tissue implies that no space or undue pressure on the host tissue surface will be created following the graft insertion. If even microscopic gaps between the host tissue and the graft develop, it is unlikely that axons will traverse this space when growing across the bridge area. While it is possible for axons to grow in many different types of media, they do not appear to grow across a space or fluid-filled tissue microcyst (Kiernan 1979; de la Torre et al. 1984). For example, we have observed catecholaminergic axons in brain and spinal cord take a winding route around a tissue microcyst following neuritic outgrowth during tissue repair. Conversely, if undue mechanical stress is exerted by the graft on the tissue surface of the host cord stump - as may occur if the graft is simply "stuffed" into gap created after trimming the cord stumps - it is likely that tissue

edema and reactive gliosis will result, with the net effect of reducing possible axonal growth across the host tissue: graft junction. Reducing mechanical stress from the area of neural tissue repair has been reported by de Medinaceli and his group (1983) to be a key factor in the successful regeneration of peripheral nerves following transection injury. This group also showed that blood, pus, fat, collagen scar fibers, and perineurium could hamper the growth of cut peripheral nerve axons. Our results on spinal cord repair appear to support their findings on peripheral nerves (de Medinaceli et al. 1983).

Besides attempting to minimize mechanical stress between graft and host tissue, it is also important to provide a relatively flat and smooth surface area at their junction. This may be achieved by reducing surface deformities of host tissue and graft to negligible limits. Our data indicate that a good surgical technique is fundamental in providing minimal stress on the graft: tissue junction. In addition, the density of the graft material may also be critical. For example, even when a physically smooth interphase is attained between tissue and implant, if enough pressure on the host tissue is exerted by a tight fitting graft, significant mechanical distortion on the tissue may occur. If the density of the graft is great enough, for example in the case of scar tissue, axons probably will not enter it due to the lack of physical space created by the tight woven fiber network (Schoenheit 1968: de la Torre et al. 1984).

Rapid vascularization of the graft after its implantation into the cord is critical since it is axiomatic that without an adequate blood supply mammalian tissue will necrotize and become functionally useless (Robbins 1974). In view of previous findings, we reasoned that axonal regeneration would not occur in a vascular (Heinecke 1978) or poorly vascularized tissue (de la Torre et al. 1984). Vascular density of the host tissue: graft junction and graft itself may be evaluated using a technique that measures tissue blood flow, such as ^{14}C-antipyrine or the hydrogen washout method. In addition, light microscopic examination of tissue sections using a fluorescent tracer, such as Evans Blue, can provide a semiquantitative estimate of the vascular density in a given region.

3) The ability of the graft to better support neuritic growth may be reflected as items 1-2 above are optimized. If our premise is valid, neuritic growth across an experimental graft will be increased by providing such axons with the most favorable physical conditions for growth. The above steps, however, do not guarantee the most optimum chemical environment to these axons. Since little is known about the latter, we propose that experiments d-e must examine the possibility of stabilizing the microchemical environment of the transected tissue.

We propose to do this by making a series of assumptions in our experimental design. These assumptions are based on previous investigations, notably that of de Medinaceli et al. (1983), who showed that transected peripheral nerve growth and function could be re-established in most of their animals following chemical and pharmcological manipulation of the fibers' stump-ends during surgical reconnection.

Since it has been previously shown that CNS trauma can alter calcium ion shifts across cellular membranes and may cause axonal (Schlaepfer 1974; Balentine and Hilton 1980), transmitter (de la Torre et al. 1984) and myelin damage (Schlaepfer 1974), this cation needs to be investigated in terms of its net activity on cord tissue repair. Moreover, calcium may play an essential role in axoplasmic transport (Ochs 1983) and in the release of adrenergic (Katz 1969) and cholinergic neurotransmitters (del Castillo and Katz 1957). Moreover, regulation of the pH, oxygen tension, and membrane electrolytes in the damaged tissue may provide a better microchemical environment for neuritic regeneration.

4) Pharmacological treatment that may accelerate axonal growth has been reported using thyroxine, NGF, insulin, cis-hydroxyproline, and estrogenic substances, to name but a few (Kiernan 1979). Treatment of the tissue with any of the above or some promising new agent may reveal whether the rate and quantity of axonal regeneration can be increased.

If reinnervation with new synapse formation is possible following spinal transection in mammals, it may well depend on the successful application of the above-mentioned factors.

Acknowledgment. This work was supported by a grant from the American Paralysis Association

References

Albin MS, White RJ, Locke, GE, Kretchner HE (1966) Spinal cord hypothermia by localized perfusion cooling. Nature 210:1059-1061

Balentine JD, Hilton CW (1980) Ultrastructural pathology of axons and myelin in calcium induced myelopathy. J Neuropathol Exp Neurol 39:339

Cotman CW, Hoff SF (1983) Synapse repair in the hippocampus: the effects of aging. Birth Defects 19:119-134

de la Torre JC (1980) An improved approach to histofluorescence using the SPG method for tissue monoamines. J Neurosci Methods 3:1-5

de la Torre JC (1984) Spinal cord injury models. Prog Neurobiol (to be published)

de la Torre JC, Boggan, JE (1980) Neurophysiological monitoring in rat spinal cord trauma. Exp Neurol 70:356-370

de la Torre JC, Hill PK, Gonzalez-Carvajal M, Parker JC (1984a) Evaluation of transected spinal cord regeneration in the rat. Exp Neurol 84:188-206

de la Torre JC, Richard MT, Ivan LP (1984b) Spinal cord catecholamines after calcium-induced myelopathy. Neurosci Soc Abstr 10: 31

del Castillo J, Katz B (1957) La base "quantale" de la transmission neuromusculaire. In: Microphysiologie comparée des elements excitables. Colloq Int Cent Natl Rech Sci 67:245-258

de Medinaceli L, Wyatt RJ, Freed WJ (1983) Peripheral nerve reconnection: mechanical, thermal and ionic conditions that promote the return of function. Exp Neurol 81:469-487

Deschner SH, Gelderd J (1984) Hypothermic treatment of spinal cord injury - the effect of treatment temperature and duration on tissue pathogenesis. Anat Rec 208:44A

Heinecke EA (1978) Vascular permeability and axonal regeneration into tissues autotransplanted into the brain. Proc Can Fed Biol Soc 21:40

Katz B (1969) The release of neural transmitter substances. Thomas, Springfield

Kiernan JA (1979) Hypotheses concerned with axonal regeneration in the mammalian nervous system. Biol Rev 54:155-197

Meacham WF, McPherson WF (1973) Local hypothermia in the treatment of acute injuries in the spinal cord. South Med J 66:95

Ochs S (1983) Calcium and mechanism of axoplasmic transport in nerve fibers. In: Kao CC, Bunge R (eds) Spinal cord reconstruction, Raven, New York, p 87

Robbins SL (1974) The pathologic basis of disease. Wilkins, Philadelphia, p 23

Schlaepfer W (1974) Calcium-induced degeneration of axoplasm in isolated segments of rat peripheral nerve. Brain Res 69:203-215

Schönheit B (1968) Weitere Untersuchungen zur Regeneration des Rueckenmarks bei Rana esculenta L. und Rana temporaria. Z Mikrosk Anat Forsch 78:577-596

Thienprasit P, Bantli H, Bloedel JR, Chou SN (1975) Effect of delayed local cooling on experimental spinal cord injury. J Neurosurg 42:150

Transplantation of Fetal Central Nervous System into the Adult Spinal Cord: A Possible Aid to Regeneration

J. J. Bernstein, D. Underberger, and D. W. Hoovler*

Veterans Administration Medical Center, 50 Irving Street, N.W., Washington, DC 20422, USA

Introduction

Transplantation of fetal central nervous system into the brain has shown that transplants can: (1) survive for years (Freed et al. 1980; Das 1974; Stenevi et al. 1976; (2) send nerve fibers to/and receive nerve fibers from the host (Björklund et al. 1976, 1979, 1983); (3) serve as bridges for host nerve fibers to cross lesions (Björklund et al. 1976; Kromer et al. 1981); (4) form electrically active synaptic complexes since stimulation of the host neurons can alter electrical activity in the transplant (Harvey et al. 1982); (5) ameliorate the symptoms of an animal model for Parkinsons (Björklund et al. 1980, 1981; Dunnett et al. 1981a, 1981b; Freed et al. 1980; Perlow et al. 1979); and (6) return brain-induced learning deficits toward normal (Björklund et al. 1983; Dunnett et al. 1982; Low et al. 1982).

Compared to the advances made in the transplantation of fetal CNS into the brain of adult hosts, fetal CNS transplantation into spinal cord is in its infancy (Bernstein 1985; Das 1981; Hallas 1982; Patel and Bernstein 1982). This is due in part to the responses of the spinal cord as a host tissue. Unlike brain, the spinal cord does not tolerate the injury induced by the implantation of the fetal CNS transplant (Bernstein et al. 1984, 1985b). Spinal cord rejects fetal CNS transplants if meninges are entrapped upon implantation (Das 1983). Transplants have been placed in pockets in the spinal cord. These pockets become lined with a fibrotic scar which effects the connectivity of the transplant with the host (Nornes et al. 1983). In addition, the mechanics of implanting the transplant result in the formation of a cyst in the host spinal cord (Bernstein et al. 1984, 1985b; Bernstein and Hoovler 1984).

In spite of these differences from brain implants, fetal CNS transplants survive for long periods in the adult spinal cord (experiment duration 4 months; Bernstein et al. 1984; Patel and Bernstein 1983), grow noradrenergic and other nerve fibers into the host spinal cord (Nornes et al. 1983), and in very preliminary studies, may aid in the amelioration of paraplegia (Shimizu 1983). However, fetal CNS transplants implanted into spinal cord injured animals have not acted as bridges for the regeneration of centrally derived spinal cord axons across the site of lesion (Bunge et al. 1983) as they do in hippocampal transplants (Kromer et al. 1981).

---

*with the technical assistance of M. Jefferson and S. Turtil

Experimental Brain Research, Suppl. 13

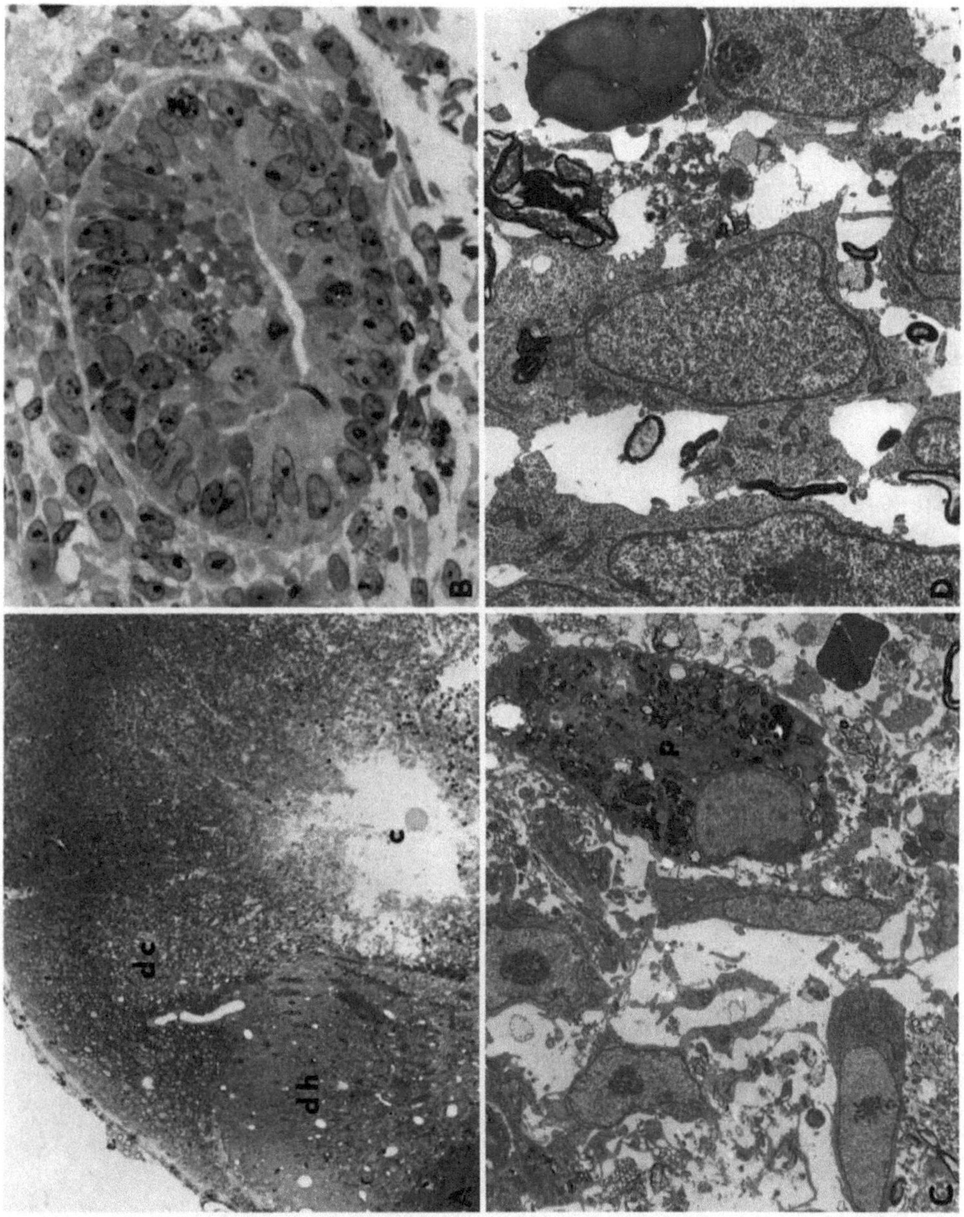

Fig. 1. A. Light micrograph of the dorsal horn (dh) and dorsal column (dc) containing a presumably fluid-filled cyst (c) in the host spinal cord 1 day after implantation. B. Light micrograph of a spherical neuroepithelium from an E11 cortical transplant (3 days postimplantation, DPI). C, D. Electron micrographs of E11 cortical transplants 3 days DPI. A neuroepithelium has formed and host phagocytes (p) remove only cellular debris, sparing the transplant (Bernstein 1985a)

Fig. 2. A. Schematic drawing of the three zones of a mature transplant: a central basal lamina (dashed line) lined zone (black), a medial zone of differentiating neurons (dark gray), and an outer zone (light gray) that was lined with basal lamina in the cyst and had overlapping neuropil with spinal gray matter. B. Central zone of an E11 spinal implant 30 DPI. Basal lamina (arrows) lined channels (C) contain peripheral nerve fibers (P). Neuroglial cells (n) are subjacent to the basal lamina (x 5600). C. E15 fetal cortical transplant 30 DPI in the medial zone of differentiation. The nuclei (N) and cytoplasm are differentiated. However, there is undifferentiated, developmental extracellular space in the neuropil (arrows; x 5000). D. E15 fetal spinal cord 60 DPI at the zone of interface between implant (I) and host (H, upper left). The neurons (N) appear to be in the transplant. The neuroglial cell (n, upper left center) is either in the transplant or host. The region between host (H) and implant (I) contained overlapping neuropil (x 2000) (Bernstein et al. 1984)

Implantation

The fetal cortex and spinal cord are derived from fetuses of time pregnant Sprague-Dawley females (11 days gestation, E11, parturition 21 days). The cortex (0.5 x 1.0 mm) is dissected free in a well dish containing Tyrode-Ringer's solution. A 1.0 mm length of spinal cord is also dissected free. These fetal tissues are implanted with a 30 gauge needle, attached to a 50-μl Hamilton syringe, between the dorsal horn and dorsal column of the host spinal cord, under the T6 vertebra, by pressure injection (see Bernstein et al. 1985a for details).

Transplant Viability and Morphology

One day after implantation of E11 Sprague Dawley rat fetal cortex and spinal cord into the thoracic spinal cord of adult Sprague-Dawley male hosts (300-350 g), the transplants are observed in a presumably fluid-filled cyst formed by rapid phagocytosis of debris by the host (Fig. 1A). At 1-3 days the transplant is in the form of multiple oval neuroepithelia (Fig. 1B). These neuroepithelia are replete with active phagocytes (Fig. 1C, D). These phagocytic cells remove the cellular debris from the host spinal cord and transplanted fetal CNS and necrotic cells from either source. However, as an indication of the immunologically privileged status of transplanted fetal CNS, viable transplanted cells are not phagocytized. On the basis of cell recognition, fetal CNS is recognized as self by the host spinal cord.

E11 and E15 cortical and spinal cord transplants will grow to fill the cyst made by the implantation process and interface with the ventral, rostral, and caudal portions of the cystic cavity. However, the most frequent interface of transplant and host is the isthmus of the spinal cord. The transplants appear to sediment to the ventral portion of the presumably fluid-filled cyst which, because of the placement between the dorsal horn and dorsal column and the degeneration of the nerve fibers, is the spinal gray at the isthmus of the host spinal cord. The transplants grow to fill the cyst and contain three concentric zones (Fig. 2A). There is a central zone of basal lamina lined channels (Fig. 2A, B) which form a syringomyelocele that may open to the surface of the spinal cord (Bernstein et al.

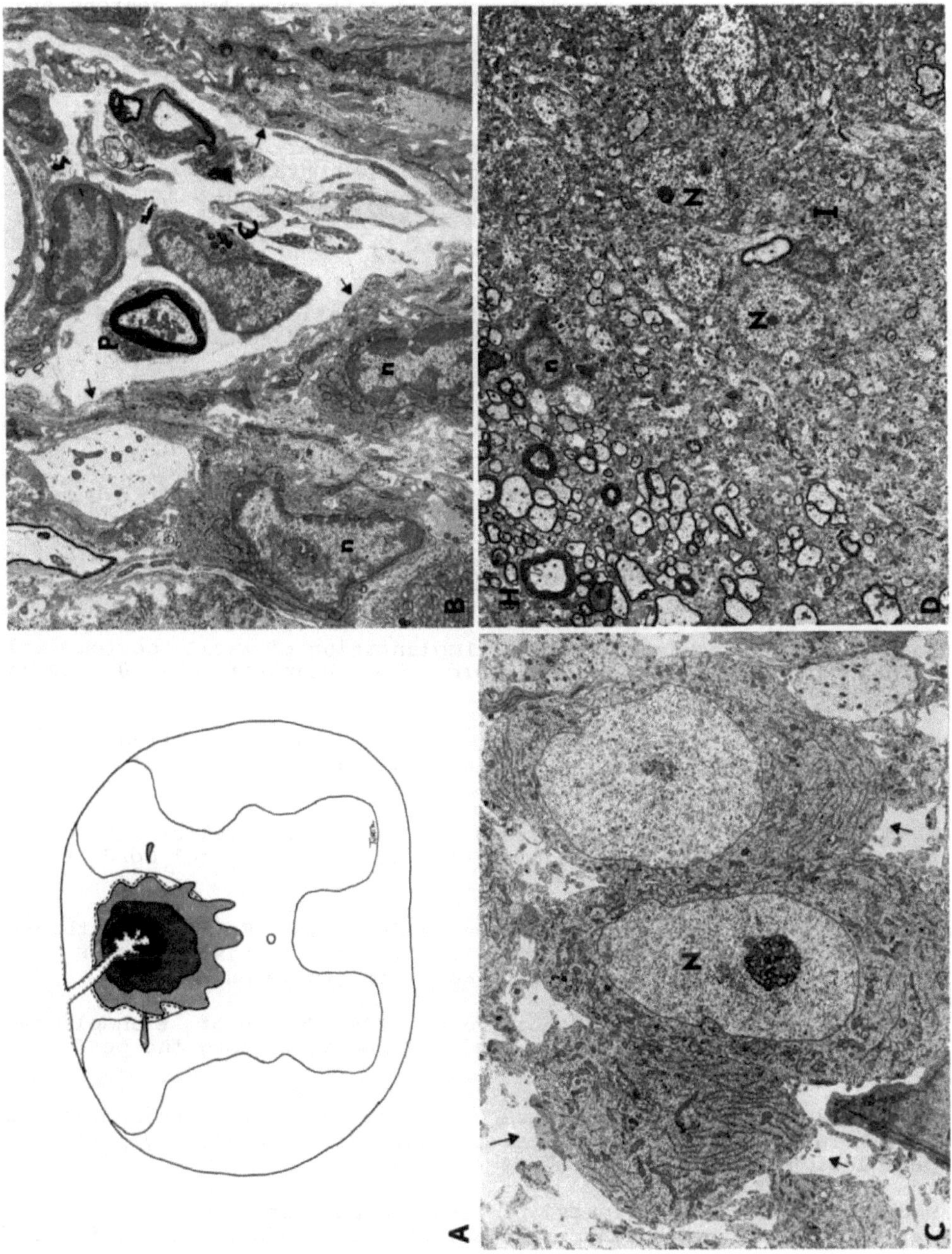
A
B
C
D

1984, 1985a). There is a medial zone of differentiating neurons and neuroglia around the central zone (Fig. 2A, C) and a third outer zone which is lined with basal lamina on its dorsal aspects (Fig. 2A). The ventral aspects of the outer zone contain areas of overlapping-coextensive neuropil with the host (Fig. 2A, D).

Nornes et al. (1983) demonstrated (fluorescence microscopy) transplanted fetal locus coeruleus noradrenaline-containing neurons with nerve fibers that extended into the host spinal cord for millimeters. It is cell process markers of this type that will be needed to extend the study of connectivity of fetal CNS transplants into the spinal cord.

These experiments have shown that transplantation of fetal CNS into the spinal cord is feasible and should be studied further as a future aid to paraplegics.

Acknowledgments. Supported by the Veterans Administration.

References

Bernstein JJ (1985) Spinal cord regeneration. In: Davidoff R (ed) Spinal cord handbook, vol IV. Dekker, New York (to be published)

Bernstein JJ, Hoovler DW (1984) Implantation of fetal cortex and spinal cord into adult spinal cord. Soc Neurosci Abs 10: (to be published)

Bernstein JJ, Patel U, Keleman M, Turtil S (1984) Ultrastructure of fetal spinal cord and cortex implants into adult rat spinal cord. J Neurosci Res 11: (to be published)

Bernstein JJ, Getz R, Jefferson M, Keleman M (1985a) Neuroglia secrete basal lamina after hemisection of the rat spinal cord. Brain Res (to be published)

Bernstein JJ, Underberger D, Hoovler DW (1985b) Fetal CNS transplants into adult spinal cord: techniques, initial effects and caveats. Central Nervous System Trauma 1: (to be published)

Björklund A, Stenevi U, Svendgaard N (1976) Growth of transplanted monoaminergic neurons into the adult hippocampus along the perforant path. Nature 262:787-790

Björklund A, Kromer LF, Stenevi U (1979c) Cholinergic reinnervation of the rat hippocampus by septal implants is stimulated by perforant path lesion. Brain Res 173:57-64

Björklund A, Segal M, Stenevi U (1979b) Functional reinnervation of rat hippocampus by locus coeruleus implants. Brain Res 170:409-426

Björklund A, Dunnett S, Stenevi U, Lewis M, Iversen S (1980) Reinnervation of the denervated striatum by substantia nigra transplants: functional consequences as revealed by pharmacological and sensorimotor testing. Brain Res 199:307-333

Björklund A, Stenevi U, Dunnett SB, Iversen SD (1981) Functional reactivation of the deafferented neostriatum by nigral transplants. Nature 289:497-499

Björklund A, Stenevi U, Dunnett S (1983) Transplantation of brainstem monoaminergic "command" systems: models for functional reactivation of damaged CNS circuitries. In: Kao C, Bunge R, Reir P (eds) Spinal cord reconstruction. Raven, New York, pp 397-415

Bunge R, Johnson M, Thuline D (1983) Spinal cord reconstruction using cultured embryonic spinal cord strips. In: Kao C, Bunge R, Reir P (eds) Spinal cord reconstruction. Raven, New York, pp 341-358

Das G (1974) Transplantation of embryonic neural tissue on the mammalian brain. 1. Growth and differentiation of neuroblasts from various regions of the embryonic brain in the cerebellum of neonate rats. Life Sci 4:93-124

Das GD (1981) Transplantation of neural tissue in the spinal cord of adult rats. Soc neurosci. Abs 7:625

Das GD (1983) Neural transplantation in the spinal cord of the adult mammal.In: Kao C, Bunge R, Reir P (eds) Spinal cord reconstruction. Raven, New Yrok, pp 367-396

Dunnett SB, Björklund A, Stenevi U, Iversen SD (1981a) Behavioral recovery following transplantation of substantia nigra in rat subjected to 6-OHDA lesions of the nigrostriastal pathway. I. Unilateral lesions. Brain Res 215:147-161

Dunnett SB, Björklund A, Stenevi U, Iversen SD (1981b) Behavioral recovery following transplantation of substantia nigra in rats subjected to 6-OHDA lesions of the nigrostriatal pathway. II. Bilateral lesions. Brain Res 229:457-470

Dunnett SB, Björklund A, Stenevi U, Iversen SD (1981c) Grafts of embryonic substantia nigra reinnervating the ventrolateral striatum ameliorate sensorimotor impairments and akinesia in rats with 6-OHDA lesions of the nigrostriatal pathway. Brain Res 229:209-217

Dunnett SB, Low WC, Iversen SD, Stenevi U, Björklund A (1982) Septal transplants restore maze learning in rats with fornix-fimbris lesions. Brain Res 48:251-335

Freed W, Perlow MJ, Karoum F, Seiger A, Olson L, Hoffer B, Wyatt RJ (1980) Restoration of dopaminergic function by grafting of fetal rat substantia nigra to the caudate nucleus: long-term behavioral, biochemical, and histochemical studies. Ann Neurol 8:510-519

Hallas BH (1982) Transplantation into the mammalian adult spinal cord. Experientia 38:699-701

Harvey AR, Golden GT, Lund RD (1982) Transplantation of tectal tissue in rats. Exp Brain Res 47:437-445

Jaeger CB, Lund RD (1981) Transplantation of embryonic occipital cortex to the tectal region of newborn rats: a light microscopic study of organization and connectivity of the transplants. J Comp Neurol 194:571-597

Kromer LF, Björklund A, Stenevi U (1981) Regeneration of the septohippocampal pathways in adult rats is promoted by utilizing embryonic hippocampal implants as bridges. Brain Res 210:173-200

Low WC, Lewis PR, Bunch ST, Dunnett SB, Thomas SR, Iversen SD, Bjorklund A, Stenevi U (1982) Function recovery following neural transplantation of embryonic septal nuclei in adult rats with septo-hippocampal lesions. Nature 300:260-262

Nornes H, Björklund A, Stenevi U (1983) Reinnervation of the denervated adult spinal cord of rats by intraspinal transplants of embryonic brain stem neurons. Cell Tiss Res 230:15-35

Patel U, Bernstein J (1982) Adult mammalian spinal cord regeneration by fetal spinal implants. Soc Neurosci Abs 8:915

Patel U, Bernstein J (1983) Growth, differentiation and viability of fetal rat cortical and spinal cord implants into adult rat spinal cord. J Neurosci Res 9:303-310

Perlow MJ, Freed WJ, Hoffer BJ, Seiger A, Olson L, Wyatt RJ (1979) Brain grafts reduce motor abnormalities produced by destruction of nigrostriatal dopamine system. Science 204:643-647

Shimizu Y (1983) Transplantation of cultured cerebellar autografts into the spinal cords of chronic paraplegic dogs. In: Kao C, Bunge R, Reir P (eds) Spinal cord reconstruction. Raven, New York, pp 359-366

Stenevi WF, Björklund A, Svendgaard N (1976) Transplantation of central and peripheral monoamine neurons to the adult rat brain: Techniques and conditions for survival. Brain Res 114:1-20

Growth and Development of Neural Transplants: Some Quantitative Parameters

G. D. Das

Department of Biological Sciences, Purdue University, West Lafayette, IN 47907, USA

Introduction

Neural transplantation has attracted both scientific and popular interest in recent years. This and other factors have influenced the lopsided growth of this field, where reports dealing with the alleviation of neurologic syndromes or functional aspects of transplants overshadow those dealing with basic morphological and physiological aspects of the neural transplants. In many instances it is uncertain whether transplantation has been successful or not; and in some cases transplants credited with the properties of alleviating neurologic and functional deficits appear to be necrotoic. However, some studies addressed to basic research on neural transplantation do provide definition and histological characteristics of successful neural transplantation, but they are qualitative in nature (Das and Altman 1972; Das 1974, 1983a; Bjoerklund et al. 1976; Lund and Hauschka 1976; Stenevi et al. 1976; Jaeger and Lund 1980). The qualitative findings are of importance insofar as they help establish a foundation for conceptualization of various aspects of neural transplantation, suggest further questions for investigation, help delineate artifacts of experimentation and observation from facts, and provide a perspective for long-range research objectives. The field of neural transplantation, when viewed comprehensively, is seen to be inherently fraught with problems. Some of the major problems are related to the facts that neural tissue from different regions of the neuraxis of the donor embryos of a given developmental stage may not have the same survival and growth rates, a given neural tissue from the embryos of different developmental stages may not have the same growth characteristics, and a given type of neural tissue from embryos of the same age may show different degrees and characteristics of growth depending upon the nature and degree of trauma caused to the host brain and degree of pathologic environment surrounding the transplant. These and other such problems must be completely understood before any statements on the functional and clinical significance of neural transplants can be made. These problems, by the very fact of their encompassing large territories, require quantitative analysis. Furthermore, there are many advantages to quantification of various characteristics of neural transplants, some of them being: it offers a basis for consensus in definition of successful and unsuccessful transplants; it aids in the replication of experiments and confirmation by independent investigators; it helps in the discrimination of significant findings from less significant ones; and it provides strong and valid support for interpretation of the findings.

Experimental Brain Research, Suppl. 13

Quantitative analysis of neural transplants can be performed at various levels, such as overall growth of the transplants, developmental changes in the transplants, cell estimates and cell density, interface areas between transplants and the host brain tissue, and magnitude of connectivity in the transplants (Das 1983a; Oblinger and Das 1982, 1983). Over the years, in our laboratory various types of embryonic neural tissue have been transplanted in different conditions and by different methods of transplantation. Transplants into the host brain have been analyzed qualitatively as well as quantitatively during their development and when fully grown. In this presentation only those data that have a bearing on neural transplantation under traumatic conditions are presented. With this restriction findings made in various other studies have been excluded.

Some General Comments on Neural Transplants and Methods

In all these studies embryonic neural tissues obtained from freshly dissected rat embryos were used. It is important to stress that these donor embryos were obtained from sperm-positive female animals: mating of the animals was restricted to a few hours only, and the day of sperm-positivity was regarded as day 1 of gestation. Other methods of mating animals without restriction for 1 day or 2, and determining developmental stages of embryos from their crown-rump length always involve the risk of underestimation of the age of the donor embryos and were therefore not used. The neural tissue used for transplants was always viable. Nonviability of neural tissue because of a partially resorbed state of the embryos or excessive damage to the embryos or exposure of the dissected neural tissues to air or application of nonsterile procedures for dissection and manipulation of tissues always results in necrosis of the entire transplant. Such transplants invariably yield negative results. Further, neural tissues were always transplanted intraparenchymally, which means the tissues were injected inside the host brain into various regions, such as forebrain or cerebellum, and were allowed to grow intraparenchymally. The tissues transplanted extraparenchymally in the subdural space or left in ventricles, or left in surgically prepared cavities in the host brain, yield highly variable results with a low survival rate. Data on such preparations are not included in this presentation. Moreover, the neural tissue was always freshly obtained, that is not frozen-thawed, and were used as solid pieces of tissue for transplantation. As solid pieces they were easily retained within the host brain parenchyma. Generally, such transplants, if manipulated carefully, do not cause problems of nonretainability of the tissue such as are encountered when cell suspensions are used. Nonretainability of neural tissues at the site of injection, an artifact of the transplantation technique, can lead to erroneous conclusions regarding survival capacity or immunological rejection of the transplants.

The host animals used in these studies were of different ages, varying from neonatal to adult animals. Various studies in our laboratory on the effects of host age on the survival and growth of neural transplants, have shown that there are no significant differences in terms of survival of the transplants but there may be some differences in the growth of the transplants with high growth potential between neonatal and adult host animals. Wherever pertinent, the ages of the host animals are presented in the different studies. Great importance attaches to the fact that all techniques of neural transplantation, by their very nature, involving surgical manipulation of the host brain, are invasive and traumatic to the host nervous system. Different techniques vary in the degree of trauma that they inflict upon the host brain. The technique of transplantation used in these and other studies has been observed to cause the minimum amount of trauma to the host brain. The details of this technique have been given elsewhere (Das 1974; Das et al. 1979; Das and Ross 1982). With other techniques excessive trauma is caused to the brain, resulting in hemorrhage, edema, extensive cell necrosis, and other pathologic changes in the host brain. When neural tissue is transplanted into such pathologic surroundings it generally does not survive well; and if it does survive it does not achieve normal growth and differentiation. More often than not such neural transplants become extraparenchymal. Furthermore, in the studies presented below the neural transplants were of known volume. Their initial volume was established with a high degree of accuracy, and the method for this has been described elsewhere (Das and Ross 1982). It has been found that determining the initial volume of the neural tissue in the calibrated glass capillary needle just before its injection into the host brain not only gives an accurate measurement but also reduces the degree of variability from one preparation to the next. This method, in comparison with others, is simple and direct. Other methods of determining the initial volume of the transplants, such as measuring linear dimensions of the dissected pieces of tissue floating in Ringer's solution in a petri dish, suffer from high degrees of inaccuracy as well as variability of the measurements. Finally, in these studies neural tissue was transplanted in the forebrain region and cerebellum of the host animals, each animal receiving only one transplant, to determine whether there was any difference in the survival and growth of the transplants in different regions of the host brain. As long as neural tissues were transplanted intraparenchymally and all other conditions of transplantation were kept constant there was no evidence to suggest that different sites of transplantation in the host brain had any appreciable effect on the survival, growth, and integration of the transplants. With this background, the findings presented below refer to neural transplants in the cerebellum of the host animals. These findings also hold true for the transplants in the forebrain region of the host animals.

Since the studies reported below mainly addressed volumetric analyses of neural transplants, it is important to comment on the methods used in these analyses. In all preparations, closely spaced serial sections were collected from one end of the transplant to the other. The stained sections were used to make drawings of the neural transplants with the aid of a Wild-M5 microscope with a drawing tube attachment. On these drawings various details of transplants, such as necrotic region, viable region, integrated interface, apposing unintegrated surface, and cavities in the transplants, were drawn. The areas and linear measurements takes from these drawings were combined to give volume and surface area of the transplants. These methods have been described in detail by Oblinger and Das (1982, 1983). In the studies on histogenesis in the transplants which were based on ^{3}H-thymidine autoradiography, analysis was performed for the labeled neurons in the transplants. In randomly selected sample areas of transplants, labeled and unlabeled cells were counted; these data were used to determine the percentage of labeled cells on different days after transplantation. These data, when plotted in the developmental sequence, yielded a comprehensive picture of histogenesis in the transplants.

Study I: Growth and Development of Different Types of Neural Transplants

The most basic issue in neural transplantation pertains to the growth of the transplants. Some neural transplants grow large, some grow less, and some do not grow at all. It appears important to establish the conditions that determine these differential growth characteristics of different neural transplants. A neural tissue used for transplantation may be defined in terms of the region of the brain from where it is obtained, and/or the age of the donor embryos. These two basic variables have a great bearing on many aspects of the growth and differentiation of neural transplants and provide a perspective for other quantitative analyses of the transplants. The main questions in this study were: Do the neural transplants from different regions of the neuraxis of donor embryos of a given age grow equally or differently? Do the transplants from a given region of the brain from embryos of different ages grow to the same extent or not? Since these two questions are complementary, they have been analyzed in this single study.

Two variables, introduced simultaneously, were different types of neural tissue, and donor embryos at different developmental stages. The tissue used for transplantation was obtained from neocortex, basal ganglia, diencephalon (mainly thalamus), tectum, brain stem, and spinal cord of embryos of 15, 16, 17, 18, 19, 20, 21, and 22 days of gestational age. In each instance the transplant was of 7.0 mm^3 in volume. Each type of neural transplant obtained from donors at different embryonic stages was transplanted in four to six host animals. These transplants were injected intraparenchymally in the vermis of the cerebellum of 30-day-old host animals. Following transplantation these animals were allowed to survive for 3 months. During this period, the transplants in the host brain had reached their fullgrowth, differentiated, and achieved their individual neural characteristics. These preparations were used for the determination of volume and survival rate of the transplants.

The findings showed that different neural transplants had grown to different extents during their development (Fig. 1). Of these transplants, those from the neocortical region seemed to show by far the highest growth. Those from the 15-day rat embryos showed a 21-fold increase in volume over their initial volume. This growth of neocortical transplants seemed to decline with increasing age of the donor embryos. The transplants from 21- and 22-day rat embryos showed rather moderate growth, progressing to less than 2-fold the original volume. It is important that all these neocortical transplants were well integrated with the host cerebellum. Areas of different types of interface could be easily identified between the transplants and the host brain, and afferents and efferents passed through them. These transplants contained apparently normal, well-differentiated neurons. There was no evidence of any persisting pathologic changes in these transplants. All the transplants were intraparenchymal. In all these cases there was evidence of loss of the host cerebellum and this loss was commensurate with the size of the transplants. In other words, the neocortical transplants from the 15-day donor embryos that had grown very large had caused the loss of a large amount of host cerebellum; and at the other extreme those from the 21- and 22-day embryos, which had not grown to such large volumes, had caused a lesser extent of loss of the host cerebellar tissue. Transplants were then performed with tissue from the basal ganglia. Transplants from the 15-day embryos showed an increase of a little more than 2-fold over the initial volume. These transplants also showed a decline in growth with increasing age of

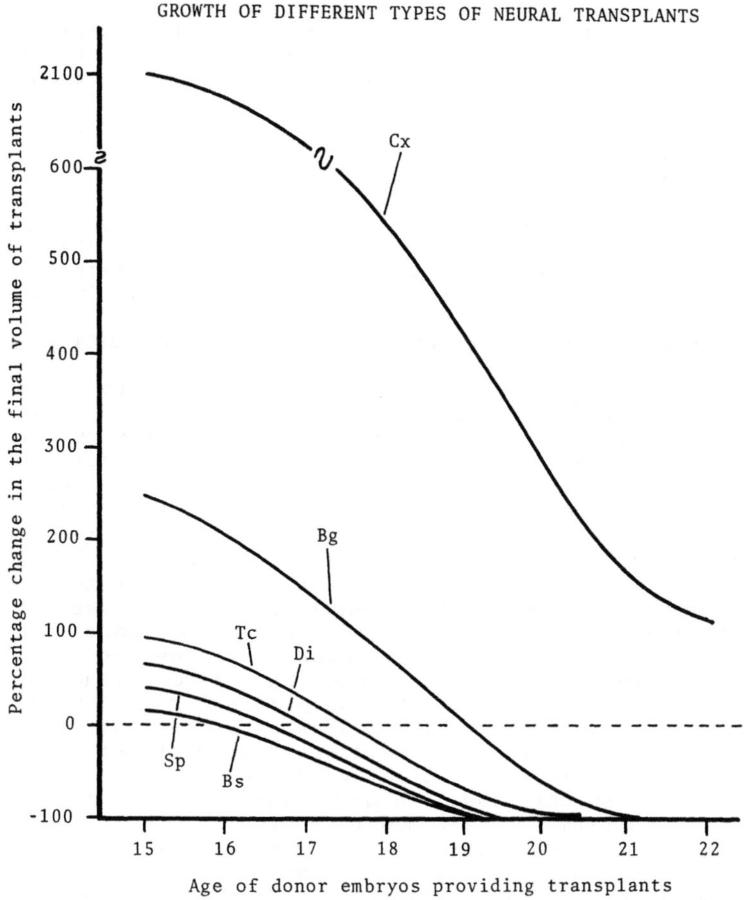

Fig. 1. Percentage increase or decrease in the volume of the fully grown neural transplants of various tissues obtained from embryos of different developmental stages. Cx, neocortex; Bg, basal ganglial; Tc, tectum; Di, diencephalon; Sp, spinal cord; Bs, brain stem

the donor embryos. In this case the transplants from the 19-day embryos did not grow at all, and those from the older embryos failed to survive. The transplants that survived without growth contained poorly differentiated, shrunken, and atrophying neurons. There were some cysts in the transplants, and a large space separated them from the host brain. In essence they were extraparenchymal transplants. Other transplants obtained from other regions of the embryonic neuraxis did not show any impressive growth. The neural transplants obtained from the tectal and diencephalic regions in 15- and 16-day embryos showed some growth, but those obtained from the older donor embryos generally did not survive at all. In the case of transplants from spinal cord and brain stem regions of the donor embryos only the transplants obtained from 15-day embryos grew at all, and those from older donor embryos failed to survive.

When these findings are viewed comprehensively the following general observations can be made. Various neural transplants can be classified qualitatively in terms of their growth potential. Neocortical transplants from 15- and 16-day embryos appear to be endowed with a high growth potential, and those from 17-, 18-, and 19-day embryos have moderate growth potential. Neural transplants from neocortical regions of 20-day embryos and from basal ganglia of 15- and 16-day donors can be considered to show moderately low growth potential. Transplants from the neocortex of 21- and 22-day embryos, basal ganglia of 17- and 18-day donors, tectum and diencephalon of 15- and 16-day embryos, and spinal cord and brain stem of 15-day embryos have a low growth potential. The remaining neural transplants that do not survive can be considered to show negative growth potential.

Differences in the growth potentials of transplants can be explained with reference to their cellular composition. From this viewpoint embryonic neural tissues can be considered to be composed of neuroepithelial cells with different degrees of potential proliferation, undifferentiated neuroblasts, partly differentiated neuroblasts, mostly differentiated neuroblasts, and fully differentiated neurons. Neocortical tissue from 15- and 16-day rat embryos contains predominantly neuroepithelial cells, which are capable of extensive proliferation following their transplantation. It is the proliferative activity that gives rise to a large number of cells and thus to a large increase in the volume of the transplants. The transplants showing moderate growth potential, such as those taken from the neocortical region in 17- and 18-day embryos, contain a mixture of neuroepithelial cells and neuroblasts. The neuroepithelial cells, which are fewer in number and which have also already gone through some mitotic activity prior to transplantation, show only a small amount of proliferative activity following transplantation. Neuroblasts in these transplants are undifferentiated or partly differentiated and do not show any mitotic activity; they simply differentiate following transplantation. These cellular characteristics of the transplants determine their moderate growth. The transplants with moderately low growth potential, such as those obtained from the neocortical region of 20-day rat embryos, contain mostly neuroblasts and a few neuroepithelial elements. In fact, the neuroblasts in these transplants are not homogenous; they are at different stages of differentiation. Many neuroblasts may be relatively undifferentiated, and some may be partly differentiated. The partly differentiated neuroblasts generally do not survive the trauma of physical manipulation involved in transplantation so well as those that are relatively undifferentiated, and many of them die following transplantation. The loss of partly differentiated neuroblasts coupled with a very low level of mitotic activity in the few neuroepithelial cells contributes to the moderate to low growth potential of these transplants. The transplants with low growth potential, such as those obtained from neocortex of 22-day rat embryos or spinal cord and brain stem of 15-day embryos, are characterized by a mixture of cells that are mostly differentiated and some that are partly differentiated. Such cellular elements are highly susceptible to the trauma of transplantation, and most of them die. The few neuroblasts that survive may be those that are partly differentiated at the time of transplantation, and they determine the low growth potential. Finally, the transplants that fail to survive or show negative growth potential are seen to be composed of mostly differentiated neuroblasts. They simply die away following their transplantation. In essence, the final growth of a neural transplant is the result of three factors, presence and degree of mitotic activity of the neuroepithelial cells, extent of the presence and loss of partially and mostly differentiated neuroblasts, and extent of the presence of relatively undifferentiated neuroblasts that seem to survive the manipulations of transplantation.

Furthermore, anatomically neural transplants with a high growth potential are always seen to be fully integrated with the host brain, to contain healthy-looking and well-differentiated neurons, and not to be affected by cysts or other pathologic changes. The transplants that have moderate and moderately low growth potential are highly variable in their anatomical characteristics. However, they are generally relatively small, show only partial integration with the host brain, contain a mixture of apparently normal and necrotic neurons, may contain cysts and small patches of gliosis reaction, and may be separated from the host brain in patches by intervening glial scar formation. The transplants with low growth potential are invariably very small, contain mostly shrunken and poorly differentiated neurons, may have very little or no anatomical integration with the host brain, and as extraparenchymal transplants are surrounded by glial scar formation and meningeal membranes. These observations are very significant, in that they demonstrate a clear-cut relationship between the growth potential of the neural transplant and the presence or absence of cysts and other pathologic changes. This relationship is determined by the developmental interaction of the growing transplants and the persisting pathologic conditions following invasive microsurgery of transplantation. In essence, neural transplants with a high growth potential, during their growth after transplantation, are able to displace the necrotic tissue from the site of transplantation, extend into the viable host brain parenchyma, and establish parenchymal attachment with it long before the processes of glial scar formation have set in. In this way such transplants become fully integrated with the host brain and do not show any pathologic changes. Other neural transplants, with moderate and low growth potential, do not show such striking developmental interaction with the pathologic reaction at the site of transplantation. In these cases, therefore, pathologic changes persist permanently.

The findings relating to the growth and development of different neural transplants acquire a greater significance when they are viewed from the perspective of the transplant survival rate (Fig. 2). Generally, neural transplants with a high growth potential also have a high survival rate; those with moderate growth potential have a moderate survival rate; and those with a low growth potential also have a low survival rate. These observations suggest that data on the survival rate and growth of the transplants should be highly correlated. Indeed, statistical analysis of the correlation between these two sets of data supports this conclusion. The concept of survival of neural transplants requires some further clarification. By strict definition the term survival should be used to mean permanent survival of the transplants for the life of the host animals. According to this definition only those neural transplants that have a high or moderate growth potential and are anatomically completely or partially integrated with the host brain are seen to survive permanently, whereas transplants with low growth potential, which generally tend to remain unintegrated with the host brain and become extraparenchymal, do not survive permanently. They may linger in the ventricles or the subdural space or in the surgically prepared cavities for some months, and may undergo a very slow process of atrophy and degeneration. They may be seen as extraparenchymal transplants 2 or 3 months after transplantation. But after a year or so they may not be present at all.

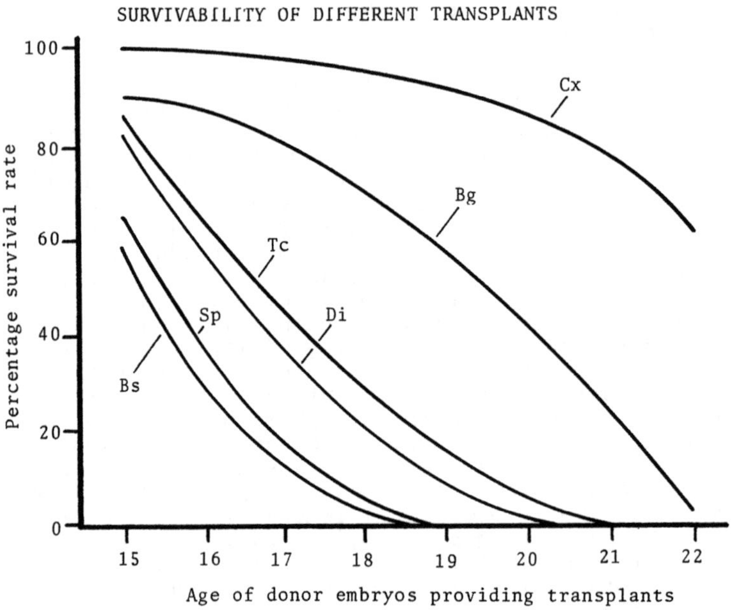

Fig. 2. Percentage survival of various neural transplants obtained from donor embryos of different ages. This graph includes all transplants, intraparenchymal, partly intraparenchymal, and extraparenchymal. Cx, neocortex; Bg, basal ganglia; Tc, tectum; Di, diencephalon; Sp, spinal cord; Bs, brain stem

Further, data on the survival rate of transplants raise some interesting questions. It is relatively easy to explain 100% and zero survival of transplants in terms of growth potential of the neural tissue and cellular characteristics of the transplants, but not the survival rates falling between these two extremes. Why, in the case of transplants with a 50% survival rate, do half the transplants fail to survive? A detailed study of this problem has shown that other factors, in addition to those of the neural transplant itself, also influence the survival or otherwise of the transplants. These are: nature of site of transplantation in the host brain, degree of pathologic reaction induced in the host brain, retainability of the transplant, and viability of the transplant. Consideration of these factors reveals that surgically preparation of cavities in the host brain for the transplants is generally followed by a lower survival rate than transplantation into the normal parenchyma. In other words, transplants injected into normal parenchyma of the host survive better than those injected into surgically prepared cavities. Different techniques of insertion of neural transplants in the host brain cause different degrees of trauma. Some cause more extensive bleeding than others. Neural transplants placed in an environment of excessive pathologic change generally do not survive. Neural transplants must be retained at the site of transplantation for them to survive. If owing to technical flaws the transplants are forcibly shot through the brain or are allowed to ooze out and settle subdurally, they do not survive at all. Finally, embryonic neural tissues at various stages of dissection and manipulation may be inadvertently exposed to air or to some nonsterile and contaminated solution or material. Owing to these uncontrolled conditions they invariably become nonviable. Such transplants are seen to degenerate very rapidly after transplantation, and also cause a pathologic reaction in the host brain tissue. Even the most sophisticated instrumentation or transplantation technique will not yield successful transplants if the tissue itself is nonviable. These are some of the important factors that contribute to variability in the results of transplantation from animal to animal. In the case of transplants with high growth potential, their overwhelming characteristic of high growth overshadows and compensates for negative influences related to techniques and procedures of transplantation. But in the case of the transplants with moderate, moderately low, and low growth potential various negative aspects of techniques of transplantation can overshadow their limited growth potential, and this may vary from animal to animal. Because of this, such transplants have a variable survival rate.

Study II: Development of Neocortical Transplants from Embryos of Different Ages

The first study described above indicated that neocortical transplants obtained from rat embryos of 15 days of gestational age onward provide material that can be used to analyze the growth characteristics of a tissue at different stages of its neurohistogenesis. They provide tissues with different cellular compositions ranging from neuroepithelial cells in the tissue obtained from the 15-day donor embryos to partly and mostly differentiated neuroblasts in the tissues obtained from 20-, 21- and 22-day embryos. Because of these properties neocortical tissue has been used extensively in our analyses of the development of neural transplants. The second study was adressed to the problem of what developmental events underlie the growth of the neocortical transplants with different cellular compositions? In this case, neocortical tissue was obtained from rat embryos of 15, 16, 17, 18, 19, 20, 21, and 22 days of gestational age. In each case the transplant was 3.5 mm^3 in initial volume. It was transplanted into the midvermis of the cerebellum of 30-day-old host rats. Following transplantation, host animals from each group were sacrificed 6 h and 1, 2, 3, 4, 6, 8, 10, 15, 30, and 60 days after the surgery. In each group there were four to six host animals. The tissue obtained from these host animals was processed for histology, and the sections were used for volumetric analysis of the necrotic mass, viable mass, and growing mass in the transplants.

The findings showed that in all cases between 1 and 2 days after transplantation about 50% of the neural tissue became necrotic (Fig. 3). The amount of tissue that survived following this initial necrotic phase was 50% or less of the initial volume of the transplant. After this phase, the transplants started to recover. In the transplants from 15-, 16- and 17-day rat embryos recovery was characterized by the formation of rosettes by the surviving neuroepithelial cells, whereas in the transplants from older embryos it was characterized by clustering of the surviving neuroblasts. The necrotic and recovery phases were completed by day 5 from transplantation in all cases. From this stage onward the transplants showed growth and differentiation. It is important to note that the end of necrosis and clearing of the debris was accelerated in the transplants from 15-, 16-, and 17-day rat embryos, but was rather slow in the transplants from older embryos. Further, although all the neocortical transplants grew following necrosis and recovery, they differed widely in the rate and extent of their increase in size.

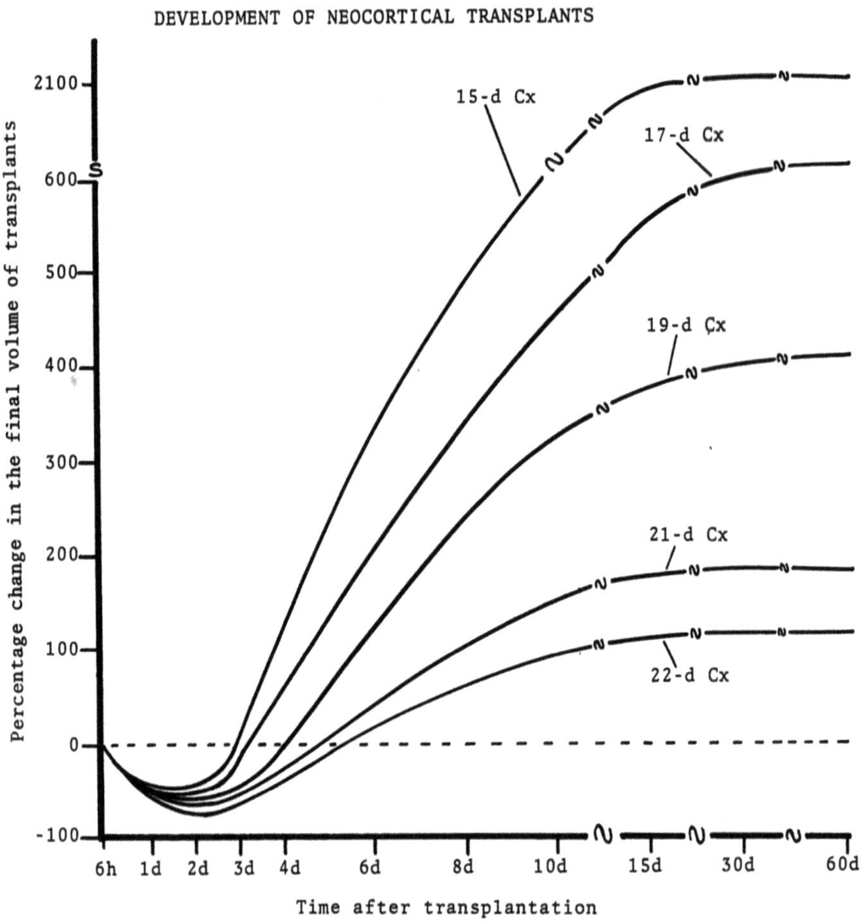

Fig. 3. Changes in the growth of neocortical transplants in relation to the age of the donor embryos. The tissue used for transplantation was obtained from embryos of all ages between days 15 and 22 of gestation, but for clarity of presentation only the data from 15-, 17-, 19-, 21-, and 22-day embryos are presented in this graph. The data on the neocortical transplants from 16-, 18-, and 20-day donor embryos, if plotted on this graph, would fit properly in relation to the rest of the data. 15-d Cx, neocortical tissue from 15-day donor embryos etc.

The transplants from the 15-day embryos showed the most rapid growth after day 3 of transplantation, and achieved most of their growth by day 10 after surgery. This rapid growth in volume was due largely to the proliferative activity in the neuroepithelial cells, a massive increase in the number of neuroblasts arising out of this proliferative activity, and active movement of neuroblasts in clusters into the surrounding host cerebellum. This particular phase of growth could thus be considered to show growth by an increase in the number of cells and by outward movement of clusters of neuroblasts. Following this, the rate of growth was seen to decline, and the transplants achieved their final volume. This phase, overlapping with the previous two phases, was seen to be characterized largely by the differentiation of neuroblasts and an increase in the neuropil. When these three phases of development are viewed comprehensively it becomes evident that there is a considerable amount of overlap from one phase to the next. They are not rigidly compartmentalized in the manner of one following the other with well-defined points of initiation and termination. The second phase, i.e., the one characterized by proliferation of neuroepithelial cells and an increase in the number of neuroblasts, is seen to have most bearing on the volumetric growth of the transplants. It is also characterized by the outward movement of clusters of undifferentiated neuroblasts, which in turn contributes further to the increase in the volume of the transplants. The third phase, that characterized by differentiation of neuroblasts into neurons and increase in the neuropil, contributes the least to the ultimate volume of the transplants.

At the other extreme are the neocortical transplants from 22-day donor embryos, which contain a mixture of undifferentiated, partly differentiated, and mostly differentiated neuroblasts formed at different times before transplantation. They do not seem to contain neuroepithelial cells. Followling necrosis and recovery in the transplants, the surviving neuroblasts simply undergo the phase of differentiation. In this case, it is only cellular differentiation along with the increase in the neuropil that determines the small amount of growth of the transplants. This increase is observed to progress at a slow rate. Other neocortical transplants between these two extremes are characterized by different cellular compositions; and they may contain neuroepithelial cells with different potencies for proliferation and a mixture of undifferentiated, partly differentiated, and mostly differentiated neuroblasts in varying proportions related to the stage of development of the donor embryos. Depending upon the magnitude of the surviving neuroepithelial and neuroblastic elements, and the proliferative potency of the neuroepithelial cells, they may show differential growth. Repeated analyses of various developmental studies have shown that these complex. factors of magnitude of neuroepithelial cells, their proliferative potency, differential numbers of neuroblasts at different stages of their differentiation, and differential susceptibility of these cells to trauma are related to the age of the donor embryos. Therefore, the degree of initial necrosis, the rate and extent of recovAry, the rate and extent of growth, and the final volume of the neocortical transplants are seen to follow an orderly gradient in relation to the age of the embryos (Das et al. 1983a).

The data have shown that immediately after transplantation all neural transplants undergo necrosis and recovery before starting their characteristic growth. Various developmental studies conducted in our laboratory, where the influence of various experimental conditions on the survival and growth of transplants has been analyzed, have shown that the initial necrosis of neural tissue is an inevitable consequence of the very procedures involved in their transplantation. However, it may vary considerably, and there are many factors that determine the degree of necrosis. Two of them are very important: nonviability of the neural tissue and the injection technique used to place the transplants. If problems of manipulation and inadequate observance of precautions cause the neural tissue to be exposed to air and they are still transplanted they undergo necrosis without any signs of recovery. These tissues are nonviable at the time of surgery and they degenerate completely. Various precautions recommended during dissection and preparation of embryos and neural tissues are aimed at preserving the viability of the transplants

(Das et al. 1979). At every step during the entire sequence of preparation and transplantation of neural tissues even slight carelessness can contribute to nonviability of the tissues. Use of nonviable tissues for transplantation can lead to serious misinterpretation of the results and unscientific claims. The technique used for injection of neural tissues has a strong bearing on the degree of initial necrosis in the transplants. These techniques vary considerably from one investigator to the other, very often with little apparent reason. With the conservative techniques of transplantation it is possible to control the trauma to the host brain and internal bleeding. If the techniques of injection of tissues cause extensive trauma to the host brain and massive hemorrhage the transplants are surrounded by blood and various other pathologic changes in the brain, and the host brain tissue in that region fails to provide a conducive milieu for their initial survival and growth. Owing to these conditions the transplants may show more than the expected amount of necrosis, prolonged persistance of the necrotic reaction, and a very slow rate of recovery.

Another important factor that seems to play a part in determining the degree of initial necrosis is the susceptibility of different types of cells in the neural tissues to the trauma of manipulation and transplantation. Various developmental studies have shown that neuroepithelial cells, which are mitotically potent, are by far the most resistant to trauma. They survive better even in a pathologic environment in the host brain. Neuroblasts in general tend to have less resistance to trauma. Among neuroblasts, those that are undifferentiated survive better than those that are partly or mostly differentiated. This factor, even when all other conditions of procedures and technique are kept under control, determines the greater loss of cells in neocortical transplants from 21- or 22-day old embryos than in tissue from 15- or 16-day donor embryos. It is this very factor that determines a greater degree of necrosis in the tissue from the spinal cord of 16-day embryos than in the neocortical tissue from the same embryo when both are transplanted under identical conditions. The former contains partly differentiated neuroblasts and the latter, mainly neuroepithelial cells.

Sometimes it is suggested that it is the lack of immediate blood supply to the embryonic neural tissue at the transplant site that contributes to the initial necrosis. This may be true, but only to a very small extent, for in all intraparenchymally injected transplants new capillaries have grown from the host into the transplanted tissue in less than 24 h. This rapid ingrowth of capillaries is seen not only in the viable portions but also in the necrotic masses of transplants. If lack of blood supply were the main cause of necrosis in the transplants, it would cease or slow down rapidly once the blood supply became available, but this is not observed to be the case. Necrosis in the transplants continues for at least 2 - 3 days, and may even continue longer in some instances. Further, in our studies on extraparenchymal transplants these transplants isolated from the host brain parenchyma have also proved to be capable of receiving a blood supply from the host brain, and still show initial necrosis followed by recovery and some degree of growth even though the penetration of capillaries may be far slower than that observed in the case of intraparenchymal transplants (Das 1982).

Other interesting observations have also been recorded. Different neural transplants show different rates of recovery following necrosis. To a great extent the rate of recovery of the surviving portion of a neural transplant is also related to the cellular composition of the neural tissue. The transplants composed of neuroepithelial cells, such as those from the neocortex of 15- or 16-day embryos, show resumption of mitotic activity soon after transplantation. A fast increase in the number of cells aids in clearance of the necrotic mass and debris, allowing the growing transplant to occupy the space available. In quantitative analysis these changes are seen as the rapid rate of recovery following initial necrosis. At the other extreme, the transplants composed mostly of differentiated neuroblasts, such as those obtained from the neocortex of 22-day embryos, do not show any growth in volume after the phase of necrosis. The debris is still present amidst the fragements of the viable transplant, and the surviving neuroblasts appear quiescent. It is the persistence of this condition that quantitative analysis reveals as a slow process of recovery from necrosis. As the surviving neuroblasts gradually recover, they show a slow process of differentiation. Other neural transplants that are composed of a mixture of neuroepithelial cells and neuroblasts in different stages of their differentiation fall between these two extremes. Their rate and degree of recovery are also related to the age of the donor embryos, that is to the cellular composition of the transplants.

Thus, various complex developmental events, including initial necrosis, recovery, and growth of surviving transplants, that are observed in neocortical transplants can also be observed in other neural transplants from different regions of the embryonic brain. The principles determining the growth of the neocortical transplants also determine the growth of other neural transplants. Further, the findings suggest that a clear understanding of survival or nonsurvival of different transplants can be better gained by thorough developmental analysis of the transplants.

Study III: Growth of Transplants Under Different Conditions of Trauma

The studies described above involved transplantation of neural tissue into normal brains. They have provided important basic information on various aspects of neural transplantation, which can be extended to transplantation under traumatic conditions. Different types of neural traumas induced in the host brain prior to neural transplantation cause different forms and degrees of pathologic reaction. Therefore, it is imperative that experimental traumas should be of such a kind that they lead to an understanding of pathogenic processes in the host brain and their interaction with neural transplants having different growth potential characteristics. For experimental investigations there are various methods of inflicting trauma to the host brain, each of which has its own advantages and limitations. Since neural transplantation involves surgical invasion of the host brain it appears appropriate to use surgical means to create trauma before injecting the neural tissue. Further, surgical trauma has the advantage that the investigator can determine the extent and magnitude of anatomical damage caused to the host brain with a high degree of precision. If there is extensive hemorrhage at the site of trauma the investigator can make an attempt to control it or at least make a note of it and attempt to understand the effect of this variable on the survival and growth of the transplants. Finally, surgically induced traumas, at least at the time of transplantation, are focal or localized rather than widespread and diffuse. This permits reliable and meaningful investigation of the interaction between histopathological changes in the host brain and the growing neural transplants. Findings made in such preparations will have a direct bearing on the acute traumatic conditions.

In this study three different conditions were used. For the first, the transplants were simply injected into the cerebellum of the host animals, and this represented the condition of minimal trauma. For the second condition, immediately before transplantation knife lesions were made in the midvermis regions of the host cerebellum, and the transplants were injected in the region of the knife lesion. In this case the trauma was greater than in normal conditions of transplantation. The third condition involved ablation of a large mass of cerebellar tissue from the host brain, followed by injection of the transplant into the parenchyma of the host cerebellum surrounding the cavity. It is very important to stress that in this condition the transplants were injected into the parenchyma and not left isolated, floating free in the cavity. The transplants used in this study were obtained from the neocortical region of donor embryos at 15, 16, 17, 18, 19, 20, 21 and 22 days of gestational age. Each transplant was 7.0 mm^3 in volume at the time of transplantation. These transplants were injected in such a way that they were retained within the host cerebellum. Care was taken to see that the transplants did not ooze out into the fourth ventricle. The host animals were allowed to survive for 3 months after surgery, when they were killed and their brains processed for histological examination. Histological preparations were used for both qualitative and quantitative analyses.

It was found that in normal conditions and after knife lesion the transplants grew normally (Fig. 4). They both showed a gradient in the size of the transplants in relation to the age of the donor embryos. The shape of gradient of growth of the neocortical transplants was similar to that observed in the studies described above, and the two sets of data were very close and followed a parallel course. These findings showed that the very procedure of injection of neural tissue amounts to some degree of trauma to the host animal, which does not differ appreciably from the trauma caused by knife lesions. By themselves the two types of traumas may be different in terms of severity, but in terms of their influence on the growth of transplants they appeared very similar. Further, in both these conditions even the transplants from 21- and 22-day embryos grew. All the transplants, irrespective of the age of the donor embryos, appeared to contain normal, well-differentiated neurons, and were anatomically integrated with the host brain. After the ablation of host cerebellum, growth of the neocortical transplants followed a different course. The neocortical transplants from 15-day embryos grew large and filled the surgically prepared cavity. Although there were a few small cysts between the transplants and the host cerebellum, the transplants appeared healthy and were integrated with the host brain wherever the two were closely apposed. The tissue from the 16-day embryos grew less than those transplanted in the other two conditions; they did not completely fill the cavity. There were large cysts between the transplant and the host cerebellum. The transplants were only partly integrated with the host brain tissue. The neocortical tissues from the 17- and 18-day embryos grew still less in volume. It appeared that in these two cases large amounts of tissue had become necrotic, resulting in a much reduced final volume of the transplants. Further, these transplants were not integrated with the host brain. They were extraparenchymal, and contained poorly differentiated and atrophied neurons. The transplants from older embryos did not survive at all.

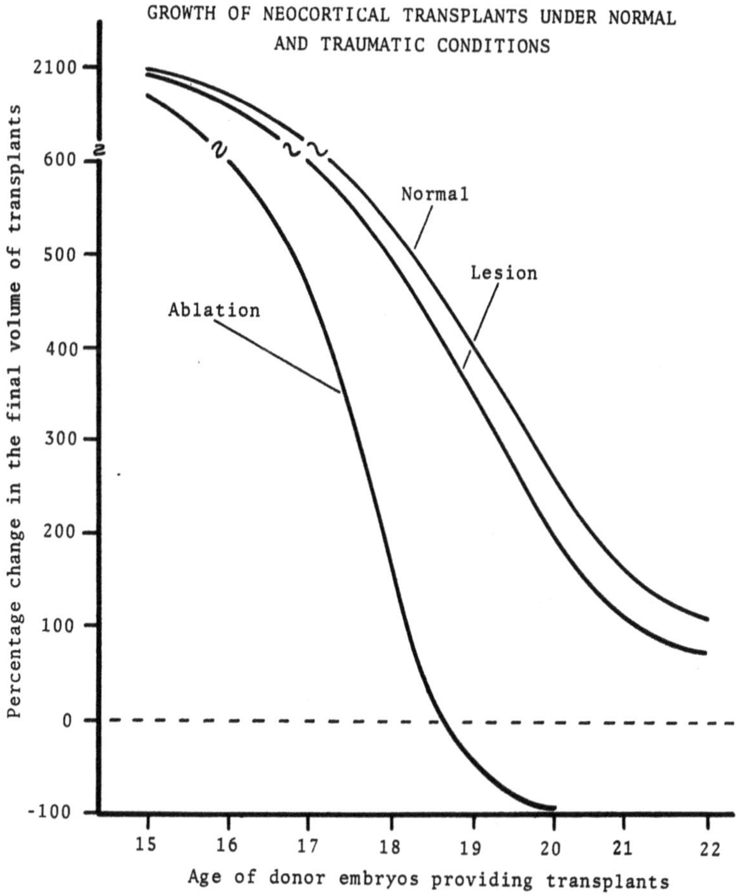

Fig. 4. Growth of neocortical transplants obtained from embryos of different ages and transplanted in three different conditions. Normal denotes transplantation into normal cerebellum without any surgical trauma. Lesion denotes data from animals subjected to knife lesions just before receiving transplants. Ablation denotes data from animals subjected to surgical ablation of a large mass of cerebellum just before receiving the transplants

When the findings described above are viewed together it becomes evident that in conditions of minimal trauma to the host brain neural transplants with high, moderate, and low growth potential can grow and differentiate, although they may have different survival rates (cf. Study I). Under excessively pathologic conditions, however, only the transplants with high growth potential, such as neocortical tissue from 15- and 16-day embryos, can grow and survive. Other types of neural tissue with moderate and low growth potential, show little or no growth and survival. Those transplants that do survive are extraparenchymal, undifferentiated, and necrotic. To understand these differences it is important to comment on developmental events in the histopathology of the host brain following lesions of different degrees of severity. When small lesions are made in the host brain, such as those caused by penetration of the needle and injection of neural tissue or knife lesions, they are restricted. Various histopathological changes, such as cell necrosis, edema, hemorrhage, and ischemia, are also restricted, and they are confined to small regions of host parenchyma immediately adjacent to the site of lesion. These reactions generally tend to subside within a few days; they do not appear to be progressive. The glial reaction surrounding this region is restricted. Such limited pathologic changes do not constitute an extensively adverse environment for the transplants to recover and grow. Hence, under these conditions even the transplants with moderate and low growth potential can grow and achieve anatomical integration with the host brain parenchyma. But when extensive lesions are made in the host brain, such as those resulting from ablation of a neural structure, pathologic changes in the host brain parenchyma are extensive. They may extend far beyond the borders of the lesion, often affecting the neighboring neural structures as well. In the case of the cerebellum, in this study ablations were made in the vermis but the pathologic reactions, in particular hemorrhage and edema, were seen to extend into the cerebellar hemispheres. As these traumatized regions of host brain become necrotic, the lesion or the cavity becomes enlarged. The glial reaction is initiated throughout the traumatized region, and glial scar formation develops along the borders of the cavity. Even when neural transplants are injected in the parenchyma of the lhost cerebellum large amounts of tissue oooze ;out and settle in the cavity. In their isolated condition, such transplants are surrounded by extensive pathologic changes, which are not conducive to their initial survival and growth. The transplants with low growth potential do not survive at all. They become necrotic, and are phagocytized immediately. Those with moderate growth potential may survive, but are unable to grow and fill the cavity. At best these become extraparenchymal and may undergo a slow process of atrophy and degeneration. Only the transplants with high growth potential, which grow at a fast rate, are able to grow and fill the cavity as it expands. These growing transplants penetrate through the necrotic tissue of the host brain surrounding the cavity and establish parenchymal integration long before host brain tissue undergoes progressive degeneration and glial scar formation sets in. These interactions between the histopathological changes and developing neural transplants are seen not only in the cerebrellum but also in the spinal cord of the host animals (Das 1983b,c). Thus, transplantation of neural tissues in traumatized regions of brain should be performed with due consideration of both the nature and severity of trauma and the nature and growth potential of the neural transplants.

This study also raises many other questions that have a bearing on neural trauma and neural transplantation. The first question is: Can successful transplantation be achieved by injecting large volumes of neural tissue with low growth potential? In another study, following ablation of cerebellum large volumes of neocortical tissue, as much as 14.0 - 16.0 mm^3, obtained from 21- and 22-day embryos were injected. The cavity created by surgical ablation was filled with the transplant, and the excess oozed back and settled in the fourth ventricle. In these animals large numbers of fragments of isolated and extraparenchymal neocortical transplants were seen in the cavity. The cavity itself was lined by dense glial scar formation. A few fragments of transplants appeared attached to the host cerebellar parenchyma. Histologically, all these fragments appeared poorly developed, contained atrophied and shrunken neurons, and had very little neuropil. It is doubtful whether such isolated and atrophied fragments of neural transplants would have survived for long, or, even if they had, served any anatomical or functional purpose. This study helped establish the importance of using neural tissue with a high growth potential when trauma of the central nervous system is present. The second question is: Why should the transplant be injected into the parenchyma and not simply left in the cavity prepared by surgical ablation? When the neural tissue is injected into the parenchyma surrounding the cavity, no doubt a considerable amount of tissue oozes back and settles in the cavity, but some tissue does remain within the parenchyma and is continuous with the tissue in the traumatized region. As the whole mass of neural tissue grows following an initial period of necrosis, the portion of tissue within the host parenchyma pushes into the host brain tissue and establishes parenchymal integration. At the same time, being continuous with the portion of tissue in the cavity, it grows in continuity with it. In the final analysis the fully grown transplant not only fills the whole cavity but also becomes anatomically integrated, if not completely at least partly, with the host brain parenchyma. This is observed only when neural tissue with a high growth potential, such as neocortical tissue from 15- or 16-day embryos, is used. These developmental changes leading to anatomical integration between transplants and host brain are not seen when neural tissue with low growth potential, such as spinal cord or brain stem tissue from 15- and 16-day embryos, is used for transplantation (Das 1983b,c). However, if neural tissue, even with a high growth potential, is simply deposited in the cavity, the important condition of neural tissue growing and pushing into the host brain parenchyma is absent. With this the favorable conditions for the establishment of anatomical integration between the transplants and the host brain are lost. The transplants may grow extensively, and may fill the cavity, but in all probability will not be able to establish anatomical integration with the host brain tissue. More often than not, the whole transplant, or most of it, will be separated from the host brain parenchyma by intervening glial scar formation. There will be no interface, and no connectivity between them. There is no evidence to suggest that such neural transplants as have grown large and filled the cavity, but are anatomically isolated from the host brain, will have any functional or clinical significance.

Concluding Comments

Neural transplantation in various conditions of neural trauma is an unexplored field. To understand the effectiveness of neural transplants in functional recovery it is imperative to study not only the problems of survival, growth, and integration of the transplants but also histopathological characteristics of the host brain in various conditions of trauma. Basic research along these various fronts and quantification of the findings are important requirements for a scientific approach to the problems of neural trauma and neural transplantation. Unconnected studies on the apparent behavioral changes in animals, without a clear understanding of neural transplantation and neural traumatic conditions, may be attractive for nonscientific purposes, but certainly do not contribute to scientific understanding of the issues or promote advances in this field. In this presentation some quantitative parameters related to the growth and survival of neural transplants are presented. They are indicative of trends underlying the dynamics of growth and development of transplants. It is possible that with other techniques or following introduction of other variables these trends may show slight shifts, but in all likelihood the overall pattern will not be much altered. More work is required to add to these findings and allow construction of a complete and comprehensive picture of the phenomenon of neural transplantation.

References

Björklund A, Stenevi U, Svendgaard NA (1976) Growth of transplanted monoaminergic neurons into the adult hippocampus along the perforant path. Nature 262:787-790

Das GD (1974) Transplantation of embryonic neural tissue in the mammalian brain. I: Growth and differentiation of neuroblasts from various regions of the embryonic brain in the cerebellum of neonate rats. T.-I.-T.J Life Sci 4:93-124

Das GD (1982) Extraparenchymal neural transplants: Their cytology and survivability. Brain Res 241:182-186

Das GD (1983a) Neural transplantation in mammalian brain: Some conceptual and technical considerations. In: Wallace RB, Das GD (eds) Neural tissue transplantation research. Springer, Berlin Heidelberg New York, p 1

Das GD (1983b) Neural transplantation in the spinal cord of the adult mammals. In: Kao CC, Bunge RP, Reier PJ (eds) Spinal cord reconstruction. Raven, New York, p 367

Das GD (1983c) Neural transplantation in the spinal cord of adult rats. Conditions, survival, cytology and connectivity of the transplants. J Neurol Sci 62:191-210

Das GD, Altman J (1972) Studies on transplantation of developing neural tissue in the mammalian brain. I: Transplantation of cerebellar slabs into the cerebellum of neonate rats. Brain Res 38: 233-249

Das GD, Ross DT (1982) Stereotaxic technique for transplantation of neural tissues in the brain of adult rats. Experientia 38:848-850

Das GD, Halls B, Das KG (1979) Transplantation of neural tissues in the brains of laboratory mammals: technical details and comments. Experientia 35:143-153

Das GD, Das KG, Brasko J, Aleman-Gomez J (1983) Neural transplants: volumetric analysis of their growth and histopathological changes. Neurosci Lett 41:73-79

Jaeger CB, Lund RD (1980) Transplantation of embryonic occipital cortex to the brain of newborn rats. An autoradiographic study of transplant histogenesis. Exp Brain Res 40:265-272

Lund RD, Hauschka SD (1976) Transplanted neural tissue develops connections with host rat brain. Science 193:582-584

Oblinger MM, Das GD (1982) Connectivity of neural transplants in adult rats: Analysis of afferents and efferents of neocortical transplants in the cerebellar hemisphere. Brain Res 249:31-49

Oblinger MM, Das GD (1983) Connectivity of transplants in the cerebellum: a model of developmental differences in neuroplasticity. In: Wallace RB, Das GD (eds) Neural tissue transplantation research. Springer, Berlin Heidelberg New York, p 105

Stenevi U, Björklund A, Svendgaard NA (1976) Transplantation of central and peripheral monoamine neurons to the adult rat brain: techniques and conditions for survival. Brain Res 114:1-20

Transplantation to the Damage Adult Hippocampal Formation

F. H. Gage and A. Björklund

Department of Histology, University of Lund, Lund, Sweden

Introduction

The cellular architectonics of the hippocampal formation (HF) were clearly described by Ramon y Cajal early this century (Ramon y Cajal 1928), and this classical knowledge has since been extended with detailed studies on the connectivity and laminar organization of its different subfields. More recently more detailed knowledge has emerged on the transmitter distribution (see Storm-Mathiesen 1979 for review), and based on these anatomical and histochemical findings the elctrophysiology of the cells within the HF and their connections has also been well characterized (see Anderson 1975).

Understanding of the HF at the behavioral level has been pursued with equal vigor in recent decades and several general hypotheses on hippocampal function have been proposed of late, all of which concur on the importance of HF in the processing of new information during learning (O'Keefe and Nadel 1978; Olton et al. 1979). Quite apart from its importance in brain function, much of the research on the HF has taken advantage of the fact that the afferent input pathways to the HF are organized to allow for selective surgical deafferentation while leaving the intrinsic anatomy essentially intact. Damage to the afferent systems of the HF results in remarkable compensatory anatomical an biochemical changes which may underlie some forms of behavioral recovery following damage.

With this background, the deafferented HF has been employed to investigate the viability of brain tissue grafting, using the anatomical, electrophysiological, biochemical, and behavioral parameters already available as indices of morphological integration, specificity of axonal ingrowth, biochemical restoration, electrophysiological function, and behavioral recovery.

In this review we will focus on the grafting of central nervous system tissue to the adult denervated HF as a model system for studies on axonal growth and regeneration in the adult mammalian CNS.

Transplantation Methods

Two principal techniques have been used to graft fetal CNS tissue to the denervated adult HF. The first involves the transplantation of solid pieces of tissue to a surgically prepared transplantation cavity. In this procedure good graft survival is ensured by preparing the cavity in such a way that the graft can be placed on a richly

Experimental Brain Research, Suppl. 13

vascularized surface (e.g., the pia in the choroidal fissure) that can serve as a "culturing bed" for the graft (Stenevi et al. 1976). Both the fimbria-fornix lesion and the perforant path aspirative lesion (Stenevi et al. 1976) provide such "culturing beds". These cavities are in direct communication with the lateral ventricle, which may allow the CSF to circulate through the graft cavity and thus possibly help the graft to survive, particularly during the early postoperative period.

The second technique used in grafting CNS fetal tissue to the denervated HF involves the implantation of dissociated cell suspensions (Schmidt et al. 1981; Björklund et al. 1983c). In this technique pieces of fetal CNS tissue are trypsinized and mechanically dissociated into a milky cell suspension. Small volumes of the suspension can then be stereotaxically injected into the desired site of the HF, using a microsyringe. The main advantage of this technique is that it allows precise and multiple placements of the cells. The technique also makes possible accurate monitoring of the number of cells injected by counting the density of cells in the suspension. For the remainder of the manuscript the first technique will be referred to as the "solid graft" technique and the second technique will be referred to as the "cell suspension" technique.

Graft Strategies

Three separate grafting strategies have used the denervated HF as a target tissue. The first strategy, that of "reafferentation", uses fetal cells from brain regions which normally send axons to the hippocampus, but which have been transected by the lesion. This strategy permits the evaluation of the extent to which donor tissue can grow axons into the host brain and accurately reestablish a new terminal network in the host hippocampus. The second strategy, that of "bridging", uses pieces of tissue placed in the cavity made by the fimbria-fornix lesion to promote regeneration of the septo-hippocampal pathway by serving as a bridge between the denervated axons passing through the septal area to the HF. The third strategy is to use the transplantation cavities as "culture beds" to investigate graft growth and development without special reference to its relationship to the HF. The results from each of these strategies will be briefly reviewed, summarizing data from solid grafts as well as cell suspensions.

Reafferentation: The idea behind this strategy is to test to what extent, and with what degree of precision, neuron transplants are able to reestablish normal terminal innervation patterns in the previously denervated HF. The HF is ideally suited for such studies for two main reasons. First, the terminal fields of the afferent inputs are, for the most part, discretely laminated, which gives each afferent system a characteristic laminar termination pattern. Second, neuronal cell groups in the brainstem and basal forebrain which send afferent fiber inputs to the HF are accessible to dissection in the fetal brain, thus making possible relatively selective grafts. In addition, the fimbria-fornix lesion transects a large portion of the afferent axons originating from these brain stem and basal forebrain nuclei. In the studies reviewed below solid grafts were obtained from 16- to 17-day-old rat fetuses, while cell suspensions were taken from 13- to 15-day-old rat fetuses. The fetal donor brain regions used were: (1) the septal-diagonal band area, containing developing cholinergic neurons that normally innervate the HF and neocortex; (2) the locus coeruleus region of the pons, containing the cells of origin for the noradrenergic innervation of the HF; (3) the pontine and mesencephalic raphe region, containing serotonergic cells which also innervate the HF.

The fetal septal-diagnoal band area has been transplanted both as solid grafts (Björklund and Stenevi 1977b; Björklund et al. 1979a) and as cell suspensions (Schmidt et al. 1981; Dunnett et al. 1982; Björklund et al. 1983a,b) to the HF in rats that had received prior fimbria-fornix lesions. They approximately doubled their initial volume and gave rise to a pattern of innervation in the HF which closely mimicked the normal cholinergic innervation [as assessed by acetylcholinesterase (AChE) staining]. Several features of this reinnervation attest to its specificity. First, when the ingrowing axons are given a choice between different denervated terminal fields, they show a clear preference for the zones denervated of the homologous fiber type. For example, septal solid grafts or cell suspensions implanted in the HF after fimbria-fornix lesion extend cholinergic axons from the graft restrictively to the normal cholinergic terminal fields, even though the lesion also denervates the extensive system of noncholinergic commissural afferents. Zones receiving dense commissural (but sparse septal) innervation, such as stratum radiatum of CA1 and CA3 and the supragranular zone of the dentate molecular layer, are largely devoid of fibers from the transplant. A second example of this specificity of growth in the cholinergic grafts is provided by taking advantage of the fact that after perforant path lesion the intrinsic septal afferents will sprout selectively into the outer molecular layer of the dentate gyrus. By lesioning both the perforant path and the fimbria-fornix, and then transplanting either solid septal grafts or cell suspensions, one observes that the grafts innervate not only the normal intrinsic terminal fields, but also the other molecular layer of the dentate gyrus (Björklund et al. 1979a; Gage et al. unpublished) (see Fig. 2). Thus, the fiber outgrowth from septal cholinergic grafts illustrates a remarkable selectivity, and yet the same plasticity as the intrinsic septal afferents.

The locus coeruleus solid grafts appear to reestablish a "coeruleo-hippocampal" noradrenergic pathway in animals initially denervated of their hippocampal noradrenaline input by a 6-hydroxydopamine lesion (Björklund et al. 1976, 1979b). The selectivity of the patterning of the ingrowing noradrenergic fibers is emphasized by the innervation of the hilar zone of the dentate gyrus, the area most densely innervated by the intrinsic noradrenergic fibers, but not certain other regions of the HF which are also denervated by the fimbria-fornix lesion and innervated by fibers from septal or raphe grafts.

The ingrowth from grafted serotonin-containing neurons, taken from the brainstem raphe region, has so far not been studied in detail. Such grafts have, however, been demonstrated at the ultrastructural level to form morphologically mature synaptic connections with dendrites in the dentate molecular layer in the adult host (Beebe et al. 1979).

From a morphological perspective the reafferentation strategy has thus demonstrated the capacity of grafted monoaminergic and cholinergic brainstem or basal forebrain neurons to reestablish innervation patterns in the previously denervated HF, leading to selective terminal patterns and synaptic contacts with the host target neurons.

Bridge: Grafted peripheral tissues have been used as bridges to promote axonal regeneration across damaged tissue or tissue defects in the nervous system (Tello 1911; Svendgaard et al. 1975; Aguayo et al. 1982). The specific issue being addressed in the bridge strategy as it relates to the denervated HF is to what extent a fetal hippocampalB implant can replace the normal hippocampus as a target for the formation of connections with the host septum, and to what extent the same implant can promote regeneration of the septo-hippocampal pathway and serve as a bridge for the regeneration of the axons from the damaged septal afferents to the HF. Previous research had shown that implants of denervated iris to the fimbria-fornix could stimulate the cholinergic septal neurons of the host to reinnervate large portions of the denervated iris in a homotypic manner (Svendgaard et al. 1976). With this same paradigm, but using fetal hippocampal tissue as a bridge between septum and hippocampus (Kromer et al. 1981a,b), AChE-positive fibers from the septal region were seen to innervate the entire implant with time. Parallel biochemical studies revealed a dramatic increase in choline acetyltransferase (ChAT) activity in the implant. In addition, by 4-6 weeks after transplantation, new AChE-positive fibers appeared in parts of the subiculum, hippocampus, and dentate gyrus, immediately bordering the implant, and at later times, new fibers from the graft had expanded 2.5 - 3.5 mm into the dorsal HF. The ingrowing regenerating host axons were seen to pattern themselves in a manner closely resembling the normal cholinergic innervation pattern. In a parallel study, Segal et al. (1981) obtained evidence that the regenerating axons establish functional synaptic contacts with neurons in the host HF (see further below). Again, these histochemical observations were confirmed by measuring a recovery in ChAT levels in the dorsal HF of up to about 50% of control. Lesions of the septum-diagonal band area in the regenerated grafted animals, and HRP injections in the host HF caudal to the hippocampal bridge graft, showed that the origin of the regenerated cholinergic axons was predominantly the ipsilateral ventral medial septum and diagonal band area. These results suggest that a piece of fetal CNS tissue, in addition to mature peripheral tissue (Svendgaard et al. 1975, 1976), can promote regeneration of axotomized central neurons both into the implant tissue and through the implant tissue to innervate the denervated host tissue in a homotypic manner.

Culture bed: The objective of this strategy is to utilize the richly vascularized pial surface provided by the hippocampal transplantation cavities to address developmental questions, similar to what has previously been done in in vitro explant or reaggregation cultures and with the intraocular grafting techniqe. The general conclusion from these studies is that the time-course of development of the grafted embryonic CNS tissue seems to follow the normal developmental sequences in its new, ectopic environment (Jaeger and Lund 1980a, b; Wells and McAllister 1982; Alvarado-Mallert and Sotelo 1982). The grafts have repeatedly been shown to develop intrinsic organization features, such as lamination and foliation, similar to the ones seen in situ (Kromer et al. 1979, 1983; Sunde and Zimmer 1981, 1983; Alvarado-Mallart and Sotelo 1982; Wells and MacAllister 1982). Interestingly, Kromer et al. (1983) have shown that, in contrast to the rudimentary development of Purkinje cells and subnormal granule cell-to-Purkinje cell ratio in in vitro explant cultures and in intraocular transplant of cerebellar primordia, transplants of early gestational cerebellar tissue to the denervated HF demonstrate a remarkably mature state of morphological differentiation in the Purkinje cells and a more normal granule cell-to-Purkinje cell ratio. This suggests that the grafts may interact with the host brain also in the development of the graft's intrinsic structure. The intracerebral grafting technique should therefore provide a useful tool for obtaining additional information concerning the cellular events which determine the organization of the developing CNS.

An additional application of the "culture bed" strategy is that which uses the denervated HF as a system that may release neurotrophic factors upon damage. Thus, by transplanting different cell types which may or may not have a dependency on neurotrophic factors, survival of grafted cells can be assessed as a quantitative in vivo bioassay of the alleged factors. Experiments with grafts of adult and neonatal sympathetic ganglia have already provided some evidence for such denervation-dependent neurotrophic effects (Björklund and Stenevi 1981; Gage et al. 1984).

Functional Effects of Intrahippocampal Neural Grafts

The survival and growth of solid grafts and cell suspensions in the denervated HF have thus been well characterized for several different neuronal cell types, and intrahippocampal grafting has received interesting applications to the understanding of neural plasticity and development in the HF. However, methods other than those required to assess survival and growth are needed to determine whether the grafted cells can function in their new environment. In addition, "function" itself can be assessed at several different levels, each of which provides important information about the quality and quantity of information that is communicated between the graft and the host tissue. At present three general types of functional procedures have been applied to this issue: (1) biochemical, (2) electrophysiological, and (3) behavioral; each of these will be discussed separately.

Biochemical measures of graft function: The biochemical activity of the cholinergic innervation of the denervated HF, derived from solid or suspension grafts of the septal-diagonal band area, has been studied fairly extensively. Although AChE is a useful anatomical marker of the septo-hippocampal projection system, the AChE enzyme is not a specific marker for cholinergic neurons. The acetylcholine synthetic enzyme, ChAT, by contrast, is an enzyme specifically localized in cholinergic neurons and is therefore a better marker of cholinergic transmission. This enzyme has been used to measure the time-course and magnitude of fiber outgrowth from both solid and suspended septal grafts (Björklund and Stenevi 1977b); Björklund et al. 1983b). Graft-derived ChAT activity was barely detectable 10 days and 1 month in the HF regions close to the graft. By 6 months, ChAT activity was restored to near normal levels in all segments of the previously denervated HF. When comparing the total amount of ChAT activity derived from the solid grafts and the cell suspension grafts, the cell suspension grafts appeared to be about twice as effective as the solid grafts, although the amount of tissue grafted was about the same in each case.

The functional activity of the cell suspension grafts was further assessed by measurements of ^{14}C-acetylcholine synthesis from ^{14}C glucose in vitro. The overall hippocampal ^{14}C-acetylcholine synthesis was restored to normal levels in the grafted animals, and estimates of the acetylcholine turnover rate suggested that the transmitter machinery of the newly established septo-hippocampal connections operated at a rate similar to that of the intrinsic septo-hippocampal pathway. Thus, these septal cell suspensions seem capable of maintaining functin at a relatively "physiological" level despite their abnormal position.

In a more recent study, Kelly et al. (1984) investigated the magnitude of lesion-induced alterations in the HF as reflected in the local rates of ^{14}C-2-deoxyglucose (2-DG) utilization, and the degree to which the index of functional activity could be normalized following reinnervation by solid septal grafts. Six months after unilateral fimbria-fornix lesion the 2-DG utilization was markedly reduced in the HF of the nongrafted animals. Rats that received the solid grafts displayed a significant increase in 2-DG use as compared to the rats with lesion alone. In addition, the changes in 2-DG utilization were significantly correlated with the density of AChE staining in adjacent sections from the same brains, thus suggesting a relationship between the cholinergic innervation from the septal grafts and the restoration of functional glucose utilization. Though it is difficult to generalize beyond the septal grafts in the denervated hippocampus to other transmitter systems, these results at least strongly suggest that the cholinergic component of the grafts is functioning at the biochemical level and influencing, or normalizing, the function of the deafferented HF.

<u>Electrophysiological studies</u>: Several questions concerning the function of the graft can be addressed with electrophysiological methods. As previously stated, fetal locus coeruleus grafts implanted in the denervated HF have been shown to grow new noradrenergic axons from the implant to the appropriate denervated hippocampal terminal fields. In electrophysiological experiments (Björklund et al. 1979b), stimulation of the locus coeruleus implants caused an inhibition of the spontaneous activity of neurons in the host hippocampus, similar to the long latency, long duration inhibition observed after stimulation of the intrinsic locus coeruleus in the intact rat. Likewise, as in the intact rat, the evoked inhibitory responses were blocked by local or systemic application of the beta-adrenergic receptor blockers, propanolol, or sotalol. In a separate experiment (Low et al. 1982), stimulation of septal grafts evoked characteristic field potentials in the dentate gyrus, and the depth profile revealed a relatively diffuse innervation characteristic of the normal septal afferents. Furthermore, prepulse stimulation of the septal grafts were found to potentiate the field potentials evoked by perforant path stimulation. This latter effect is a characteristic response to septal cholinergic stimulation also in the intact HF. In these two latter studies the reafferentation strategy was used to assess the graft-to-host interaction. However, using the bridge strategy, Segal et al. (1981) have studied whether the graft could act as a conduit for electrophysiological impulses within the denervated septo-hippocampal system. Fetal hippocampal tissue was placed in the fimbria-fornix cavity to promote regeneration of the septal fibers through the graft to the denervated HF, as described above. By stimulating the host septal area and recording in either the host HF or the grafted hippocampus, evidence was provided that the AChE-stained axons were able to reestablish atropine-sensitive excitatory synaptic connections in the grafted hippocampus as well as in the dorsal part of the initially denervated host HF. Together these results demonstrate that the grafts can establish functional synapses in the host brain, that they can receive functional synapses from the host brain, and that they can act as a conduit for functional synapses between two disconnected brain structures.

<u>Behavioral studies</u>: Bilateral fimbria-fornix lesions in rats are known to result in severe impairments in working memory (Olton et al. 1979) or spatial memory (O'Keefe and Nadel 1978). In addition,

these same types of impairment can be obtained by pharmacological blockade of cholinergic transmission in a variety of tests (Sutherland et al. 1982; Eckerman et al. 1980). Using several standard behavioral tests of learning and memory, researchers have demonstrated the ability of the fetal septal grafts to reverse these impairments. In the eight-arm radial maze (Low et al. 1982) rats with septal grafts showed a positive linear trend in maze performance over days of testing, but did not differ significantly from nongrafted rats with lesions overall. However, potentiation of the cholinergic transmission by pretreatment with physostigmine produced a significant enhancement of maze performance in the grafted group, but not in the lesioned control group, and in some cases the grafted rats performed as well as the nonlesioned control animals. In another study (Dunnett et al. 1982) using a T-maze forced-choice alternation test seven of nine rats with solid septal grafts and four of five rats with septal cell suspensions learned the task, some of them up to the level of the control rats. The remaining rats with septal grafts, and a separate group of rats with locus coeruleus grafts, performed at chance level, similar to the rats that only received the fimbria-fornix lesion. In this study there was a positive and significant correlation between performance of the grafted rats and the amount of graft-derived AChE-positive staining in the previously denervated HF. Together these two studies strongly suggest not only that the grafts can partially ameliorate deficits induced in learning that result from FF lesions, but that the amelioration shows some specificity for the septal grafts which provide a cholinergic reinnervation of the deafferented HF.

Conclusion

Transplantation of fetal CNS tissue to the denervated adult HF has thus far provided useful information relating to the survival and outgrowth as well as functional capabilities of intracerebral grafting. The well-described anatomy of the HF and its efferent and afferent pathways has enabled researchers to obtain clear answers relating to the specificity of different grafts for their respective homotypic terminal fields. The selectivity of the denervating lesions, and the favorable conditions for graft survival in the lesion cavities, have allowed for reliable and reproducible results. The dramatic and longlasting functional functional deficits observed following bilateral fimbria-fornix lesions have provided a valuable starting point to test the capabilities of the grafts to reinstate or restore lost functions due to the damage.

In addition to grafting to the adult HF, some experimenters have used neonatal rats as hosts and have reported great success at the anatomical level (Sunde and Zimmer 1983). Furthermore, peripheral tissues, such as sympathetic ganglia and iris, have been successfully grafted to the denervated hippocampus and have been used successfully as targets for sprouting CNS fibers (Svendgaard et al. 1975, 1976; Björklund and Stenevi 1977b, 1981; Schonfeld and Katzman 1983). The limitations of the hippocampal model system for some questions are the advantages of the system when asking other questions. For example, the fimbria-fornix lesion is large and it results not only in a deafferentation of the entire dorsal hippocampus, but also in a de-efferentation. However, when testing for potential recovery following hippocampal damage it is essential to have a complete lesion or else the propensity of compensatory sprouting to occur in the partially denervated HF will make interpretations of graft-induced effects difficult.

The reinnervation of the previously deafferented HF by grafted cholinergic and monoaminergic neurons is extensive and exhibits a high degree of specificity. This has suggested to us that those mechanisms which normally operate during ontogenesis to direct, guide, and pattern intracerebral connections are retained also in the fully mature brain, at least with respect to the types of nonmyelinated and weakly myelinated regulatory systems we have studied so far. The functional capacity of the reconstructed "septo-hippocampal" connections from the grafted septal neurons now seems well established. Future studies of this experimental model system should therefore provide new insights into the mechanism(s) of functional recovery in the damaged CNS.

Acknowledgments. Our own reserach summarized here was supported by grants from the Swedish MRC and from the National Institutes of Aging.

References

Aguayo A, Richardson PM, David S, Benfrey M (1982) Transplantation of neurons and sheath cells - a tool for the study of regeneration. In: Nicholls JG (ed) Repair and regeneration of the nervous system. Springer, Berlin Heidelberg New York, pp 91-105

Alvarado-Mallart RM, Sotelo C (1982) Differentiation of cerebellar anlage heterotypically transplanted to adult rat brain: a light and electron microscopic study. J Comp Neurol 212:247-267

Anderson P (1975) Organization of hippocampal neurons and their interconnections. In: Isaacson RL, Primbram KH (eds). The hippocampus, vol 1. Plenum, New York, p 155

Beebe BK, Møllgård K, Björklund A, Stenevi U (1979) Ultrastructural evidence of synaptogenesis in the adult rat dentate gyrus from brain stem implants. Brain Res 167:391-395

Björklund A, Stenevi U (1977a) Experimental reinnervation of the rat hippocampus by grafted sympathetic ganglia: 1. Axonal regeneration along the hippocampal fimbria. Brain Res 138:259-270

Björklund A, Stenevi U (1977b) Reformation of the severed septohippocampal cholinergic pathway in the adult rat by transplanted septal neurones. Cell Tissue Res 185:289-302

Björklund A, Stenevi U (1981) In vivo evidence for a hippocampal adrenergic neurotrophic factor specifically released on septal deafferentation. Brain Res 229:403-428

Björklund A, Stenevi U, Svendgaard N-A (1976) Growth of transplanted monoaminergic neurones into the adult hippocampus along the perforant path. Nature 262:787-790

Björklund A, Kromer LF, Stenevi U (1979a) Cholinergic reinnervation of the rat hippocampus by septal implants is stimulated by perforant path lesion. Brain Res 173:57-64

Björklund A, Segal M, Stenevi U (1979b) Functional reinnervation of rat hippocampus by locus coeruleus implants. Brain Res 170:409-426

Björklund A, Gage FH, Stenevi U, Dunnett SB (1983a) Intracerebral grafting of neuronal cell suspensions. VI. Survival and growth of intrahippocampal implants of septal cell suspensions. Acta Physiol Scand [Suppl] 522:49-58

Björklund A, Gage FH, Schmidt RH, Stenevi U, Dunnett SB (1983b) Intracerebral grafting of neuronal cell suspensions: VII. Recovery of choline acetyltransferase activity and acetylcholine synthesis in the denervated hippocampus reinervated by septal suspension implants. Acta Scand Physiol[Suppl] 522:59-66

Björklund A, Stenevi U, Schmidt RH, Dunnett SB, Gage FH (1983c) Intracerebral grafting of neuronal cell suspensions: I. Introduction and general methods of preparation. Acta Physiol Scand [Suppl] 522:1-10

Dunnett SB, Low WC, Iversen SD, Stenevi U, Bjöerklund A (1982) Septal transplants restore maze learning in rats with fornix-fimbria lesions. Brain Res 251:335-348

Eckerman DA, Gordon WA, Edwards JD, MacPhail RC, Gage MI (1980) Effects of scopolomine, pentobarbital and amphetamine on radial arm maze performance in the rat. Pharm Biochem Behav 12:595-602

Gage FH, Björklund A, Stenevi U (1984) Denervation releases a neuronal survival factor in adult rat hippocampus. Nature 308:637-639

Jaeger CB, Lund RD (1980a) Transplantation of embryonic occipital cortex to the tectal region of newborn rats: a light microscopic study of organization and connectivity of the transplants. J Comp Neurol 194:571-597

Jaeger CB, Lund RD (1980b) Transplantation of embryonic occipital cortex to the brain of newborn rats. An autoradiographic study of transplant histogenesis. Exp Brain Res 40:265-272

Kelly PAT, Gage FH, Ingvar M, Lindvall O, Stenevi U, Björklund A (1984) Functional reactivation of the deafferented hippocampus by embryonic septal grafts as assessed by measurements of local glucose utilization. Exp Brain Res (to be published)

Kromer LF, Björklund A, Stenevi U (1979) Intracephalic implants: a technique for studying neuronal interactions. Science 204:1117-1119

Kromer LF, Björklund A, Stenevi U (1981a) Innervation of embryonic hippocampal implants by regenerating axons of cholinergic septal neurons in the adult rat. Brain Res 210:153-171

Kromer LF, Björklund A, Stenevi U (1981b) Regeneration of the septohippocampal pathway in adult rats is promoted by utilizing embryonic hippocampal implants as bridges. Brain Res 210:173-200

Kromer LF, Björklund A, Stenevi U (1983) Intracephalic neural implants in the adult rat brain. I. Growth and mature organization of brain stem, cerebellar and hippocampal implants. J Comp Neurol 218:433-459

Low WC, Lewis PR, Bunch ST, Dunnett SB, Thomas SR, Iversen SD, Bjorklund A, Stenevi U (1982) Functional recovery following neural transplantation of embryonic septal nuclei in adult rats with septohippocampal lesions. Nature 300:260-262

Low WC, Lewis PR, Bunch ST (1983) Embryonic neural transplants across a major histocompatibility barrier: survival and specificity of innervation. Brain Res 262:328-333

O'Keefe J, Nadel L (1978) The hippocampus as a cognitive map. Clarenden, Oxford

Olton DS, Becker JT, Handelman GE (1979) Hippocampus, space and memory. Behav Brain Sci 2:313-365

Ramon y Cajal S (1968) The structure of Ammon's horn, transplanted by L.M. Kraft. Thomas, Springfield

Schmidt RH, Björklund A, Stenevi U (1981) Intracerebral grafting of dissociated CNS tissue suspensions: a new approach for neuronal transplantation to deep brain sites. Brain Res 218:347-356

Schonfeld AR, Katzman R (1983) In vivo cholinotrophic activities. In: Katzman R (ed) Biological aspects of Alzheimer's disease. Branbury Report 15. Cold Spring Harbor Laboratories, New York, p 269

Segal M, Stenevi U, Björklund A (1981) Reformation in adult rats of functional septohippocampal connections by septal neurons regenerating across an embryonic hippocampal tissue bridge. Neurosci Lett 27:7-12

Stenevi U, Björklund A, Svendgaard N-A (1976) Transplantation of central and peripheral monoamine neurons to the adult rat brain: techniques and conditions for survival. Brain Res 114:1-20

Storm-Mathisen J (1979) Localization of transmitter candidates in the brain: the hippocampal formation as a model. Prog Neurobiol 8:381-388

Sunde NAa, Zimmer J (1981) Transplantation of central nervous tissue. An introduction with results and implications. Acta Neurol Scand 63:323-335

Sunde NAa, Zimmer J (1983) Cellular histochemical and connective organization of the hippocampus and fascia dentata transplanted to different regions of immature and adult rat brains. Dev Brain Res 8:165-191

Sutherland RJ, Whishaw IQ, Regeher JC (1982) Cholinergic receptor blockade impairs spatial localization using distal cues in the rat. J Comp Physiol Psychol 96:563-573

Svendgaard NAa, Björklund A, Stenevi U (1975) Regenerative properties of central monoamine neurons as revealed in studies using iris transplants as targets. Adv Anat Embryol Cell Biol 51:1-77

Svendgaard NAa, Björklund A, Stenevi U (1976) Regeneration of central cholinergic neurons in the adult rat brain. Brain Res 102:1-22

Tello F (1911) La influencia del neurotropismo en la regeneracion de los centros nerviosos. Trab Lab Invest Biol 9:123-159

Wells J, McAllister JP (1982) The development of cerebellar primordia transplanted to the neocortex of the rat. Dev Brain Res 4:167-179

Concluding Remarks

G. W. Kreutzberg

Max-Planck-Institut für Psychiatrie, Am Klopferspitz 18A, 8033 Martinsried, FRG

This meeting has covered a broad range of topics and work done in many different fields using a wide spectrum of methods such as electrophysiology, biochemistry, morphology or behavioural analysis. At this point it might therefore be useful to summarize the proceeding which are the outcome of 4 days of working on processes of recovery from neural trauma.

By going through the contributions presented at the meeting and to a large extent reported in these proceedings, it appears that the problems have necessarily been approached at several levels:

1. At the very basic level we are dealing with phenomena which could almost be concerned with the biophysical properties of membranes, such as conductivities, fluxes, channels. During the meeting Itzak Parnas has reminded us several times that we have to consider very basic electrical phenomena and that, e.g., synaptic efficacy undergoes changes as a consequence of injury and local synaptic loss. Bob Werman showed us the importance of considering ionic fluxes with regard to the survival of the vagal neurons following axotomy. This mimics the phenomenology of a Mg^{++} poisoning of the neuron with the possibility of a blockage of Ca^{++}, causing excess accumulation of intraneuronal Ca^{++} which may be deleterious to the cell. Electrotonic coupling produced as a result of dendrotomy as demonstrated by dye coupling (Gutnick), could turn out to be an important factor in the causal genesis of posttraumatic epilepsy.

Only a few of the innumerable electrical events have been touched upon which occur in connection with trauma, axotomy, deafferentation, etc. The link between the electrophysiological and cell biological changes is not yet established. A possible exception is that we are beginning to understand that Ca^{++} is sometimes involved as a second messenger, sometimes helpful sometimes not. A search for injury induced triggering signals is needed. Are electrical events involved in initiating injury response in the nerve cell? What tells surviving intact axons to grow and take over vacated territories or to strengthen synaptic efficacy?

2. The second level in the hierarchy of complexity might be called the molecular level. It is the field of factorology, receptorology, and molecular biology. What are the mechanisms that turn on the genom in the response to trauma, as part of a growth or regenerative process?

Experimental Brain Research, Suppl. 13

Dr. Freeman reported on proteins which are almost specifically synthesised during the process of regeneration. The discovery of growth associated proteins (GAP) is probably one of the most exciting new things in the field and we are looking forward to learn more on their function soon. It was interesting to hear that there are also new data on old molecules such as the beta-tubulin now found in the particulate fraction of axonal transport, meaning that we probably have to revise the dogma that microtubular proteins are strictly in the slow component (SCa) of axonal transport (Grafstein). Growth factors and extracellular matrix molecules such as fibrin, fibronectin, collagen and laminin have turned out to be important partners for the nerve cell during growth and regeneration as demonstrated here by S. Varon, M. Schwab and others. Natural constituents of the cell membrane, e.g., the GM1 gangliosides have an astonishing effect on the injured neurons and help to speed up functional restitution (Gorio, Jonson, Shahar and Leon). It is interesting to see that gangliosides have a noticeable effect on neurons only when they are injured.

An attractive idea seems to be that of signal producing cells: a hypothesis that attempts to explain why mammalian CNS neurons do not express regenerative responses (M. Schwartz). Is there a lack of specific molecules produced by the non-neuronal cellular environment? Y. Misrachi in Dr. Schwartz's group has obviously found a factor which acts as a substratum for neurite extension (SNEF). It is a glycoprotein isolated from goldfish retinae following nerve crush.

The molecular mechanisms involved in turning on the regeneration program of neurons are still unclear; growth factors may play a role and one of the early steps might be under the control of polyamines. The early increase of ornithine decarboxylase (ODC) as the key enzyme of the biosynthesis of this class of molecules may show us a line for future research, especially since application of polyamines seems to speed up regeneration and may be a factor improving survival of neurons as shown by G. Gilad. The search for substances to promote axonal growth such as thyroxin, testosterone etc. will continue (L. Guth). The complexity of intercellular relationships is shown in a frequently used model: Interruption of innervation causes denervation of muscle demonstrating control of the neuron over target tissue, e.g., the stability of acetylcholine receptors in the postsynaptic site (M. Salpeter). To elucidate further the function of receptors and ion channels the toxicoimmunological approach as used by Meiri could prove to become a powerful tool.

3. At the cellular level, we attempt to analyse the cell biology of the traumatised neurons.

Some of the mechanisms of this level have already been mentioned in the molecular section. Axonal transport is an important vehicle for axon elongation and changes in it are well documented. Very little is known on dendritic changes following injury except a retraction and extension of the arborization which could serve to compensate for changes in axonal transport rates. Such a change in the dendritic field as well as the synaptic stripping occurring on the soma and the stem dendrites cannot be without consequences for input of the neurons and hence for its function. The observation of a polyribosomal organelle in dendrites especially at the base of spines during synaptogenesis or growth is very exciting. The idea of specific local protein synthesis as put forward by Oswald

Steward seems to me to be very attractive, especially if this idea is used to isolate ribosomes and get this molecule into the test tube.

Formation of sprouts, of course is a cellular event, although it takes place often at a considerable distance from the cell body. Invertebrate systems have been demonstrated here by I. Parnass and M. Spira. The growth of the giant axon is most impressive. How is the decision made to elongate the axon rather than inducing branching? The dependence of the local environment is obvious, the factors unknown. There is a necessity for a specific cellular terrain for the regeneration of axons as L. Guth pointed out. This local milieu might be decisive for spinal cord axons to grow. M. Berry's reconnection of the optic nerve to a peripheral nerve transplant shows again in principal that CNS neurons are capable of regeneration. However, the specificity of some neurons to react with growth or the selectivity of the death of large neurons confronts us with a new and unsolved problem.

4. The local tissue response to trauma deserves our special attention. I think that Dr. Feinsod's seminar illustrating clinical aspects of head injuries has impressed all of us. There is no question that our research could and should have consequences in this field. One has to realize that the most severe changes in the brain after head injury are of a secondary nature. The primary tissue damage is irreversible but the deficits resulting from bleeding, ischemia, edema, inflammation could probably be avoided if we would know more about the pathophysiology. We heard from Drs De La Torre, E.D. Means and E. Hall about the enormous complexity of these events. However, I think there is hope since the most noxious factors, e.g., free radicals, high intracellular Ca++, oxygen and substrate deficit caused by local ischemia have now been recognized and could be attacked. Thus the outcome of the recovery process from CNS trauma depends very much on our ability to suppress the secondary lesions in order to save as much brain tissue as possible.

The role of sprouting, new synapse formation or synapse reactivation in the process of rehabilitation of function has served widely as a justification for our experimental work. And indeed some remarkable results presented here may help to understand these processes. Models such as the motorsystem with a contralateral signal transfer inducing sprouting at the neuromuscular junction could be a useful basis for further discussions (S. Rothshenker).

5. Here we are already confronted with the specificity of systems. The central monoaminergic system behave in many ways like PNS. Sprouting, axon elongation, reinsertion of synaptic terminals take place in the CNS and have been discussed here by Azmitia, Zigmond and others. A remarkable report by F. Gage showed the reconstruction of the delicate acetylcholinesterase system in the hippocampus. Similar recoveries may occur in the dorsal horn of the spinal cord (M. Devor) or a replacement of vacant synaptic sites takes place by a new population of terminals (Goldberger). This must all involve cellular interaction and molecular communication. A system to study this further might be the "brain in the eye" preparation of Lars Olsson demonstrated here by Ake Seiger.

The successful development of techniques for transplantation and grafting in the CNS has opened a new field in recent years. Sometimes it seems debatable where this is leading to. Although we have seen remarkable experiments demonstrating the capacity of the central neuron for regeneration we must also admit that the finer control of a directed growth of regenerating axons and a functional reconstitution is not yet in our hands. Painstaking and critical experiments reported here show how much work has still to be done (Bernstein, Das, Gage). Nevertheless the perfection of many transplantation experiments can only be admired.

6. The critical analysis of the functional outcome from the transplantation experiments shows very clearly how much good behaviour studies are needed. A transfer of animal experimentation into clinical application will depend widely from such studies. Analysis of functions can reach from simple motor performances (Tsukahara) to the most complex behaviour such as learning. It is clear that regeneration research urgently needs experimental psychologists to develop adequate behaviour model to quantify results.

Subject Index

MIX
Papier aus verantwortungsvollen Quellen
Paper from responsible sources
FSC® C105338

If you have any concerns about our products,
you can contact us on
ProductSafety@springernature.com

In case Publisher is established outside the EU,
the EU authorized representative is:
Springer Nature Customer Service Center GmbH
Europaplatz 3, 69115 Heidelberg, Germany

Printed by Libri Plureos GmbH
in Hamburg, Germany